GLOBAL CORAL REEF
MONITORING NETWORK

Coral Reefs of the World: 2000

Edited by Clive Wilkinson

AUSTRALIAN INSTITUTE
OF MARINE SCIENCE

Dedication

This book is dedicated to the many people who are collectively working to conserve the vast heritage of coral reef ecosystems throughout the world – we thank them for their efforts. It is also dedicated to the partners of the International Coral Reef Initiative and to agencies of the Government of the United States operating through the US Coral Reef Task Force. Of particular mention is the support to the GCRMN from the US Department of State and the US National Oceanic and Atmospheric Administration.

This report has been produced for the sole use of the party who requested it. The application or use of this report and of any data or information (including results of experiments, conclusions, and recommendations) contained within it shall be at the sole risk and responsibility of that party. AIMS does not provide any warranty or assurance as to the accuracy or suitability of the whole or any part of the report, for any particular purpose or application. Subject only to any contrary non-excludable statutory obligations neither AIMS nor its personnel will be responsible to the party requesting the report, or any other person claiming through that party, for any consequences of its use or application (whether in whole or part).

Cover images courtesy of Sandy Tudhope and Lida Pet-Soede.

© Australian Institute of Marine Science, 2000
Cape Ferguson, Queensland, and Dampier, Western Australia

Postal addresses:

PMB No 3   PO Box 264
Townsville MC QLD 4810   Dampier WA 6713
Australia   Australia

Telephone: (07) 4753 4444   (08) 9183 1122
Facsimile: (07) 4772 5852   (08) 9183 1085
World Wide Web: www.aims.gov.au

National Library of Australia Cataloguing-in-Publication data:

Status of coral reefs of the world: 2000.

ISBN 0 642 32209 0.

1. Coral reefs and islands. I. Wilkinson, Clive R.
II. Australian Institute of Marine Science.

578.7789

FOREWORD

Over the past few years, I have become increasingly concerned about the continued degradation of coral reefs, some of the most valuable and spectacular places on earth. Coral reefs are important to us in so many ways. They provide food and livelihoods for millions of people; they help protect coastal communities from tropical storms; and they support rich communities of marine life that rival even rainforests in their biological diversity. These 'rainforests of the sea' have already begun to provide new medicines and other compounds to combat diseases and solve other human needs. And we've just begun to explore the rich life of our coral reefs.

Today, millions of people visit coral reef ecosystems every year to experience the beauty and bounty of healthy coral reefs. In the United States alone, coral reefs support millions of jobs and billions of dollars from tourism related activities. These are just a few of the reasons why coral reefs are so valuable, and why it is so important for us to protect the world's vulnerable coral reefs.

In 1997 and 1998, coral reefs began bleaching and dying across huge regions of the world in association with changes in ocean and climate conditions. In many areas, coral reef ecosystems were devastated, along with the human communities that depended on them. The impacts were on a global scale never before recorded. Scientists now believe that coral reefs may be the first natural ecosystem to clearly show the potential impacts of global climate change.

The struggle to conserve the coral reefs is now at a critical stage. While changes in climate conditions may continue to impact coral reefs for some time, coral reefs are losing an even greater battle with water and air pollution, sediment runoff caused by deforestation and coastal development, and over-exploitation, including the insidious practices of dynamite and cyanide fishing. In many areas, the living coral reefs may soon be gone. If coral reefs are lost, many coastal populations will lose their primary source of food, jobs, cultural heritage and long-term prosperity.

To conserve these natural treasures, we must reduce human impacts on coral reefs by immediately controlling pollution, reducing over-fishing and increasing protection and sustainable use of our valuable coral reef resources. By working together – from local communities to regions and internationally – I believe we can, and must, reverse the tide of destruction and conserve the world's precious coral reefs.

This report is a tremendous example of what partnerships can do to protect and sustain coral reefs worldwide. This historic document brings together scientific information from around the world to provide the most comprehensive status report to date on the health of the world's coral reefs. The report is made possible by the efforts of nations, communities and individuals from all over the world to monitor the health of coral reef ecosystems. Without these monitoring programs, this report would not be possible, and we would be severely limited in our ability to address the coral reef crisis. And these efforts are just the beginning. We need more monitoring, research, management and education to adequately track, and improve, the health of coral reefs around the world.

This report documents the continuing and disturbing decline in the health of the coral reefs worldwide, which was initially expounded in the first report from the Global Coral Reef Monitoring Network in 1998. Unfortunately, the report provides powerful new evidence that coral reefs continue to be destroyed and degraded in many areas, threatening the very existence of these valuable resources. Now the question is: what are we willing to do about it?

I am very pleased that the United States has been able to join with so many international partners in efforts to protect coral reef ecosystems. I sincerely commend the International Coral Reef Initiative (ICRI), the Global Coral Reef Monitoring Network (GCRMN), the sponsors of this report, and all those who have supported efforts to monitor, manage and conserve the world's reefs.

The children of tomorrow have the right to experience the beauty and wonder of coral reef ecosystems. I call on all of us – nations, societies and individuals – to act now to reduce the threats to these remarkable ecosystems. We must ensure that this report marks the beginning of a powerful new age of coral reef protection, not the sad ending to their very existence.

Al Gore
Vice President
United States of America

Contents

Countries, States and Territories vii
Acknowledgements xi
Introduction 1
Executive Summary 7
   Clive Wilkinson

1. The 1997-98 Mass Coral Bleaching and Mortality Event: 2 years on
Clive Wilkinson 21

2. Regional Status of Coral Reefs in the Red Sea and the Gulf of Aden
Nicolas Pilcher and Abdullah Alsuhaibany 35

3. Status of Coral Reefs in the Arabian/Persian Gulf and Arabian Sea Region (Middle East)
Nicolas J. Pilcher, Simon Wilson, Shaker H. Alhazeem and Mohammad Reza Shokri 55

4. Status of Coral Reefs in East Africa: Kenya, Mozambique, South Africa and Tanzania
David Obura, Mohammed Suleiman, Helena Motta, and Michael Schleyer 65

5. Status of Coral Reefs of the Southern Indian Ocean: the Indian Ocean Commission Node for Comoros, Madagascar, Mauritius, Reunion and Seychelles
Lionel Bigot, Loic Charpy, Jean Maharavo, Fouad Abdou Rabi, Naidoo Ppaupiah, Riaz Aumeeruddy, Christian Viledieu and Anne Lieutaud 77

6. Status of Coral Reefs in South Asia: Bangladesh, India, Maldives and Sri Lanka
Arjan Rajasuriya, Hussein Zahir, E.V. Muley, B.R. Subramanian, K. Venkataraman, M.V.M. Wafar, S.M. Munjurul, Hannan Khan and Emma Whittingham 95

7. Southeast Asian Reefs - Status Update: Cambodia, Indonesia, Malaysia, Philippines, Singapore, Thailand and Vietnam
Loke Ming Chou 117

8. Status of Coral Reefs of East and North Asia: China, Japan and Taiwan
Shuichi Fujiwara, Takuro Shibuno, K. Mito, Tatsuo Nakai, Yasunori Sasaki, Dai Chang-feng and Chen Gang 131

9. Status of Coral Reefs of Australasia: Australia and Papua New Guinea
Thomas Maniwavie, Hugh Sweatman, Paul Marshall, Phil Munday and Vagi Rei 141

10. STATUS OF CORAL REEFS IN THE SOUTHWEST PACIFIC: FIJI, NAURU, NEW CALEDONIA, SAMOA, SOLOMON ISLANDS, TUVALU AND VANUATU
   Robin South and Posa Skelton 159

11. STATUS OF SOUTHEAST AND CENTRAL PACIFIC CORAL REEFS IN 'POLYNESIA MANA NODE': COOK ISLANDS, FRENCH POLYNESIA, KIRIBATI, NIUE, TOKELAU, TONGA, WALLIS AND FUTUNA
   Bernard Salvat 181

12. STATUS OF CORAL REEFS OF AMERICAN SAMOA AND MICRONESIA: US-AFFILIATED AND FREELY ASSOCIATED ISLANDS OF THE PACIFIC
   C.E. Birkeland, P. Craig, G. Davis, A. Edward, Y. Golbuu, J. Higgins, J. Gutierrez, N. Idechong, J. Maragos, K. Miller, G. Paulay, R. Richmond, A. Tafileichig and D. Turgeon 199

13. STATUS OF CORAL REEFS IN THE HAWAIIAN ARCHIPELAGO
   David Gulko, James Maragos, Alan Friedlander, Cynthia Hunter and Russell Brainard 219

14. STATUS OF CORAL REEFS IN THE US CARIBBEAN AND GULF OF MEXICO: FLORIDA, TEXAS, PUERTO RICO, US VIRGIN ISLANDS AND NAVASSA
   Billy Causey, Joanne Delaney, Ernesto Diaz, Dick Dodge, Jorge R. Garcia, Jamie Higgins, Walter Jaap, Cruz A. Matos, George P. Schmahl, Caroline Rogers, Margaret W. Miller and Donna D. Turgeon 239

15. STATUS OF CORAL REEFS IN THE NORTHERN CARIBBEAN AND WESTERN ATLANTIC
   Jeremy Woodley, Pedro Alcolado, Timothy Austin, John Barnes, Rodolfo Claro-Madruga, Gina Ebanks-Petrie, Reynaldo Estrada, Francisco Geraldes, Anne Glasspool, Floyd Homer, Brian Luckhurst, Eleanor Phillips, David Shim, Robbie Smith, Kathleen Sullivan Sealey, Mónica Vega, Jack Ward and Jean Wiener. 261

16. STATUS OF CORAL REEFS OF NORTHERN CENTRAL AMERICA: MEXICO, BELIZE, GUATEMALA, HONDURAS, NICARAGUA AND EL SALVADOR
   Philip Kramer, Patricia Richards Kramer, Ernesto Arias-Gonzalez and Melanie McField 287

17. STATUS OF CORAL REEFS IN THE EASTERN CARIBBEAN: THE OECS, TRINIDAD AND TOBAGO, BARBADOS, THE NETHERLANDS ANTILLES AND THE FRENCH CARIBBEAN
   Allan H. Smith, Mark Archibald, Trish Bailey, Claude Bouchon, Angelique Brathwaite, Ruleta Comacho, Sarah George, Harold Guiste, Mark Hastings, Philmore James, Cheryl Jeffrey-Appleton, Kalli De Meyer, Andre Miller, Leonard Nurse, Clive Petrovic and Paul Phillip. 315

18. STATUS OF CORAL REEFS IN SOUTHERN TROPICAL AMERICA: BRAZIL, COLOMBIA, COSTA RICA, PANAMA AND VENEZUELA
   Jaime Garzón-Ferreira, Jorge Cortés, Aldo Croquer, Héctor Guzmán, Zelinda Leao and Alberto Rodríguez-Ramírez 331

19. SPONSORING ORGANISATIONS, CORAL REEF PROGRAMMES AND MONITORING NETWORKS 349

SUGGESTED READING 359

LIST OF ACRONYMS 363

Countries, States and Territories

American Samoa	Chapter 12	199
Anguila	Chapter 17	315
Antigua and Barbuda	Chapter 17	315
Australia	Chapter 9	141
Bahamas	Chapter 15	261
Bahrain	Chapter 3	55
Bangladesh	Chapter 6	95
Barbados	Chapter 17	315
Belize	Chapter 16	287
Bermuda	Chapter 15	261
Bonaire	Chapter 17	315
Brazil	Chapter 18	331
British Virgin Islands	Chapter 17	315
Cambodia	Chapter 7	117
Cayman Islands	Chapter 15	261
Chagos	Chapter 6	95
China	Chapter 8	131
Colombia	Chapter 18	331
Comores	Chapter 5	77
Cook Islands	Chapter 11	181
Costa Rica	Chapter 18	331
Cuba	Chapter 15	261
Curaçao	Chapter 17	315
Djibouti	Chapter 2	35
Dominica	Chapter 17	315
Dominican Republic	Chapter 15	261
Egypt	Chapter 2	35
El Salvador	Chapter 16	287
Eritrea	Chapter 2	35
Federated States of Micronesia	Chapter 2	35
Fiji	Chapter 10	159
French Polynesia	Chapter 11	181
Grenada	Chapter 17	315
Guadeloupe	Chapter 17	315
Guam	Chapter 12	199
Guatemala	Chapter 16	287

Haiti	Chapter 15	261
Hawaii	Chapter 13	219
Honduras	Chapter 16	287
Iran	Chapter 3	55
India	Chapter 6	95
Indonesia	Chapter 7	117
Israel	Chapter 2	35
Jamaica	Chapter 15	261
Japan	Chapter 8	131
Jordan	Chapter 2	35
Kenya	Chapter 5	77
Kiribati	Chapter 11	181
Kuwait	Chapter 3	55
Madagascar	Chapter 5	77
Maldives	Chapter 6	95
Malaysia	Chapter 7	117
Martinique	Chapter 17	315
Mauritius	Chapter 5	77
Mexico	Chapter 16	287
Montserrat	Chapter 17	315
Mozambique	Chapter 4	65
Nauru	Chapter 10	159
New Caledonia	Chapter 10	159
Nicaragua	Chapter 16	287
Niue	Chapter 11	181
Northern Marianas	Chapter 12	199
Oman	Chapter 3	55
Pakistan	Chapter 6	95
Palau	Chapter 12	199
Panama	Chapter 18	331
Papua New Guinea	Chapter 9	141
Philippines	Chapter 7	117
Puerto Rico	Chapter 14	239
Qatar	Chapter 3	55
Reunion	Chapter 5	77
Samoa	Chapter 10	159
Saudi Arabia	Chap 2 & 3	35, 55
Seychelles	Chapter 5	77
Singapore	Chapter 7	117
Solomon Islands	Chapter 10	159
Somalia	Chapter 2	35
South Africa	Chapter 4	65
Sri Lanka	Chapter 6	95
St. Kitts and Nevis	Chapter 17	315
St. Lucia	Chapter 17	315
St. Vincents and Grenadines	Chapter 17	315
Sudan	Chapter 2	35

Taiwan	Chapter 8	131
Tanzania	Chapter 4	65
Thailand	Chapter 7	117
Trinidad and Tobago	Chapter 17	315
Tokelau	Chapter 11	181
Tonga	Chapter 11	181
Turks and Caicos	Chapter 15	261
Tuvalu	Chapter 10	159
United Arab Emirates	Chapter 3	55
USA (Florida, Gulf of Mexico)	Chapter 14	239
US Virgin Islands	Chapter 14	239
Vanuatu	Chapter 10	159
Venezuela	Chapter 18	331
Vietnam	Chapter 7	117
Wallis and Futuna	Chapter 11	181
Yemen	Chapter 2	35

Acknowledgements

This book results from the hard work of many people monitoring coral reefs, assembling data, writing reports, and running projects. The process started with requests in 1999 to produce National status of coral reefs and monitoring reports – about 75 reports were produced which form the basis for the regional summaries in this global report. The national and regional authors are specifically thanked (their names are given in the chapters to which they contributed). Many of these authors are also acknowledged as national or regional coordinators of GCRMN Nodes throughout the world. I also wish to thank Meriwether Wilson for her careful editorial contributions in the final phases of preparing this global report.

The concept for and driving force behind the GCRMN comes through the International Coral Reef Initiative (ICRI) and its members who have provided considerable moral and financial support. Special thanks go to those running the ICRI Secretariat from France: Bernard Salvat, Francis Staub and Genevieve Verbrugge.

The GCRMN is now a flourishing network of regional nodes and regional partnerships involving several hundred people, some working under other banners, such as Reef Check, CORDIO and partners like ReefBase, WCMC, Reefs @ Risk, CARICOMP and AGRRA. A special thanks goes to Gregor Hodgson and his Reef Check volunteers for providing considerable data for this report and stimulating the global network to spread its coverage to remote parts of the world.

The Co-sponsors of the GCRMN have provided substantial assistance, advice and support. The Intergovernmental Oceanographic Commission of UNESCO, the United Nations Environment Programme (UNEP), IUCN the World Conservation Union, the World Bank, AIMS and ICLARM come together on an opportunistic basis or provide advice electronically as the GCRMN Management Group, and all of this is coordinated through Ned Cyr (and previously George Grice) of IOC-UNESCO and I owe them special thanks.

The GCRM primarily operates though an evolving web and partnerships of Regional Nodes. The Department for International Development (DFID) of the UK Government, especially John Tarbit and Chris Price, have actively supported the South Asia Node and assisted with contacts in the Caribbean. The Swedish International Development Agency has provided considerable support to the Eastern African Node and to the CORDIO programme, with Olof Linden and Petra Lundgren as the key links. Other financial support and advice has come from the World Bank, directed through Marea Hatziolos and Andy Hooten, and the

Global Environment Facility. The Government of France is thanked for assisting with Nodes in the Pacific and Indian Oceans and for global advice. Additional support has come from the David and Lucile Packard Foundation, and the United Nations Foundation.

Existing regional programmes throughout the world have provided essential support to help catalyse the 'Node' process and ensure linkages across other relevant national and regional programmes. The GCRMN is grateful for the considerable support from the regional offices of the UNEP network and similar agencies coordinated via Agneta Nilsson; specifically the Caribbean Environment Programme of UNEP in Jamaica, the South Pacific Regional Environment Programme (SPREP) in Samoa, the East Asian Seas Regional Coordinating Unit of UNEP in Thailand, South Asian Cooperative Environment Programme (SACEP) in Sri Lanka, the Indian Ocean Commission in Mauritius, and PERSGA, the Programme for the Environment in the Red Sea and Gulf of Aden in Saudi Arabia. UNEP is now catalysing several Nodes in the Caribbean with financial and technical assistance.

In addition to political and institutional support noted above, there are many sponsors throughout the world providing direct financial support for the GCRMN and Reef Check networks. A special mention is for the support from the Government of the USA, without which there would be no global coordination and no global status report in 1998 or 2000. I give special thanks to the US Department of State (through Jamie Reaser, Peter Thomas and Brooks Yeager), and the National Oceanic and Atmospheric Administration (via Arthur Paterson and Charles Ehler) which have been the major supporters of the GCRMN and assisted funding central coordination, in association with the Australian Institute of Marine Science. These US agencies are partners in the US Coral Reef Task Force, which is continuing the lead role of the USA in actions across many fields to conserve coral reefs.

About 30 leading scientists and resource managers assist voluntarily as members of the GCRMN Scientific and Technical Advisory Committee, and it is Edgardo Gomez who coordinates their input into the network. Special thanks are due to them and to Ed for his valuable advice and support.

The GCRMN benefits greatly by being associated with the Australian Institute of Marine Science and the staff provide a wide range of day to day assistance from funds management, computer support, and library services, in addition to assistance with ongoing advice on coral reef science, monitoring and database management. I owe all staff at AIMS special thanks. Finally this book was prepared under an extremely tight schedule and would not been realised without the always professional and friendly Science Communication team at AIMS, Steve Clarke, Wendy Ellery and Liz Tynan – thanks team.

Funds to print this book have come from the Government of the USA (Department of State and NOAA), SIDA of Sweden, the World Bank, the Great Barrier Reef Research Foundation and ICLARM (The World Fish Center).

INTRODUCTION

We are pleased to support the Second Edition of the Status of Coral Reefs of the World, particularly as the target audiences are decision makers, major donors and national and international agencies who are requested to take urgent action to conserve these valuable resources. It is also targeted at the informed public, who continually put pressure on decision makers to act and devote resources for reef conservation and management. The report was written to provide an overview, without the use of scientific jargon, of the status of coral reefs and causes of reef decline, with recommendations for action by national authorities. The chapters are expanded executive summaries of the National and Regional reports that form the feedstock for this report. We refer the reader to these sources listed after each chapter for the data that support the statements made in this summary report.

This second edition of the Status of Coral Reefs of the World is a major expansion and improvement on the first edition written in 1998. The improvements reflect encouraging development and maturation of the Global Coral Reef Monitoring Network (GCRMN) with National Status of Coral Reef Reports submitted by 97 authors from 86 countries or states. Many of these reports constitute baseline information for large areas of the world's coral reefs, and are often the first summaries written by countries on the status of their reefs. When this process is repeated in 2002, these countries will be asked to look at the 2000 report and assess the progress of efforts to manage reef resources. Similar success was evident when many Pacific countries presented their first Status of Reefs Reports at the International Coral Reef Initiative (ICRI) regional Pacific meeting in Noumea in May, 2000.

The 1998 report was released at the ICRI International Tropical Marine Ecosystems Management Symposium (ITMEMS) in Townsville in November, 1998. Material for that report arose out of papers presented at the 8th International Coral Reef Symposium in Panama City in June 1996 when leading coral reef scientists and managers were invited to present summaries at a reef status symposium organised by Clive Wilkinson and Bernard Salvat. Those summaries were updated for the GCRMN status report of 1998, which was the first review of the status of coral reefs status the ground breaking 3 volume series edited by Susan Wells for UNEP and the IUCN in 1988 (references at the back).

An examination of the 1998 and this report shows that the regions and groupings of countries have changed, which reflects the development of Nodes of the GCRMN. These Nodes are groupings of countries around the world which cooperate to monitor their reefs, usually with the assistance of one country with the capacity to assist in training, monitoring and data analysis. The first Node, formed in South Asia for India, Maldives and Sri Lanka, has been functioning since 1997 under financial assistance from DFID UK. Another Node was formed in the Indian Ocean for the island states of Comoros, Madagascar, Mauritius, Reunion and Seychelles through the Indian Ocean Commission, with initial support from the European Union and recently from the Global Environment Facility. The countries of Eastern Africa are supported under the CORDIO programme (COral Reef Degredation in Indian Ocean) including Kenya, Mozambique, South Africa, Tanzania and eventually Somalia with financial and logistical assistance coming from Sweden (SIDA) and the World Bank.

Five Countries of southeast Asia have all requested to remain as independent country nodes (Indonesia, Malaysia, Philippines, Singapore and Thailand), whereas mechanisms are being sought to provide assistance for Burma/Myanmar, Cambodia and Vietnam as a Node. The Government of Japan has formed a Node for East Asian countries to assist China, Korea and Taiwan from the International Coral Reef Research and Monitoring Center at Ishigaki Island, southern Japan.

Nodes in the Pacific are consolidating. The IOI Pacific Islands Node assists Nauru, New Caledonia, Samoa, Solomon Islands, Tuvalu and Vanuatu from the University of the South Pacific in Fiji with support from the International Ocean Institute. The Polynesia Mana Node includes the Cook Islands, Kiribati, Niue, Tokelau, Tonga, and Wallis and Futuna with coordination provided by the University of Perpignan marine station on French Polynesia and financial support from France. The Micronesian countries (Federated States of Micronesia, Guam, Marshall Islands, Northern Marianas and Palau) under the MAREPAC consortium is functioning out of the Palau International Coral Reef Center. The Hawaiian Islands and other US territories in the Pacific are members of the US Coral Reef Task Force and are coordinated out of Hololulu.

Nodes in the Arabian Region remain to be resolved, but they were the topic of a meeting in Saudi Arabia in February, 2000. There are two coordinating bodies – PERSGA (Programme for the Environment in the Red Sea and Gulf of Aden) and ROPME (Regional Organisation for the Protection of the Marine Environment) which may take eventual responsibility for coordination.

The US Coral Reef Task Force is coordinating and financing US interests in the Caribbean (Florida and other southern states, Puerto Rico, and US Virgin Islands). The World Bank coordinates the Mesoamerican Barrier Reef Project, which is effectively the GCRMN Node for Belize, Guatemala, Honduras and Mexico for reefs on the Caribbean side. The Government of Colombia is assisting in the formation of a Node for Brazil, Costa Rica, Panama and Venezuela with Ecuador likely to join in the future. UNEP in the Caribbean has provided start up assistance. Many of the remaining countries and states are being assisted through centres of the CARICOMP network, including their database node in Jamaica at the University of the West Indies. DFID UK is assisting in coordination of many of the small islands states, likewise France is assisting with her territories as is the Netherlands.

Better understanding of the oceans and their resources to facilitate improved ocean management and sustainable development will not be possible without the establishment of routine ocean observing systems. This is the premise behind the Global Ocean Observing System which is co-sponsored by IOC (Intergovernmental Oceanographic Commission), WMO (World Meteorological Organisation) UNEP (United Nations

Environment Programme) and IUCN (The World Conservation Union), with assistance of FAO (Food and Agriculture Organisation). The GCRMN is a critical component of this system as monitoring of coral reefs is particularly important because of the apparent links between global climate change and coral bleaching illustrated in this report. Coral reefs may yet prove to be the first major marine ecosystem to show significant impacts from global climate change.

A major focus of UNEP is to provide accurate and accessible information on the state of our environment for informed decision-making, adaptive management action and policy-setting. The partnerships and active networks, collaborating through the Regional Seas Programme of UNEP, are working with the GCRMN to promote and coordinate monitoring of coral reefs throughout the tropics. Assessments of social, cultural and economic values of reefs serve to highlight the close relationship between reefs and coastal communities. UNEP is a co-sponsor of the GCRMN, and a member of ICRI since it was launched in 1994. UNEP is a partner in the International Coral Reef Action Network (ICRAN), an operational unit under the ICRI umbrella, with the goal of catalysing concerted action for protection of coral reef resources.

Monitoring by non-governmental organisations is a major component of the GCRMN, both using advanced methods, or the more basic ones of Reef Check. IUCN interacts with 900 members in nearly 140 countries and some of these are active in coral reef monitoring and conservation. These linkages are provided to the GCRMN to broaden its influence and encourage and assist societies throughout the world to conserve the integrity and diversity of coral reefs. Another critical process is addressing the links between coral reef status and global climate change via interactions with the Convention on Biological Diversity, the United Nations Framework Convention on Climate Change. IUCN helps provide these links by being an active member of the International Coral Reef Initiative (ICRI) and a co-sponsor of the GCRMN.

Healthy coral reefs are critical to the livelihoods and cultures of millions of people in tropical coastal environments, as well as forming part of the crucial life support system of the biosphere. A major theme of the World Bank is alleviating poverty, therefore it is essential to understand the fundamental nature of factors that determine the productivity, diversity, and resilience of coral reefs. Monitoring and assessment are keys to such understanding. Therefore the World Bank is partnering with the GCRMN in four of the five coral reef regions of the world to promote monitoring and information for decision making as part of its expanding program of national and regional initiatives to improve the management of coral reefs. The contribution of the GCRMN via this latest edition of the Status of Coral Reefs will be a vital handbook for managers and serve as an important benchmark for our efforts to reverse degradation of coral reefs and measure progress toward their conservation around the world. The GCRMN also makes a significant contribution by bringing people and organisations at all levels into the process.

Unfortunately, many coral reefs are overfished and so damaged that they no longer provide sufficient income and food for people in developing countries. ICLARM is examining how marine protected areas can serve as fisheries reserves, and how to restore or increase productivity. The potential methods include restocking wild populations, developing sustainable low impact aquaculture of valuable species low on the food chain, and training in coastal management. ICLARM places great importance on regular monitoring and the dissemination of status reports, and has developed ReefBase as a repository for data and information on coral reefs as a key link for the GCRMN. ICLARM relies on information generated by the GCRMN to identify reefs that are not yielding potential harvests, and the causes for declines. ICLARM will continue to work with the partners in expanding global monitoring and making the data accessible through ReefBase.

Environmental monitoring is fundamental to informed management and policy making, and it also functions to raise public consciousness on conservation. The research priorities at the Australian Institute of Marine Science recognise the importance of monitoring the status of reefs in Australia and the region for two decades, and significant monitoring programmes on the Great Barrier Reef and on Western Australian reefs are maintained. Also considerable scientific effort has gone into developing and formalising monitoring techniques, with the Survey Manual for Tropical Marine Resources (English et al. 1997 in suggested reading) as a major product that the GCRMN has adopted as a recommended set of methods. Many coral reefs of the world are under greater pressures than those in Australia, and the GCRMN enhances the value of AIMS efforts by putting this monitoring in a global context and broadening the application methods to reefs worldwide.

Since the last report, the GCRMN and Reef Check have formed a strong strategic partnership, with the GCRMN focussing on implementing monitoring in nations and states, whereas Reef Check assists communities and volunteers establish monitoring projects. This is facilitated because similar methods are used and the protocols recommend that training start with Reef Check methods and then progress to more detailed GCRMN methods.

This report demonstrates that there has been progress in implementing global monitoring networks and providing sound information on the status of reefs of the world. But a close examination will show that there are large regions of the world where the baseline data are insufficient to determine the status of many coral reefs. This was particularly evident after 1997-98 when it was not possible to determine the impacts of a major bleaching event because of a lack of such baseline data. Monitoring now has a new role. In places where the reefs were laid bare and lost much of the previous coral cover, there is a need to first assess what are the probable losses of coral cover, and then determine the amount and extent of any recovery, either through the settlement of new corals or regrowth of the few survivors. A lack of recovery will indicate to reef managers that factors preventing recovery must be removed, or rehabilitation must be considered.

Coral reef monitoring should never exist on its own – 'monitoring for monitoring sake'. It should always be a component of coral reef management, providing the data to determine suitable areas for protection and then assist management by assessing effectiveness of management decisions. This will enable resource managers to continually adjust regulations to ensure maximum sustainability without excessive interference in the lives of local communities.

Therefore, there is a need to enhance this network in those areas where monitoring is progressing, and introduce monitoring into areas without effective coverage. Moreover, it is essential that the process of monitoring be made sustainable with adequate financial and personnel resources. The monitoring programmes should soon become routine tasks for governments and communities to assess their coral reefs as part of wider state of the environment monitoring. This will require continued coordination, which must come from the regions to be fully effective and sustainable.

Patricio Bernal
Executive Secretary, Intergovernmental Oceanographic Commission

Klaus Töpfer
United Nations Under-Secretary General and Executive Director,
United Nations Environment Programme

John Waugh
Marine Programme Coordinator, IUCN – The World Conservation Union

Robert Watson
Chief Scientist, The World Bank

John Bell
Acting Director, Australian Institute of Marine Science

Meryl Williams,
Director General, ICLARM – The World Fish Center

Executive Summary

Clive Wilkinson

The Changing Agenda

A major conclusion in the 'Status of Coral Reefs of the World: 1998' report was of two simultaneous, but contradictory, global phenomena occurring on coral reefs: there is an increasing rate of degradation of many coral reefs due to direct human activities; while there is greatly enhanced awareness by people everywhere about the problems facing reefs and actions are being catalysed to conserve them.

The paradox continues and coral reefs continue to deteriorate in all areas where human activities are concentrated, notably along the coast of eastern Africa, all of continental South Asia, throughout Southeast and East Asia and across the wider Caribbean region. In addition to the human activities and associated reef degradation observed over the past four decades, an increase in sea surface temperatures associated with the major El Niño–La Niña change in climate in 1997-98 resulted in extensive coral bleaching and mortality over large parts of the Indian Ocean and Southeast and East Asia. On some reefs, there were mortality levels greater than 90% leaving these reefs almost bare of corals and with early indications of major shifts in the population structures of these reefs. The critical feature of the 1998 bleaching events was that areas were struck indiscriminately, irrespective of the status of reef health; impacts were equally severe on pristine, remote reefs as on reefs already under major human stresses. It was also very disappointing for those people managing reefs from human impacts to observe their reefs die from stress factors outside their control. We are being forced to recognise that human reliance on burning fossil fuels and clearing rainforests is leading to changes in global climate, and that events like the extensive coral mortality in 1998 may occur more frequently and devastatingly in the future; and not just to coral reefs.

It is pertinent to revisit the report of the UNEP-IOC-ASPEI-IUCN Global Task Team on the Implications of Global Climate Change on Coral Reefs compiled by Clive Wilkinson and Bob Buddemeier from the deliberations of a 15 person team of experts in 1994. The report concluded that the major problems for coral reefs were direct anthropogenic stresses of nutrient pollution, excessive sediments and over-exploitation acting at many local sites near concentrations of people, and that global climate change was not yet an issue. To quote "Our major finding is that human pressures pose a far greater immediate threat to coral reefs than climate change, which may only threaten reefs in the distant future"; and "Climate change by itself is unlikely to eliminate coral reefs ...". Yet within 4 years of that report, both authors have become convinced that evidence points to global climate change posing an equal or even greater threat to coral reefs than direct anthropogenic impacts. Coral bleaching and mortality due to elevated sea surface temperatures is one of several major climate-related threats to coral reefs. Possibly the most alarming is evidence that increasing concentrations of CO_2 in seawater will reduce the rate of coral calcification, thus slowing growth and increasing their fragility and capacity to build reefs. A major concern is that the location and timing of the next bleaching event is unpredictable, and climate induced bleaching can devastate pristine coral reefs as well as reefs which are effectively managed.

Thus, the agenda for coral reef conservation changed radically after the 1997-98 coral mortality events. Until then, the major action to conserve coral reefs was to reduce direct human impacts of land-based pollution and sediment releases, and over-exploitation, by establishing Marine Protected Areas (MPAs). Now the fight has shifted to two fronts: the need to increase management to abate direct anthropogenic impacts at all scales; and action to study the impacts of global climate change on coral reefs and reduce global emissions of greenhouse gases.

CORAL BLEACHING AND MORTALITY IN 1997-98

There is both encouraging and dismal news in the following summary reports on the coral bleaching event of 1997-98 (and early 2000). The mortality after the bleaching was varied in scope and scale with large parts of the Indian Ocean, Southeast Asia and the far western Pacific being the most dramatically affected. In the wider Caribbean and parts of the Great Barrier Reef, mortality was minimal even after extensive bleaching, as many severely bleached reefs recovered almost fully. Over vast areas of the Pacific, there was no bleaching (summarised in Chapter 1).

There are now encouraging reports of new recruitment on coral reefs in Eastern Africa, the Seychelles, the Maldives and Palau indicating that sufficient parent corals survived to provide some larvae, if not for the 1999 season, then for the 2000 spawning season. However the down side is that many of the major reef building corals, such as *Acropora* and *Pocillopora* are either recruiting in very low numbers or not at all. This will mean that many of the affected reefs will be dominated by slow growing massive species that provide poor habitat for fishes and are less attractive for diving tourists. Many reefs that were dominated by large table and branching corals, now have a smaller, low profile corals.

It may be several years before we can state that the reefs will recover, or whether there will be local losses of species, including the extinction of rare endemic species. Reef recovery is dependent on few or no repeats of the extreme event of 1997-98, at least not within the next 20 to 50 years which will be the time required for many of the reefs to recover to structures resembling those before the bleaching. Recovery, in many cases, will also depend on the reduction of human pressures through the application of sound management.

Coral Bleaching in 2000 - Early Reports

A worrying sign that severe bleaching may become more regular was seen in parts of the southwest Pacific between February and April 2000. Satellite images from the National Oceanic and Atmospheric Administration of USA showed a developing 'HotSpot' of increased seawater temperatures in the region in February and soon after severe bleaching often involving around 80% of the corals was seen in Fiji, and the Solomon Islands. Fortunately the 'HotSpot' dissipated rapidly and many corals appear to have recovered, although there were losses of up to 90% of corals in parts of Fiji. These affected areas are being followed by GCRMN and Reef Check teams. Coral bleaching has also been reported from some locations in the Western Atlantic.

Coral Reef Diseases in the Tropical Americas

In the last 20 years, 2 of the 3 major reef building corals in the Caribbean have succumbed to white-band disease, and are now scarce. An important algal grazing sea-urchin also suffered mass mortality, facilitating the overgrowth of reefs by macroalgae. Now, other diseases are appearing and massive corals are declining. We know very little about these apparently natural factors, but the worrying concept is that the increased frequency may be linked to anthropogenic disturbance.

Global Progress Towards Conserving Coral Reefs

There has been a major expansion in international efforts to monitor, research, manage and conserve coral reefs during the last few years. The International Coral Reef Initiative (ICRI), formed in 1994, has continued to expand with the major operational unit, the Global Coral Reef Monitoring Network (GCRMN) also growing. Many coral reef nations participated in two ICRI global meetings (1995 and ITMEMS 1998) to review ICRI implementation. They determined a set of priority steps for coral reef conservation: the Renewed Call to Action, addressed to the world's governments, following the 1995 Call to Action and Framework for Action (www.environnement.gouv.fr/icri). During the last 2 years, the Secretariat of ICRI has operated through the environment department of the Government of France and next year, coordination of ICRI will pass to the Philippines in partnership with Sweden. Two new ICRI operational units are been formed. The International Coral Reef Information Network (ICRIN) was established in 1999 to raise awareness about coral reefs, particularly targeting senior decision-makers. Another network being formed is the International Coral Reef Action Network (ICRAN) which has initial funding from the UN Foundation to establish demonstration sites around the world to showcase successful MPA conservation projects and serve as major training facilities.

President Bill Clinton introduced Executive Order 13089 in June 1998 urging all arms of the Government of the USA to do their utmost to map, document, research and conserve the coral reefs under US. jurisdiction as well as assist other international agencies, including ICRI and the GCRMN. This Order established the US Coral Reef Task Force (USCRTF) which is comprised of many government agencies committed to carrying out this order with increased budgetary allocations. This has resulted in a massive increase in US sponsored coral reef action, including assistance to the GCRMN.

The Governments of the UK, through the Department for International Development (DFID), and Sweden, through the Swedish International Development Agency (SIDA) have made major contributions to reef conservation. DFID continues to support monitoring and conservation activities in South Asia and Sweden, in conjunction with the World Bank, formed CORDIO (COral Reef Degradation in the Indian Ocean) as a direct response to the massive bleaching in 1998, and it operates to assess the damage and seek corrective measures for all countries of the Indian Ocean. The World Bank also supports coral reef projects around the world worth many millions of dollars. The Global Environment Facility has provided funds for the Indian Ocean Commission to facilitate monitoring in the southern Indian Ocean over the next 3 years. France is supporting activities in both the Pacific and Indian Oceans.

The volunteer network of reef monitors, Reef Check, continues to expand and has assessed reefs in 40 countries during 1999. The AGRRA network, operating out of the University of Miami has undertaken large scale assessment of coral reefs in the Caribbean, and CARICOMP continues to assess reefs and other coastal systems there and maintains a database in Jamaica. The global database, ReefBase within ICLARM continues to assemble data on coral reefs and assist the Reefs at Risk project of the World Resources Institute and the World Conservation Monitoring Network in the production of regional assessments of the risks facing coral reefs and an atlas of reefs and resources around the world. The DIVERSITAS program of UNESCO will have a focus project in 2001, the International Biodiversity Observation Year, on assessing the impacts of the 1998 bleaching event to assess the probability of recovery of reef biodiversity.

This is the second global report of the GCRMN. The process for writing this report varies from the last, in that it is primarily derived from national status reports written by governments and nationals from 86 countries. The majority of these countries are now constituted into GCRMN Nodes (regional groupings of countries forming networks for monitoring, usually with a country with more capacity acting as the coordinator). These Nodes and other regional groupings combined these national reports into a series of regional reports that form the chapters that follow. The original national and regional reports contain the data and other information that form the basis for this report.

STATUS OF CORAL REEFS OF THE WORLD

Arabian Region

Nearshore reefs in the Arabian/Persian Gulf were virtually obliterated by severe coral bleaching in 1996 and again in 1998, so that most do not have any living corals. These reefs had survived many years of development of the petroleum industry, yet were virtually obliterated by a climate induced impact. Offshore reefs were less affected and still have some healthy corals. Major coral bleaching is now occurring in the Northern Gulf in September 2000. Red Sea reefs remain predominantly healthy with few localised anthropogenic stresses although there was some bleaching damage during 1998 to the reefs in the southern sector. Monitoring and management capacity is low, but is improving with plans established for regional monitoring and management networks. Coordinated action is particularly urgent as this region is developing rapidly through growth in tourism and shipping, especially associated with the petroleum industry. Most countries lack the capacity or legal structures to manage these developments.

South Asia
Most reefs in this region were severely affected and some were almost totally devastated during the extreme climate events of 1998. Reefs on the Maldives, Sri Lanka and parts of western India lost much of their coral cover, whereas those on some parts of the east coast of Sri Lanka and most of the Andaman Islands were not seriously affected. These losses have added to major anthropogenic damage to reefs off the mainland of India and Sri Lanka, particularly from coral mining, over-fishing and pollution. These pressures will continue as populations grow. There is an active regional network to assess coral reef health and socioeconomic aspects of coral reef use, which is raising awareness about the need to establish more protected areas.

Eastern Africa
All reefs in this region are close to the large continental land mass of Africa, and therefore experience all the impacts from land: significant levels of sediment runoff; nutrient pollution; and definite over-exploitation of reef resources from growing populations. Developing national programmes to assess and manage coral reefs for both sustainable fishing by local communities and the rapidly expanding tourism industry were given a major sense of urgency in 1998. There was massive coral bleaching and mortality associated with the El Niño climate switch with some large areas losing up to 80% of their live corals (especially parts of Kenya and Tanzania). There is now new coral recruitment, which is an encouraging sign for recovery, although it is probable that the reefs will have a different composition in the immediate term. A call has been made for improved regional cooperation to build national capacity and to manage transboundary problems.

Southern Indian Ocean
The reefs of the northern part of the region suffered severe damage during 1998 as a result of the El Niño bleaching, with losses of 80 to 90% of the corals in parts of the Comoros and Seychelles. Reefs around the southern islands were only slightly affected, but the reefs of Madagascar continue to be under very high human pressures. Prior to the recent formation of regional monitoring, there was little capacity and limited baseline data. Now there are 44 permanent monitoring sites in these states, however the capacity for the management of coral reefs remains well behind many other parts of the world. A substantial Node of the GCRMN was formed with European funding, and has received substantial support from the Global Environment Facility to expand in the future.

Southeast Asia
This region has both the largest area of coral reefs in the world with the highest biodiversity, while the reefs are probably under the greatest threats from anthropogenic activities. Simultaneously there is rapidly expanding capacity to monitor, research and manage the coral reef resources, which must accelerate as both populations and the economies expand and put greater demand on reef resources. The extensive network of reef monitoring is documenting an inexorable decline in the status of the coral reefs, some of which were also damaged by the 1998 bleaching event (but not as severely as those in the Indian Ocean). The area is still the centre of the reef live fish trade, worth over US$1billion per year, with virtually all reefs being hit by roving cyanide and blast fishers. Co-management with local communities is emerging as the best method to implement sustainable management of these high diversity resources.

East Asia
The reefs of southern Japan and Taiwan were severely affected by coral bleaching and mortality during the second half of 1998, coincident with the La Niña climate switch. Bleaching ended when the first typhoon of the season came in late September. There are many reports of coral losses of 30-60%, with some losses as high as 80-90%. Some localised extinctions of prominent corals have been reported. China has established one MPA in Hainan and has commenced coral reef monitoring. The Japanese government has established an international coral reef centre on Ishigaki Island (the southern islands of Okinawa) to facilitate coral reef conservation in the region and assist the GCRMN with monitoring.

Australia and Papua New Guinea
Australian reefs continue to be the best managed in the world with the lowest levels of anthropogenic impacts of any continental reefs. In general, they are in good to excellent condition, however there are problems on the Great Barrier Reef (GBR) with sediment and nutrient runoff from the lands, particularly from over-grazed range lands after long dry spells, and increasing professional and recreational fishing. Coral bleaching was intense, but only on inshore reefs and the bulk of the GBR was not affected, although crown-of-thorns starfish are attacking the offshore reefs in 2000. Management of the GBR is also assisted by the largest capacity of any region for coral reef research, monitoring, teaching and management, which flows onto strong public awareness and acceptance of and demand for effective reef conservation. Similar levels of research, monitoring and conservation initiatives are being implemented for reefs off Western Australia. During 1998 there were markedly different affects of the coral bleaching; Scott Reef way offshore was nearly devastated, whereas the nearby Ashmore reef escaped damage (See Chapter 1). Similarly most of the reefs of PNG are generally in very good condition, except for localised areas of damage from excessive logging and increasing levels of exploitation on nearshore reefs. These reefs have very high biodiversity and a wide range of structures, however, there is virtually no management capacity or commitment for management in government, and only weak capacity to assess and monitor these resources.

Micronesia
There is now greater coordination amongst Micronesian countries with the formation of a consortium of major coral reef institutes and agencies, and strong recognition from the United States Government. This is resulting in enhanced monitoring and marine resource management training in these countries. The reefs are predominantly in good to excellent condition, although there is damage resulting from development activities on the high islands and over-fishing around centres of population. Some MPAs are protecting fish populations, but there are insufficient on the islands and poor enforcement of existing reserves. Most of the region escaped damage from the 1997-98 bleaching event, however, there were unprecedented losses of corals around Palau. Reports coming through in mid-2000 are of some recovery and encouraging levels of new coral recruitment. Tourism is the major reef activity on the Northern Marianas, Guam and Palau, and is developing in the Federated States of Micronesia.

Southwest Pacific
While this region escaped major bleaching during 1997-98, it was damaged by relatively severe coral bleaching between February and April 2000 with extensive mortality in some

parts of Fiji and the Solomon Islands. Human impacts on these reefs are steadily increasing but still concentrated at a few sites per country, mainly around the capital cities and in lagoons. However, most reefs are still healthy with few real problems, except for some over-fishing for subsistence and small-scale commercial activities. Most reefs are protected by the influence of the Pacific Ocean, which both dilutes nutrient and toxic pollution and removes sediments. Recent political instability in the Solomon Islands and Fiji, has delayed conservation efforts for coral reefs e.g. the special marine science unit of the University of the South Pacific, and the ICLARM aquaculture centre have moved off the Solomon Islands awaiting developments. An active coral reef monitoring network is being established at the University of the South Pacific in Suva and in smaller campuses in other countries, but most nations lack the capacity to implement long-term monitoring. Most nations have not declared key areas as marine protected areas, and have little capacity for integrated coral reef management.

Southeast Pacific

This region also escaped major climate related coral bleaching in 1997-98 and 2000, such that most of the reefs remain healthy with few anthropogenic threats. Similar to other areas in the Pacific, most human impacts are concentrated around the major centres of population and within enclosed lagoons. Considerable shoreline modification on these islands has resulted in damage to the nearshore reefs, but the outer reefs facing the ocean show no real impacts. Fishing pressures are increasing, threatening some coral reefs, and there are increasing conflicts between fishers and tourist operators. Integrated management planning is needed to improve the design, location and management of tourist resorts and to manage the rapid growth of the black pearl culture in some lagoons of the Cook Islands and French Polynesia. While there has been active coral reef monitoring in French Polynesia, such efforts are only beginning in the other countries. The whole region will benefit from a programme of training and monitoring at the marine station on Moorea with assistance from France.

Northeast (American) Pacific

This region contains the bulk of coral reefs under US control with 2 distinct regions: the main islands of Hawaii; and scattered islands and atolls throughout the Pacific. Strong population and economic growth in the Hawaiian islands is resulting in considerable local damage to reefs around the major centres and tourist operations, whereas all reefs are subjected to high and increasing fishing pressures from both indigenous and tourist fishers. Collecting for the aquarium trade has cause major depletion of stocks of high value species. There are currently active measures to monitor the reefs throughout the main islands, and plans are being developed for considerably improved management of coral reef resources, however, local development pressures will make it difficult to achieve a target of 20% of areas under protection. In contrast, the scattered islands are under minimal anthropogenic pressures and none experienced climate related bleaching in 1998. The prognosis for these reefs into the immediate future is good, and it would be considerably enhanced by a greater awareness of sustainable management by policy makers.

The American Caribbean

The increase in activity catalysed by the formation of the US Coral Reef Task Force means that awareness has been raised in the islands of Puerto Rico and the US Virgin Islands to overcome serious problems of over-fishing and damage to coastal nursery areas of mangrove forests and seagrass beds. Similarly, there is a greater urgency to conserve reefs off the mainland of the USA, with an ambitious target of having 20% of these reef resources managed as no-take reserves within the next few decades. The major threats to reefs off Florida are pollution from massive agriculture in Florida and the growing populations of people wanting to enjoy coral reefs, and over-fishing of key target species. There is now a concerted programme to improve research and management on these reefs and all others under US jurisdiction.

Northern Caribbean and Western Atlantic

There is similar reef deterioration in all countries, with over-fishing and direct pollution being the most critical problem for reefs that are close to land (Jamaica, Haiti and the Dominican Republic). Whereas, exploitation is much less over the broad shelves and fish stocks are in better shape (Cuba, Bahamas, Turks and Caicos Islands). Finally, tourists on Bermuda and Cayman Islands are demanding healthy fish populations, which is resulting in reduced fishing pressures. Coral cover on most of the islands has dropped because white-band disease killed off the large stands of *Acropora* spp., and reefs close to land still show low cover e.g. coral cover in northern Jamaica dropped from 52% in the 70s to 3% in the early 90s, but is gradually recovering (currently 10-15% now). Bleaching in 1998 was severe in places, but there was little or no mortality. Much of the tourism development based on the coral reefs is poorly planned and results in sediment run-off and nutrient pollution, thereby damaging the reefs. While there is increasing awareness of the need for conservation and the establishment of Marine Protected Areas, there is little capacity or planning for management or enforcement of conservation laws.

Central America

Many of the reefs in this region escaped the damage that was occurring in the rest of the Caribbean until there was mass coral bleaching and mortality in 1995 and 1998, and extensive damage from the intense Hurricane Mitch, also in 1998. These events impacted heavily on reefs from the Mexican Yucatan to Nicaragua, causing losses in coral cover of 15-20% across the region with some losses as high as 75% in parts of Belize. Throughout large parts of the region there are intense fishing pressures (Honduras and Nicaragua, and Veracruz and Campeche in Mexico), and major damage to reefs from sediment runoff because of poor land-use and as a result of Hurricane Mitch. Capacity to monitor and manage coral reefs varies enormously in the region, from advanced to virtually non-existent. Now, countries of the region are cooperating to conserve and manage their reefs, and resolve cross-boundary problems. This has a real sense of urgency following recent damage to the reefs.

The Eastern Antilles

Many of these islands scattered over the eastern Caribbean face similar problems of over-exploitation, sedimentation and nutrient pollution damage to their narrow fringes of coral reefs. They also suffer from having limited capacity to conserve and manage the reef resources in the face of rising populations. Capacity is variable; some countries are well advanced in implementing community-based, or tourism-funded, management and monitoring, whereas others lag well behind. A common problem is a lack of trained staff to monitor reefs, hence the introduction of basic methods like Reef Check have helped expand the GCRMN network throughout the islands. This network is now assisting with reef management and monitoring via coordination from St. Lucia and assistance from the regional office of UNEP. Coral cover on some islands has dropped recently due to the passage of Hurricanes and coral bleaching e.g. on St. Lucia, cover declined from 50% to 25% at 3m depth and from 35% to 17% at 10m.

South America

Reefs in this region were seriously degraded throughout the 1980s and early 1990s due to a range of natural and anthropogenic stresses. Throughout this time the major natural stress has been repeated coral bleaching episodes with cumulative mortalities, however, there were only minor impacts during the 1997-98 bleaching event on both Pacific and Caribbean reefs. The major anthropogenic stresses are from increased sediment and nutrient pollution on the nearshore reefs because of deforestation, poor agricultural practices and diversion of rivers. Offshore reefs are being increasingly over-exploited for fisheries, coral rock and sand, resulting in distinct declines of coral cover and fish populations. There is strong capacity and apparent willingness to monitor reefs in the region, with the exception of Brazil, which is yet to establish a national monitoring programme. The countries are now cooperating within a regional node of the GCRMN to introduce monitoring in some countries and enhance it in others. As noted in other regions, obtaining funding for monitoring at both national and regional levels is a major challenge (especially in Columbia).

REEFS AS MODELS FOR ENVIRONMENTAL CONSERVATION

Where do we go from here? Coral reefs should be regarded as ideal models for the management and conservation of marine ecosystems, and possibly other ecosystems. Many coral reefs are degrading from stresses that can be managed, moreover reefs are often discrete entities with water barriers between them, and the land or other major ecosystems. In addition, many reefs are sufficiently remote from land to receive only marginal impacts from land based pollution and exploitation. Therefore many of those reefs that are being degraded by human activities would befit from holistic and integrated approaches to ecosystem management and conservation.

There is almost no other ecosystem that has the 'charismatic appeal' of coral reefs and they are clearly high on the public agenda as systems for conservation. It is difficult to pass a week without seeing a television documentary or an educational film on coral reefs. Importantly, there are no large economic or political concerns directly lobbying against coral reef conservation. In contrast, the massive tourism and transport industries are strongly in favour of coral reef conservation, because reef-based tourism is the fastest growing sector of the ecotourism and activity holiday market. Reef fisheries are generally subsistence or small scale commercial operations, except for cyanide fishing for the live fish trade which is conservatively estimated to be worth US$1 billion per year through Hong Kong. Major operators in the aquarium trade are actively working to make this 'fishing' industry sustainable, supported by most hobbyists who prefer 'green labelled' products.

Reefs are also strategically important. There are more than 20 countries in the United Nations with few natural resources other than coral reefs; and another 70 countries or states have substantial areas of ecologically and economically important coral reefs. These states exert considerable influence within the UN and also have guardianship over vast expanses of the oceans as Exclusive Economic Zones. Coral reefs are becoming major discussion topics in international forums such as the Conventions on Sustainable Development and Biological Diversity, as well as the Asia Pacific Economic Cooperation.

Thus there are powerful, positive indications that we should be able to succeed in conserving large areas of coral reefs and their contained heritage of biodiversity. Such success would provide models for the conservation of other ecosystems, for if we cannot succeed in conserving coral reefs, what other ecosystems will we be able to conserve?

CALLS FOR ASSISTANCE

Similar calls for assistance and recommendations for action came from many countries and regions. The specific recommendations are listed at the end of each chapter, but following is a summary of the major requests:

Monitoring efforts should be expanded in scope and scale, as there are many countries without the capacity in trained staff, logistics, and ongoing funding to monitor reefs. There are no data in this report for large areas of the world, some of which were severely damaged during the 1998 bleaching event. Therefore, there is a need for operational support to train staff in assessing reefs underwater, and archiving and analysing the data using databases. Moreover, there is a need to ensure that these people have guaranteed employment and are able to interact with colleagues in neighbouring countries; the GCRMN seeks to provide this coordination facility.

Monitoring of coral reefs should be conducted as a multi-scaled and multi-level process, with all scales and levels being equally important. Communities and volunteers are requested to monitor their 'home' reefs at a basic level to enhance involvement and ownership; governments require reef monitoring as a contribution to 'state of the environment' reporting and to provide information for management of these critical resources; and scientific monitoring is required as quality control for other monitoring and to answer specific management and research questions (e.g. oceanographic and terrestrial influences on coral reefs; synergistic coupling of anthropogenic and climate change impacts). Thus, greater cooperation and coordination is required among groups working on coral reefs to ensure that data and information are delivered in a timely manner to the world.

There is a need for greater involvement of communities in both monitoring and local management of coral reefs. Involvement in community based monitoring is a powerful tool for raising awareness both within the community and with decision makers; such involvement should lead to successful co-management of resources e.g. recently in Samoa.

Although small marine protected areas are often successful, there is a need to link them into networks to avoid the scene of devastation around a few oases of success. Alternatively, MPAs could be made larger to incorporate a multitude of uses and communities, with zoning ranging from fishing and tourism development, to total protection of fish stocks (the target suggested is 20% of the area as 'no take reserves'). MPAs at this scale can address catchment areas and trans-boundary coastal and marine influences, and accommodate industrial and tourism development along with traditional uses through different 'tenure' and zoning measures. Either way, there is a strong need for coordination and cooperation between operational and funding agencies to expand the area of reefs under protective management.

Coral reefs are generally sustainable, self-repairing systems that recover when conditions are suitable and there is an adequate supply of larvae. However, there may be instances when rehabilitation is warranted. Consideration should be given to applying practical and low-cost rehabilitation methods where recovery is not proceeding normally, but such methods must be effective at the scale of the damage, and not local 'band-aid' gimmicks.

The traditional rights and management practices, where they are still practised or remain in the memory, need to be recognised and incorporated into the codified laws of states and regions. This would be assisted by scientific validation and re-drafting of state laws to include co-management provisions over coastal and marine areas. Currently many valuable traditional practices for conservation of coral reef resources are being eroded under the combined force of codified state and international law. For example in Samoa, traditional rights and customs have been written into the fisheries regulations providing village communities, through their chiefs, the right to manage their coral reef resources and apply punishment to offenders who violate village areas and their proclaimed MPAs.

There is an ongoing need for the provision of legal advice and assistance as part of developing economic policy to ensure a balance between conservation and development. Many laws in most countries were developed during colonial times and focussed on sectoral rather than integrated management, and the short term rather than the long-term e.g. optimising harvests of fishery and forestry resources. Calls have been made to assist in redrafting national and state statutes to remove considerable multi-sectoral overlaps in jurisdiction over coastal resources and ensure sustainable use of these resources, including establishment, control and enforcement of MPAs.

There was strong general concern that global climate change had the potential to destroy the coral reefs of many countries of the world, which are making minuscule contributions to the emissions of greenhouse gases. All countries requested assistance in assessing the potential future impacts of GCC on their coral reefs as well as energy alternative programmes. Many made pleas to the larger developed countries to curb their emissions, so that their coral reefs and often the countries will survive into the future.

FUTURE PREDICTIONS FOR CORAL REEFS

Predictions were made in 1992 that in 10 to 20 years, another 30% of the world's coral reefs could be added to the 10% that were effectively destroyed, if urgent management action was not implemented. While these figures appeared alarmist, recent events show that they may be conservative. In the early 1990s, there was little organised monitoring to observe reef decline and to determine the causal relationships. Estimates suggesting that 30% of the world's reefs were not under anthropogenic stress were an underestimate, as large areas of deeper coral reefs have been identified that were not visible in satellite images of the Indian, and Pacific Oceans and South China Sea. Those dire predictions appeared extreme until the 1998 mass bleaching event struck; an event not anticipated in

1992. Many remote pristine reefs were devastated from shallow waters down to depths exceeding 30 metres. Hopefully these reefs are not destroyed and will recover if severe bleaching events do not become regular occurrences.

The authors of the regional reports were asked to examine those 1992 estimates. The table below presents a grim picture in the Wider Indian Ocean, in Southeast and East Asia and the Caribbean/Atlantic regions. However, vast expanses of reefs in the Pacific and off the coast of Australia are in reasonably good health with a positive prognosis for the future; unless global climate change events re-occur like those in 1998 (and in 2000 near Fiji and the Solomon Islands). The bleaching was caused by the combination of extremely calm conditions during the 1997-98 El Niño-La Niña events, coupled with steadily rising baseline of sea surface temperatures in the tropics (increasingly attributed to greenhouse warming). These two factors drove temperatures in parts of the tropics to higher values than have been recorded in the last 150 years.

Year of Prediction	Reef % destroyed pre 1998	Reef % destroyed in 1998	Reef % at critical stage, loss 2-10yr	More reef % threatened, loss 10-30 yr
1992 - Guam	10	-	30	30
1998 - Reefs@Risk	-	-	27*	31*
2000				
Arabian Region	2	33	6	6
Wider Indian Ocean	13	46	12	11
Australia, PNG	1	3	3	6
South & East Asia	16	18	24	30
Wider Pacific Ocean	4	5	9	14
Caribbean Atlantic	21	1	11	22
Status 2000 Global	**11**	**16**	**14**	**18**

*The R@R process determined these numbers by statistically assessing predicted threats to existing reefs, and did not include coral bleaching as a threat.

This is a compilation of current losses and predictions for the regions as mean values, after weighting for reef areas detailed in the Reefs at Risk report of 1998. The losses in 1998 should be regarded as temporary, as many of these reefs should recover, provided that major bleaching stresses are not repeated frequently. The 2-10 and 10-20 year periods for the predictions are to complement the 1992 predictions that reefs classed as either critical or threatened would be irreparably degraded unless the stresses are removed and relatively large areas are set aside as MPAs.

These grim estimates for large areas of the world indicate that efforts to conserve coral reefs should to be increased, most probably through the establishment of a network of Marine Protected Areas covering many coral reefs with full community and government co-management involvement. Conservation efforts in small areas are being undermined by damaging activities in adjacent un-managed oceanic or terrestrial areas. The major human threats to coral reefs can be managed by providing alternative livelihoods and education to people on the adjacent land. Unless these threats are managed and mitigated, reefs will continue to degrade with the inexorable increase in human populations and pressures. If predicted rates of greenhouse gas emissions continue and these are confirmed as the trigger for global warming, shifts in global climate like the El Niño-La Niña phenomenon of 1997-98 will recur with increased severity and frequency. Coral reefs which are recovering, will be set back by recurring bouts of coral bleaching and mortality. An additional climate problem is looming as corals will probably grow more slowly with more fragile skeletons due to increases in CO_2 concentrations in sea water.

Physiological restrictions will prevent corals escaping global warming by retreating into deeper water or occupying higher latitudes when tropical waters become too warm. The time required for recovery from the recent extreme mortality event to form reefs with similar levels of high coral diversity will be long because few adult colonies remain to provide larvae for new settlement. Poor management of human activities on reefs will slow this recovery. Over-fished reefs are frequently overgrown with large fleshy algae that prevent coral recruitment. Moreover, the recovered reefs will have different coral community structures, dominated initially by slow growing massive corals. Repeat bleaching events within the next 20 to 50 years will reverse any such recovery.

Already the world has lost 11% of coral reefs and a further 16% are not fully functional. Most of these should recover, except possibly for those in already stressed areas. Coral reefs will not become extinct in the immediate future, but there are likely to be major changes in the composition of coral communities and reductions in harvestable products. Corals have survived for millions of years of major climate change events, meteor strikes and changes in solar activity and somehow recovered. However, the time frames for those recoveries were long; often many thousands of years and well outside short-term human interests. Reefs will probably recover somewhat from the current bouts of anthropogenic and climate change degradation, but it is likely that worse is yet to come and we will probably experience significant reductions in the cover and health of coral reefs, and major losses in biodiversity. It is therefore imperative that globally coordinated actions are implemented soon to manage and conserve these ecosystems of global heritage value for present and future generations, and for peoples and cultures whose livelihoods depend on reef resources.

1. The 1997-98 Mass Coral Bleaching and Mortality Event: 2 Years On

Clive Wilkinson

Many coral reef scientists and resource managers were considerably shocked and depressed during 1998 when there was massive coral bleaching and mortality of corals over large reef areas in many parts of the world. This caused a major paradigm shift in concepts about the degradation of coral reefs and mechanisms for management. Previously it was accepted that direct anthropogenic impacts of nutrient and sediment pollution, over-exploitation (especially with damaging methods) and destruction of reefs during development should be ameliorated for successful conservation of reef resources. Suddenly, a new and comparable threat to coral reefs was posed by Global Climate Change related to indirect effects of increasing levels of greenhouse gases causing major switches in climate and warming surface waters, and the direct chemical imbalances in the seawater caused by increases in CO_2 concentrations. There are no easy management approaches to these threats, other than seeking global reductions in the emissions of greenhouse gases.

Global Climate Change and Coral Reef Damage – Probable Links

The world experienced the most dramatic changes in climate on record when a major El Niño event started in early 1997 (lasting for 12 months), and was followed by a rapid switch to a La Niña in mid 1998 (lasting for 11 months). It was only in mid-2000 that the La Niña diminished to 'normal' weather patterns.

There was major coral bleaching and mortality on many reefs around the world during these major switches in climate. This was the most severe bleaching ever witnessed and suddenly alerted the world to the potential impacts of Global Climate Change (GCC) on coral reefs. At the Convention on Sustainable Development meeting in Montreal, February 2000, a resolution was passed asserting that the damage to reefs across the world in 1998 was directly associated with global climate change. This resolution was argued by the Seychelles, Jamaica, Dominican Republic, Antigua and Barbuda, and Cuba; whose reefs suffered from coral bleaching.

When the first Status of Coral Reefs of the World: 1998 was written, the initial impression was of major catastrophes in many parts of the world, particularly to Indian Ocean reefs, those of Southeast and East Asia, Palau and large parts of the Caribbean. The anecdotal impressions were that we had witnessed the localised extinction of many reefs with large-scale losses of species and major economic crises for the people dependent on those reefs. Now, 2 years later, a re-examination of the evidence shows that there are both causes for pessimism and many reasons for optimism.

First, the good news. Many reefs that experienced major coral mortality are now showing encouraging signs of recovery following the recruitment of new corals, the survival of many juvenile corals during 1998, and the recovery of seemingly dead corals from tissue buried deep inside the skeletons. For example, there has been encouraging new recruitment in the Maldives, Kenya, Seychelles, southern Japan and Palau originating from a few surviving adult colonies in deep passes to the outside cooler waters. While large areas of reefs in Southeast Asia, the Great Barrier Reef (GBR) and the Caribbean suffered extensive bleaching, there was little resultant mortality, even though some corals remained bleached for up to 10 months (Case Studies 6 and 7).

But there is also bad news. In late 1998, there were either no reports or incorrect ones about many reefs which were subsequently shown to have had major levels of coral mortality and low rates of new recruitment. The reefs of the Chagos plateau in the Indian Ocean were reported as healthy in 1998, but Case Study 1 shows a different story. The reefs of southern Japan have apparently lost their major reef building corals (Case Study 4). Large areas of Kenya, the Seychelles, and southern Japan have not recovered and large corals, many hundreds of years old, have not recovered (e.g. Vietnam, Western Australia, GBR).

The 1998 report was a compilation of anecdotal and early reports obtained immediately after (and in some cases, during) the 1998 major bleaching event, but few of these had substantiating data. This report focuses on a few key case studies, with data on the status of the reefs, and comparisons with the reef status prior to 1998.

Arabian Region

Bleaching coincided with the massive La Niña (1998) that was also causing bleaching in Southeast and East Asia and the Caribbean. The duration of bleaching was short, but many corals that had survived previous bleaching in 1996 in the Arabian/Persian Gulf were killed. Most shallow water reefs on the southwestern side of the Gulf were almost completely destroyed with 80 to 100% of corals killed in August and September, 1998. Most corals that had survived a similar massive bleaching in 1996, were killed during 1998, and coral cover is almost universally less than 10%, and as little as 1% on many reefs. An example of the mortality was the destruction of coral reefs near the oil installations of the United Arab Emirates. These had been monitored for 2 decades during major developments on land and had survived. Reefs to the east on the coast of Iran suffered only minor impacts.

Most of the Red Sea was not affected in 1998, with very few reports of bleaching in Egypt, Israel, Jordan, and Sudan. There was some bleaching in Eritrea around Massawa and Green Island but most corals recovered after the temperatures dropped. No bleaching was seen on the Assab Islands. There was wide-scale coral mortality on the fringing reefs and banks off Rabigh (22°33' N to 22°58' N), Saudi Arabia, with almost all hard (including *Millepora*) and soft corals killed (more than 95%) from the reef-flat to 15m depth on the reef slope. *Millepora* species had dominated extensive lengths of the reef crest in 1997 but suffered mortality approaching 100%. Mortality was about 70% on the Rabigh reef banks 25km offshore with some corals still bleached 3 months after the onset. (Yusef Fadlallah, e-mail: yfadlal@kfupm.edu.sa).

Arabian Sea

There was major bleaching around Mirbat, southern Oman, in May 1998 with 75 to 95% of abundant *Stylophora* bleached, and 50% of large *Porites* colonies were partially bleached. No bleaching was observed at Sudh, 40km east of Mirbat, nor in the Muscat Area, Gulf of Oman at temperatures around 30.5°C. Upwelling during the southeast monsoon normally drops temperatures to 19°C. No recovery of bleached colonies was seen in mid-October when temperatures increased to 25°C after the summer upwelling. There was also extensive coral bleaching on the Socotra Archipelago (see chapter 2).

Indian Ocean

Probably the most extensive coral bleaching ever witnessed occurred in the central to northern Indian Ocean during the first 6 months of 1998, with devastation of coral reefs along Eastern Africa, in the Seychelles, Chagos Archipelago, the Maldives, Sri Lanka and India and on reefs off Western Australia. The normal monsoon and trade winds stopped and surface waters warmed during this severe El Niño, and the warm water 'followed' the sun from south to north between January and June. When the El Niño switched to an equally strong La Niña in June 1998, winds and currents started again and bleaching stopped.

Eastern Africa

Coral bleaching and mortality were severe on the coasts of Kenya and Tanzania from March to May, 1998. In Kenya, bleaching varied from 50% to 90% on most reefs, but less on reefs below 10 m depth (about 50%). Coral cover in many sites has dropped from a mean of 30% to 5–11%, irrespective of whether or not the reefs were protected in a protected area. New coral recruitment is very low, so recovery will be prolonged. Bleaching was similar or worse in Tanzania, particularly in the northern half with 60-90% losses of corals at Tutia (Mafia Island Marine Park) and Misali (west coast of Pemba). There was less bleaching on some reefs e.g. 10% or less on Unguja Island, Zanzibar. Reefs in the northern part of Mozambique with up to 99% mortality e.g. on some patch reefs at Inhaca Island. Inshore areas, which experience greater variations in temperatures experienced less bleaching. Only a few corals in South Africa were affected, probably because most coral growth is deeper than 10 m.

Southern Indian Ocean

Bleaching impacts were far more severe in the north (Comoros and Seychelles) than in the south (Madagascar, Mauritius and Reunion) where it was minimal. Mortality on the Comoros reef flats and slopes was approximately 40% to 50%. Most reefs of the inner Seychelles have less than 10% coral cover and on some it is as below 1%. On Madagascar, there was 30% bleaching of corals at Belo sur Mer (mid-west coast) and similar bleaching at Antananbe, Toliara, Nosy Bé, and Mitsio archipelago. At Mananara-Nord (northeast coast, 15°S), 40-80% of corals bleached with high mortality, and 10-40% of mixed species corals bleached in deeper water. A similar pattern was reported on Mayotte, except on the exposed southern end and the lagoon, which receives cooler water from the north.

South Asia

The most severely impacted reefs in the world during the 1998 bleaching event were probably those of Chagos, Sri Lanka, India and the Maldives. In this region, and on nearby reefs off the Seychelles and Western Australia (Case Study 3), there were losses of 80% of

CASE STUDY 1: CHAGOS ARCHIPELAGO, CENTRAL INDIAN OCEAN

An error was made in the 1998 report with no bleaching reported on Chagos. In fact these remote reefs were severely damaged like others in the Indian Ocean. The 1998 El Niño resulted in a major losses in coral cover and abundance across the Archipelago. The El Niño came at the end of a chain of events that had caused hard coral cover to decline from 50-75% (plus 10-20% soft coral) to only 12% live cover over the last 20 years. These are very remote reefs with high biodiversity (220 and 770 fish species) which experience virtually no human impacts. The 1998 'Hotspot' sat over Chagos for 2 months, so that in early 1999 seaward reefs of the 6 Chagos atolls had 40% cover of recently dead coral and 40% bare rock. The 50% of corals lost in the lagoon has turned into loose rubble. It is probable that earlier El Niño events (1982-83 and 1987-88) resulted in conversion of much live coral to rock and rubble, but the 1998 event caused a major sudden loss of corals. Meteorological records show a 1-2°C rise in mean air temperature, lower annual atmospheric pressure, reduced cloud cover, and more variable winds during the last 25 years; strong evidence for global climate change.

In the 1970s, seaward and lagoonal reefs had high coral cover and diversity, with many large table *Acropora* spp. in 5–15m depth, and *A. palifera* in the turbulent shallows. By February 1999, most of these corals and soft corals were dead with only a few large *Porites* heads having more living than dead tissue. There were also losses of calcareous red algae, *Millepora,* and blue *Heliopora* corals from shallow areas. Soft corals were almost totally eliminated from seaward slopes after 1998, resulting in large areas of bare substrate. There was much lower coral mortality on lagoon reefs, probably because these experience large normal fluctuations in temperature and the corals presumably have adapted to higher temperatures. The massive *Porites* in lagoons survived better than those on ocean facing reefs.

The effects on Chagos have been catastrophic, particularly for the dominant genus *Acropora.* In early 1999, it was possible to snorkel for 15 minutes in areas of north-western Salomon and western Blenheim and not see any live coral amongst the still-standing coral skeletons. Whereas before then, there was 50-75% cover. There were only occasional sites with more than 10% of colonies were alive, with *Porites lutea* and other similar large massive *Porites* species being the best survivors (average cover of 8% on all reefs). The immediate consequences will be a reduction in reef growth to at least 20m depth, and delayed recovery of corals and the coral community. Erosion of the coral skeletons will produce unstable rubble and sand, which will prevent the settling of new larvae from the few living parent colonies. The scale of the losses, including coral colonies that were 200-300 years old, indicates that partial recovery will take several decades at least, and may be a few hundred years before a similar community of corals is re-established. (Charles R.C. Sheppard, 1999. Coral decline and weather patterns over 20 years in the Chagos Archipelago, Central Indian Ocean. Ambio, 28, 472-478, E-mail: sh@dna.bio.warwick.ac.uk.)

the living corals or more. These now appear as virtually barren reefs with just a few percent of coral cover. The critical problem for these reefs is that there are few remaining adult colonies to provide coral larvae, thus it is probable that many reefs will have lower coral diversity for the next 20 to 40 years. Bleaching throughout the Maldives was extreme, with many reefs now having less than 5% live coral cover. There are, however, encouraging signs of recovery with many small corals surviving after the bleaching and many new recruits appearing. The 'new' reefs, however, will have a very different appearance with dominance by slow growing massive corals, instead of branching and plate *Acropora*.

Southeast and East Asia

Bleaching was variable across the region, with most countries reporting severe bleaching, but generally low levels of coral mortality. The exceptions were parts of Vietnam and the northern Philippines where up to 80% corals bleached, and mortality was high. Singapore reported the first major bleaching in the island's history.

The large La Niña event that started in June 1998 occurred during the northern summer and resulted in serious bleaching in the region, particularly in Taiwan and Japan. This area is frequently impacted by typhoons travelling to the northwest, which mix the waters and maintain them below 30°C, but the summer of '98 was particularly calm with the first typhoon being delayed until September. By then there had been massive damage to the coral reefs. Bleaching was variable with less than 20% in the north around Kyushu and Shikoku, 30–40% on the eastern Hachijo Islands (maximum of 80 – 90% in places), and

CASE STUDY 2: BLEACHING OF REEFS IN THE LAKSHADWEEP ISLANDS, INDIA

First reports from the Lakshadweep Islands off the west coast of India were of catastrophic bleaching, but follow-up surveys in mid-2000 showed that bleaching mortality was very variable between reefs, but more uniform within a reef. After the 1998 El Niño bleaching, live coral cover on Kadmat Island was 2.4% and 6.7% on Agatti, whereas on Kavaratti there was 14.3% cover. Turf and coralline algae growing on dead coral were the dominant cover at all sites, and there was clear evidence at Agatti and Kadmat that the dead *Acropora* skeletons were breaking down. An encouraging sign was that there were many small coral colonies (<8cm), up to 3.6 individuals m-2, with many *Acropora*. Most of these are probably new recruits since 1998, and also there were more small colonies growing on the top surfaces of table *Acropora* killed in 1998 (average of 13.7 colonies m^{-2} of which 60% were juvenile *Acropora*). This indicates that parent colonies occur nearby, but few were seen. Unfortunately, there was no pre-bleaching information on these reefs, therefore it is not possible to compare the new coral communities with the old, but these reefs previously had high coral cover; 56% coral cover on reef fronts and 37.2% on shallow back-reef lagoonal sites, and it had already dropped to 8.1% a year later. From Rohan Arthur, Centre for Ecological Research and Conservation, India, and James Cook University, Australia (Rohan.Arthur@jcu.edu.au)

CASE STUDY 3: BLEACHING AND RECOVERY OF WESTERN AUSTRALIAN CORAL REEFS POST 1998

A complex pattern of bleaching and mortality occurred on remote and nearshore coral reefs off the coast of Western Australia during the first half of 1998. These reefs receive very minor land runoff and minimal anthropogenic damage, but the patterns were very different on adjacent reefs:

- Ashmore Reef (S 12°16' E 123°00') only minor bleaching seen in the lagoon of these atoll-like reefs.
- Seringapatam Reef (S 13° 39' E 122° 02'); extensive bleaching with reductions in hard coral cover from 45% to 5%, and soft coral cover from 10% to 2% on the outer slopes. There was almost 100% mortality of mostly Acropora corals in the relatively shallow lagoon, small patch reefs have suffered 90 to 100% mortality and a very old single colony of *Pavona minuta* over 100 m diameter and 6m high was almost entirely killed with 0.5m2 still alive. Many species have become locally extinct.
- Scott Reef (S 14°11' E 121°48'; 50 x 35km) was the most seriously bleached of these isolated reefs, with extensive coral mortality at all depths in April 1998. There was almost 100% mortality of large *Acropora* beds with no survivors on most lagoon patch reefs. Corals on the reef crests were devastated, with few survivors. Data along 9m depth transects showed all animals with symbiotic algae were wholly or partially bleached, with variable recovery. The majority of branching *Acropora* were 80 to 100% killed, and 100% of *Millepora* were dead at all sites except one. Porites cover was reduced by 50%, with many colonies having only small patches of live tissue a year later. The once common Pocilloporids and *Acropora bruggemanni* are either extremely rare or locally extinct. Soft corals suffered at all sites with between 50% and 80% of soft corals dying, depending on the site. Even at 30m depth, 80% of the corals were bleached. The coral community has changed dramatically. Recruitment studies between 1995 and 1999 reveal that recruitment in 1998 and 1999 was less than 1% of the pre-1998 levels.
- Rowley Shoals (including Mermaid, Clerke and Imperiuse Reefs Reef between (S 17°04' and 17°33'; E 114°38' and 119°23') showed only minor bleaching in 1998, with a few Pocilloporids and branching *Porites cyclindrica* bleached. Mortality was minor with no changes in coral cover since the first surveys in 1994.
- Dampier Archipelago (S 20°38' E 116°39') showed severe bleaching in March-April, 1998 on the fringing reefs of the inner islands, but the impacts were patchy. Most hard corals were partially bleached, while soft corals, anemones and zoanthids also suffered varying degrees of bleaching. Recovery was highly variable, with most Pocilloporids and Fungids dying, while many massive corals (Turbinia, Mussids, Pectinids) recovered. Bleaching on the outer island areas was almost non-existent.

> - Onslow - Mangrove Islands (S 21°27' E 115°22') also showed extensive bleaching of 80% of corals on the inshore reefs in late March 1998, whereas there was little bleaching in the outer areas.
> - Ningaloo (S 22°13' E 113°49') showed no measurable impact of bleaching on this 220km long fringing reef, but there were some anecdotal reports of minor bleaching in some areas from February to April 1998.
> - Abrolhos Islands (S 28°41' E 113°50') much further South showed no coral bleaching in 1998.
>
> Reports from Luke Smith l.smith@aims.gov.au and Andrew Halford a.halford@aims.gov.au, Australian Institute of Marine Science, Western Australia.

40–60% on islands west of Kyushu. There was extensive bleaching (30-40%) around Penghu Islands Taiwan in June, and similar impacts (>80%) around Posunotao, southeast Taiwan in August with water temperatures as high as 32°C at 25m depth.

Australia and Great Barrier Reef

There was some bleaching along the Australian Great Barrier Reef (GBR) in early 1998 with extensive coral bleaching and mortality on some inshore reefs, while bleaching was relatively minor on middle- and outer-shelf reefs. Aerial surveys of 654 reefs between March and April showed that 87% of inshore reefs (25% of reef with severe bleaching, >60%; 30% high, 30–60% bleached; 12% mild, 10–30% bleached) had some bleaching, compared to 28% of offshore reefs (no severe bleaching; only 5% more than 30% bleached corals). Impacts were patchy: on Orpheus Island (an inshore reef of the Central GBR), 84-87% of corals bleached, but 5 weeks later, mortality was 2.5-17%; 10km away on Pandora there was almost 100% coral mortality, including some large *Porites* colonies that were centuries old. Soft corals which often dominate inner-shelf reefs were extensively bleached, with almost 100% mortality in some areas, whereas in other areas there was major recovery.

Southwest and Southeast Pacific Ocean

No bleaching associated with the El Niño-La Niña climate shift was reported in the southwest of the region, although there were reports of bleaching on shallow reef flats around Samoa associated with extreme low tides. In most places water temperatures were cooler than normal. There was some bleaching in early 1998 in French Polynesia e.g. 20% coral cover was reduced to 12% on Takapoto; there were reports of mortality from other islands e.g. Rangiroa and Manihi; and minimal bleaching on Moorea, Bora Bora and Tikehau. There was, however, severe bleaching between February and April, 2000 on Fiji and the Solomon Islands (Chapter 10).

Northwest Pacific Ocean

The islands of Palau have never experienced coral bleaching like the events of September 1998, with an estimated one third of all corals affected. The major effects were on outer ocean facing slopes where 90-99% of *Acropora* species were destroyed, whereas the

> ## CASE STUDY 4: BLEACHING ON THE REEFS OF SOUTHERN JAPAN
>
> There was unprecedented mass coral mortality in Okinawa, Japan, with the local extinction of at least 4 coral species and a loss of hard and soft coral cover from 70% to 10%. A strong La Niña coincided with mid summer, no typhoons and low cloud cover and this raised water temperatures around Okinawa to 2.8°C above the 10 year average. Bleaching started in mid-July and ended with the first typhoon of the season in September. Locals reported seeing nothing like this during the 35 years of coral reef research, and elderly Okinawans said they "have never seen such beautiful white corals". Coral cover virtually did not change between 1995 and 1997 at Sesoko Island (26°38'N, 127°52'E), but after the bleaching, mean coral cover decreased by 73%, the number of species decreased by 61% and number of colonies per m^2 dropped to less than half. Furthermore the living cover of soft corals decreased by 99%. Branching *Acropora* species which were once the most prolific corals on Okinawan reefs, were most severely affected; 3 species of branching corals species *(Seriatopora hystrix, S. caliendrum, Stylophora pistillata* and *Pocillopora damicornis)* and the branched fire coral *Millepora intricata* are now locally extinct and no new recruits were found around Okinawa in 1999. Curiously, many juvenile Acropora colonies (< 5cm in diameter) survived on the reef flats, where they are likely to be killed when exposed to air and high irradiance during midday low tides. Bleaching also affected other common corals, whereas many of the less common species survived and are now the most abundant e.g. massive and encrusting colonies of *Porites, Goniastrea, Leptastrea, Platygyra, Favia* and *Favites*. The chances for recovery of the missing species depends on coral populations on distant offshore reefs (15-30km from Okinawa) which may provide larvae for re-settlement, provided currents are favourable.
>
> Report from Yossi Loya Department of Zoology, Tel-Aviv University, Israel 69978, e-mail: yosiloya@post.tau.ac.il

impacts were often minor on inshore areas which experience natural variations in temperature and water clarity. On the northern Ngeruangel atolls, bleaching deaths varied between 10% and 70% at all depths, with a mean of 53.4% of corals at 3–5m, 68.9% at 10–12m, and 70% at 30m (Peleliu, the Blue Corner, the Big Drop-off, Iwayama Bay - Rock Islands). Severe bleaching occurred in Arakabasan and Cemetary island. The major damage was to corals in oceanic waters, with much less bleaching and mortality in shallow areas and close to the more turbid and stressed habitats near the shore. The least effect was to corals around the sewage outfall near Koror (John Bruno, e-mail: John_Bruno@Brown.edu). All adult jellyfish in the famous marine lake (Ongeim'l Tketau) were killed, but live polyps were found in 1999.

About 20% of corals, including a wide variety of hard and soft coral species, bleached to 20m depth on the north of Yap, **Federated States of Micronesia** in September 1998,. Other parts of Micronesia reported only minor bleaching.

Northeast Pacific Ocean
No bleaching was reported in 1997-98, and water temperatures were often colder than usual.

Pacific Coast of the Americas
When the pool of warm water associated with the El Niño banked up on the Pacific coast of central America, bleaching soon followed in Colombia, Mexico and Panama between May and September 1997. However, the bleaching was usually partial and only occasional complete bleaching on some corals. The resultant mortality was usually less than 5% or not detectable. This episode of bleaching was very minor compared to the disastrous events of 1982-83 when there was 50-100% coral mortality on most reefs, which have yet to recover from those losses. When this pool of warm water expanded to reach the Galapagos Islands in mid-December 1997 there was some bleaching of the coral communities that had recovered after the major bleaching of 1982-83. Bleaching continued affecting most corals by March 1998.

CASE STUDY 5: GBR BLEACHING UPDATE - SEPTEMBER 2000

Around 28% of offshore reefs, which make up the bulk of GBR reefs, were bleached in early 1998, but the majority showed full recovery by the end of 1998. The inshore reefs bleached more extensively (87% of the reefs with some bleaching), with 55% showing high to extreme bleaching and recovery was patchy. Reefs in the Keppel Island group (inshore southern GBR) recovered well, while reefs on the Palm Island group (inshore central GBR) suffered high mortality. Overall, hard coral cover at Orpheus and Pelorus Islands (northern Palm group) declined from an average of 13.1% to 5% at 4 sites, while soft coral cover declined only marginally from 18.4 to 16%. The biggest impact was on two hard coral families: the Acroporidae, which suffered a 91% relative decline and the Milleporidae, which suffered >99.99% decline. Attempts to find living colonies of *Millepora tenella* (a previously dominant species), failed to find any living colonies around Orpheus Island, however, there was an anecdotal report of a single small colony in deep water. Two years after the bleaching, recovery of Acropora species is slow in the Palm Islands. At Northeast Reef on Orpheus Island, staghorn coral patches *(A. nobilis)*, which were initially thought dead, are showing signs of recovery with many new branches (10-15cm) growing out of the dead skeleton. These are not new recruits, and they presumably derive from residual live tissue within the skeleton or underneath the colony. Other staghorn species *(A. formosa* and *A. grandis)* do not show this recovery pattern. Plating Acropora species (particularly *A. hyacinthus* and *A. cytherea)* suffered near total mortality and have not recovered, and recruitement of all *Acropora* species is very low. Thick tissued and massive coral species (poritids, favids, mussids, fungiids) generally recovered well, and pocilloporid recovery is intermediate and patchy.

From: Ray Berkelmans, Australian Institute of Marine Science, Townsville (r.berkelmans@aims.gov.au)

Case Study 6: 1998 Coral Bleaching in the Mesoamerican Barrier Reef System (MBRS)

This region experienced fewer large-scale bleaching events compared to other areas in the Western Atlantic until recently. While coral bleaching was reported for much of the Caribbean during 1983 and 1987, the first well-documented mass bleaching event in Belize occurred in 1995 where 52% of coral colonies bleached, although only 10% had partial mortality with a loss of 10-13% of coral cover. The 1995 bleaching also affected Cayos Cochinos, Honduras, where 73% of scleractinian corals and 92% of hydrocorals bleached and slightly higher mortality was reported. A less severe bleaching event was reported in 1997, with little reported damage. However, during 1998, high sea-surface temperatures first appeared in the region during August and intensified during September. Reports soon followed of intense bleaching (more than 50% of colonies) in the Yucatan in August-September, and then in Belize (September) and Honduras (September-October). The first coral mortality was seen in the Yucatan in early October, particularly on *Agaricia tenuifolia* colonies, where large scale mortality of *A. tenuifolia* and *Millepora* spp. was seen in the central and southern Belize barrier reef. Water temperatures decreased with the passage of Hurricane Mitch in late October, and some branching corals started to recover, although massive corals remained bleached into 1999. Extensive surveys showed that the 1998 bleaching event was more severe than that in 1995. Shallow reef corals either died immediately or recovered more rapidly compared to deeper depths, where recovery from bleaching was slow, and significant remnant bleaching was observed up to 10 months later. Specific findings from this study showed:

- An average coral mortality of 18% on shallow reefs and 14% on fore reefs across the region;
- Up to 75% recent coral mortality occurred on localised patch and barrier reefs in southern Belize;
- The highest mortality was on *A. tenuifolia* (>35%), *M. complanata* (28%), and *Montastraea annularis* complex (25-50%)
- There was high recent mortality and disease on *Montastraea annularis* in the region;
- There were low to moderate levels of bleaching mortality in *Acropora palmata*;
- Remnant bleaching was still evident on fore reefs 10 months later (up to 44% of corals bleached); and
- A high incidence of coral disease afterwards on Belize shallow reefs (black band) and Honduras and Belize fore reefs (white plague).

Caribbean and Atlantic Ocean

Bleaching in the Caribbean did not follow a clear pattern and varied across the region. Even in areas where bleaching was severe, mortality was generally low, except when Hurricane

Mitch hit some areas. Most of the bleaching was associated with the La Niña event of the latter half of 1998. This contrasts with the Pacific coasts which were bleached in late 1997 associated with El Niño conditions. For example, early reports in **Belize** and **Puerto Rico** were of massive bleaching, but mortality was eventually low in most places (Case Studies 6 and 7).

In August 1998, more than 60% of all head corals to a depth of 15m bleached around Walker's Cay, New Providence Island, Little Inagua, Sweetings Cay, Chubb Cay, Little San Salvador, San Salvador and Egg Is, **Bahamas** and up to 80% between 15-20m depth. Samana Cay was less effected with *Montastrea cavernosa* not bleached, and *Acropora palmata* bleached on only the upper sides. A large area of the Carribee bank, **Barbados** apparently bleached in September 1998, whereas on Bermuda bleaching started in August 1998 and continued into October, when surface temperatures rose to 30°C. There was 2-3% bleaching of the 25% coral cover at 8m on rim reefs, 5-10% bleaching of the 40% coral cover at 15m on offshore terrace reefs, and 10-15% of the 15-20% lagoon coral cover. Mortality was low, perhaps 1-2% of affected colonies.

There was mass bleaching on reefs off Bahia State (12°S; 38°W) **Brazil** in April 1998, with 60% of *Mussismilia hispida* (endemic coral), 80% of *Agaricia agaricites*, and 79% of *Siderastrea stellata* (endemic) bleached. Similar bleaching was reported on the Abrolhos Reefs (18°S; 40°W). By October 1998, all colonies have recovered.

Conclusions

The bleaching in 1997-98 was the most intense on record with damage recorded in all oceans. The major causal factor was increased seawater temperatures during the extreme El Niño and La Niña events. Bleaching predominantly occurred on southern reefs when the El Niño event coincided with the summer, whereas severe bleaching occurred in the northern hemisphere when the strong La Niña coincided with summer.

Coral bleaching started with the build up of El Niño associated warm waters in the far eastern Pacific between May and December 1997 when the Southern Oscillation Index (SOI) averaged around minus 16, in: Colombia starting May; Mexico from July-September; Panama in September; and the Galapagos Islands in December. Most of this bleaching was relatively mild, with little resulting mortality.

The most severe impacts occurred in the southern Indian Ocean between March and June, 1998 and during the El Niño when the SOI averaged below minus 23 for 4 months: in Kenya and Tanzania in March–May; Maldives and Sri Lanka in April to May; Western Australian reefs from April to June; India from May to June; Oman and Socotra, Yemen in May.

There was also some bleaching in Southeast Asia, Australia and the southern Atlantic Ocean between January and May, 1998, in: Indonesia from January to April; Cambodia, Thailand and East Malaysia during April–May; Eastern Australia and the Great Barrier Reef in January and February, 1998; and the Southern Atlantic Ocean off Brazil in April, 1998. These events were of modest severity, with some incidences of high mortality.

> **CASE STUDY 7: CORAL BLEACHING AND RECOVERY IN PUERTO RICO**
>
> Wide-spread and intense bleaching events usually result in mass coral mortalities. During the 1998-99 La Niña bleaching event, many reefs in the Caribbean were affected by wide-spread and intense bleaching, however, there were generally low levels of mortality, especially in Puerto Rico. Here, 386 colonies of 18 reef-building, hard coral species were tagged in August of 1998, when they started to bleach, on an offshore bank reef, an inshore patch reef, and a fringing reef off the southwest coast of Puerto Rico. These colonies were monitored for 7 months in 1998-99 and 5 months in 1999-2000. Many colonies were 100% bleached for more than 150 days before recovering completely; only 3 small colonies died (0.8%), 14 suffered partial tissue mortality (3.6%), 357 recovered by February 1999 (92.4%), and 12 remained pale until March, 1999 (3.1%). Fifty nine tagged colonies (15%) bleached again in the second week in September, 1999 at slightly lower water temperatures than 1998, but bleaching was less intense and all recovered after 5 months. This illustrates that bleached corals need to be observed for recovery before they are declared as dead. Recovery after 5-7 months of bleaching is possible, but it will usually depend on whether other stress factors intervene, or on the severity of the original temperature stress. Many Caribbean reefs experienced similar situations in the big bleaching events of 1997-98, and the event of 99, and recovered (with the exception of some reefs in Belize), whereas reefs in the Indian Ocean were devastated. Recent observations and surveys of the area indicate no wide-spread bleaching, and no tagged colonies had bleached by September, 2000. There are a few colonies show some slight paling over small areas.
>
> Ernesto Weil. Department of Marine Sciences, University of Puerto Rico, Lajas PR 00667, e-mail: eweil@caribe.net.

The El Niño dissipated rapidly in May 1998 and between June 1998 and April 1999 there was a major La Niña event with a SOI of about plus 12 or higher for these 10 months. From June to September 1998, this coincided with the northern summer with bleaching reported in South East and East Asia from July to October 1998, in: Singapore, Sumatra, Indonesia and Vietnam in July in July; Philippines from July to September; and Japan and Taiwan from July to September.

Simultaneously, there was bleaching in the Arabian/Persian Gulf and Red Sea from August to October 1998, in: Bahrain, Qatar and UAE in August–September; and Eritrea and Saudi Arabia (Red Sea) in August–September.

There was also bleaching throughout the Caribbean Sea and Atlantic Ocean from August to October 1998, in: Florida from July to September; Bahamas, Bonaire, Bermuda in August – September; Barbados, BVI, Caymans, Colombia, Honduras, Jamaica and Mexico in September. However, most of this bleaching resulted in minimal coral mortality.

Bleaching then followed the path of the solar zenith in the far West Pacific from September to November 1998, in: the Federated States of Micronesia in September; and Palau in September to November when there was unprecedented coral mortality.

A consistent feature of the weather during all these bleaching events was the coincidence of a major shift of the SOI away from zero, which apparently induced particularly calm weather. For example, in the Indian Ocean, the monsoon winds stopped almost completely, resulting in doldrum-calm conditions and weak currents. Under these conditions sea surface waters warmed above the temperature tolerance limits for corals and other symbiotic animals, without mixing with cooler oceanic and deeper waters. Warm waters often extended down as far as 30m or more. Coral bleaching consistently started about 4 to 6 weeks after the passage of the solar zenith.

A critical question remains; was the 1997-98 event a one-in-a-thousand year event killing off large old corals, or will bleaching events like 1997-98 become more frequent and severe in the future? Coral bleaching and mortality levels were closely associated with the El Niño and La Niña switches in climate that started in 1997 and ended in early 1999, with large deviations in the SOI. Many scientists now predict that major climate change events will become more frequent and severe, resulting in greater bleaching. It is apparent that when a strong La Niña (positive SOI) coincides with the northern summer, bleaching will follow; likewise the major trigger for bleaching in the southern hemisphere is when a strong El Niño (negative SOI) occurs in summer. Reefs closer to the equator will probably experience bleaching under either El Niño or La Niña conditions.

Thus, these predicted scenarios mean that coral reefs will change in structure over the next few decades with a major reduction in the branching and plate forming species that are the fastest growers and major contributors to coral reef accretion. In their place, reefs will have a lower profile of slow growing, massive species, with lower coral cover and lower diversity in the medium term. This will probably reduce the capacity of reefs to act as breakwaters for fragile shorelines and provide sand and rock as building materials on low islands. Fish populations will probably drop as there will be reduced structural complexity to shelter juveniles and the many small species, but changes will be gradual. The impacts on tourism are harder to predict. Coral reef tourism, one of the fastest growing sectors of the market, and growth will continue via the many tourists visiting coral reef destinations for the first time. The only adverse affects may be felt with those few tourists who revisit coral reefs for diving, and become disillusioned with large areas of dead corals. However, these tourists will probably seek new experiences and go to the many areas that will escape damage in the immediate future. Surveys of many tourists indicate that many are unaware that the corals are dead, as they focus on fish and above water activities in the tourist resorts.

ISRAEL CORAL REEFS OF THE GULF OF AQABA (EILAT)

Israel has approximately 12km of coastline in the northern Gulf of Aqaba, between Jordan and Egypt, with one Marine Protected Area (155ha), and one Costal Protected Area (16ha). The reserves are governed by the regulations that concern all nature reserves in Israel. These are among the most northerly coral reefs in the world, which grow on a narrow shelf before it drops to 400-700m. Hard coral diversity is relatively high with over 100 species, as well as about 350 species of reef fishes; there is a high proportion of endemic species. Anthropogenic stresses from intensive tourism activity and poor water quality from sewage discharges, mariculture effluents, flood waters, ballast and bilge water from various boat activities, and discharges of fuel, oil, detergents, phosphates, pesticides, anti-fouling compounds are damaging the reefs. In addition there has been extensive sand nourishment of beaches, and solid waste disposal at sea and along the shore. Hard coral mortality during 1999 was 7-28%. In the 'Japanese Gardens', which has a few divers, live coral cover has dropped from 70% in 1996 to 30%. Coral recruitment has been declining steadily by 53-96% since 1997. The release of planulae from one of the reef building coral, *Stylophora pistillata*, dropped steeply between 1975 and 2000. The highest rates of coral mortality and low recruitment have occurred in sites which have none, or little recreational activities. Bacterial diseases of fishes are on the increase, in parallel with increases of disease in the mariculture farms. During the last two years deep water (below 500m) nutrient concentrations have almost doubled, probably due to discharges from the mariculture industry that is located at the tip of the Gulf. During winter and spring, nutrient-rich deep water rises cause seasonal blooms of algae, which can smother corals and block light penetration. About 20% of shallow water corals died during a severe upwelling in 1992. These coral reefs are among the most heavily used in the world by recreational divers, with more than 200,000 dives per year, with most dives on the fringing reefs in the Coral Beach Nature Reserve. During 1996, the damage to corals was measured at a rate of 66% at sites with high diving pressure compared to 8% at low impact sites. In the year 2000, these levels have dropped following diver education programmes, but at 23% to 4%, they are still high by world standards. From: David Zakai, Israel Nature & National Parks Protection Authority Eilat, Israel (dudu.zakai@nature-parks.org.il)

2. Regional Status of Coral Reefs in the Red Sea and the Gulf of Aden

Nicolas Pilcher and Abdullah Alsuhaibany

Abstract

The coral reefs in Djibouti, Eritrea, Egypt, Jordan, Saudi Arabia, Somalia, Sudan, and Yemen are generally in good and often pristine condition. Reefs in this region are primarily fringing reefs along the mainland and island coastlines, along with barrier reefs, pinnacles and atolls which receive only minor land influences. Many reefs have 30% to 50% live coral cover at most locations. Coral bleaching caused extensive die-off in the Gulf region and the southern Red Sea in 1998, and on the Sudanese coasts a red algal film was present over most shallow reefs. Diversity of corals and other reef fauna is among the highest in the Indian Ocean region. Major threats to coral reefs include: land-filling and dredging for coastal expansion; destructive fishing methods; damage by the recreational diving industry; shipping and maritime activities; sewage and other pollution; lack of public awareness; and insufficient implementation of legal instruments on reef conservation. A number of international, regional, bilateral and multilateral agreements and other legal instruments have been adopted by the countries in this region, and each possesses a relatively complete set of national Laws and Regulations. However, the implementation of these remains generally poor and in some cases there is no implementation and enforcement whatsoever. For coral reef conservation to improve and be effective, there is a need for increased public awareness, increased enforcement and implementation of national and international legal instruments, and the implementation of coastal management plans that integrate coastal development, and control industrial effluents, tourism, for the maintenance of environmental quality of marine habitats.

Introduction

This report summarises the status of coral reefs in the countries bordering the Red Sea and the Gulf of Aden. Most of the countries along the Red Sea (Djibouti, Egypt, Jordan, Saudi Arabia, Somalia, Sudan, Yemen) coordinate conservation and management of coral reefs in this region through the Regional Organisation for the Conservation of the Environment of the Red Sea and Gulf of Aden (PERSGA). In the Arabian/Persian Gulf region coral reef conservation is coordinated by the Regional Organisation for the Protection of the Marine Environment (ROPME). The two institutions cooperate closely under a Memorandum of Understanding. This report summarises information from publications and reports provided by people from these countries and the key agencies operating in the region (some are listed at the end of this chapter). PERSGA is currently developing a Regional Action Plan for the conservation of reef resources through as part of its Strategic Action Plan. It is envisaged that the implementation of the Regional Action Plan will address current threats and assist the RSGA countries in developing Marine Protected Areas. Eritrea operates independently of PERSGA, not being a member of the Arab League.

The Red Sea and Gulf of Aden contain complex and unique tropical marine ecosystems, especially coral reefs, with high biological diversity and many endemic species. The reefs are surrounded by some of the driest parts of the world, such that continental influences are limited, but these narrow waters are also major shipping lanes to the Suez Canal and major petroleum industry activities. While large parts of the region are still in a pristine state, environmental threats, notably from habitat destruction, over-exploitation and pollution, are increasing rapidly, requiring immediate action to protect the coastal and marine environment. Humans have used these reef resources in this region for thousands of years, although most inhabitants are not 'sea peoples'. The coral reefs of the region contain more than 250 species of hard (scleractinian) corals, being the highest diversity in any part of the Indian Ocean. The stable warm waters and lack of major fresh water runoff provide ideal conditions for coral reef formation along the coasts.

The Gulf of Aden is influenced by the seasonal upwellings of cool, nutrient rich waters from the Indian Ocean, limiting coral reef development and promoting planktonic and large algae. Despite this, there are diverse and complex reefs and non-reef assemblages in the Gulf of Aden. Some of the sandy beaches are major nesting sites for sea turtles, and the Socotra archipelago is being considered as World Biosphere Reserve site. The Red Sea is an isolated reservoir of marine biodiversity of global importance. A wide range of ecosystems have developed with high biodiversity and endemism, particularly for reef fishes and other organisms. There is an almost continuous band of coral reef along the northern Red Sea and Gulf of Aqaba, which physically protects the shoreline. Coral reefs in the shallow Gulf of Suez are less well developed as this area is usually quite turbid and has extremes of temperature and salinity. Further south, the shelf becomes much broader and shallower and the fringing reefs gradually disappear, to be replaced with shallow, muddy shorelines. But many coral reefs grow on offshore pinnacles and around islands off the coast.

Although most reefs in the region are still pristine in most areas, threats are increasing rapidly and the reefs are being damaged by coastal development and other human activities. Major threats include: landfilling and dredging for coastal expansion; destructive fishing methods; damage by scuba divers, shipping and maritime activities, sewage and other pollution discharges; lack of public awareness; and insufficient implementation of the laws for reef conservation.

GEOGRAPHICAL REEF COVERAGE AND EXTENT

Reefs in the Red Sea are particularly well developed, with predominantly fringing reefs in the north and platform, pinnacle and atoll reefs offshore in the south, with sandy lagoon type environments towards the shore. The coast of Gulf of Aden supports only patchy reef distribution, but the Socotra archipelago is fringed by extensive reefs.

Djibouti
The coastline is fringed by extensive coral reefs in places; the north coast is generally shallow and sandy, with a few coral outcrops, but the Sawabi archipelago has fringing

coral reefs. The southern coast is shallow with poorly developed reefs, affected by the cold water upwelling from the Indian Ocean. The Gulf of Tadjoura contains low diversity coral reefs, and the Mousha and Maskali islands at the mouth of the Gulf are surrounded by extensive coral reefs.

Egypt

Fringing reefs have formed on the Red Sea coast, as well as along the Gulfs of Suez and Aqaba, and there are also several submerged reefs and fringing reefs surrounding 35 small islands. In the northern Gulf of Suez, there are mainly small, shallow patch reefs, whereas on the western side of Gulf, the reefs are more developed, forming a fringing reef from 50km south of Suez to Ain-Sukhna. The most extensive reefs are around the Sinai peninsula at Ras Mohammed in the southern Gulf of Suez, and surrounding the Ashrafi islands near the western shores of the Gulf. In the Gulf of Aqaba, there are narrow fringing reefs along the steep cliffs of both shores. In the Red Sea, almost continuous fringing reefs extend from Gubal in the north to Halaib, on the border with Sudan.

Eritrea

The newly declared country of Eritrea is a neighbour of Sudan and Djibouti on the eastern Red Sea, and has a coastline of 1,200km, with 59% of national area as a coastal plain. There are over 350 islands, including approximately 210 islands in the Dahlak Archipelago, which sits on a relict Pleistocene platform. There are poorly developed fringing reefs along 18% of the Red Sea coast (continental coast), but the reefs on the islands are a globally significant reservoir of marine biodiversity and relatively pristine reefs, especially the Dahlak Archipelago which was identified as a potential marine protected area in 1968. Coastal resources are largely untapped and the population density along the coast and on islands is low. The climate is conducive for reef growth: generally warm waters, with low rainfall. The extensive corals, seagrass and mangroves include over 250 species of reef fish, as well as large populations of turtles and dugongs. Corals around Massawa are patchy and separated by sand, or along fringing reefs in about 6m depth. There over 17 genera of hard corals, with a mean live coral cover from 16% to 37%, and dead coral from 16-29%. But these are up to 30% lower than records from 1996 which indicates a rapid decline. Sponges increased between both studies, suggesting that they were overgrowing dead coral areas. Bleaching in 1997 was lower than 1%, and there are few crown-of-thorns starfish. The most common genera were Porites, Montipora, Stylophora and Platygyra. No complete fish lists are available, but the aquarium trade exports at least 75 species.

The Ministry of Fisheries (MOF) is responsible for managing fisheries, and coral reef research and monitoring, and the National Environmental Management Plan and a Framework Marine Conservation Strategy is guiding activities. Current conservation efforts include the Coastal Marine and Island Biodiversity Project, a 5 year (1999-2004) GEF-funded programme, executed through the MOF to ensure the conservation and sustainable use of the globally important biodiversity of Eritrea's coastal, marine and island ecosystems, while permitting sustainable use by Eritrean people. Many coral reefs were relatively pristine away from human or economic impacts, but they are increasingly threatened by rapid, and largely uncontrolled, development of fisheries, tourism and oil exploration activities. There is an urgent need to document the resources and this is a function of the CMIBP, along with implementing management of a diversity of marine and coastal habitats, and providing training at government and community levels.

There are no declared MPAs in Eritrea, but 4 areas were proposed during GEF project preparation, including large parts of the Dahlak Islands, Dur Gaam and Dur Gella Islands near Massawa, the Fatuma island group near Aseb and the Museri island group offshore from Massawa. Biodiversity is being assessed through the GEF project to support proposals for protection, especially in the Dahlak archipelago.

Since 1990, the coastal population has increased slightly and the fisheries sector is also developing, most fishing is non-destructive and commercial trawlers operate in deeper waters away from reefs. The fisheries are predominantly artisanal targeting finfish, gastropods, sea cucumbers and pearl oysters. Threats to Eritrean reefs include coastal development, including land reclamation, primarily at Massawa which is reducing the capacity of the reefs to provide coastal erosion. In addition, there are potential threats from curio collecting, discharges from the oil terminal; sewage and solid waste disposal, cooling effluents from desalination and power plants, and sedimentation of cement dust. Recreational tourism is largely undeveloped and coral mining in uncommon. Coral collecting is prohibited, but shell and ornamental fish collection are commercial operations, with more than 100,000 fish exported between 1995 and 1997. Shipping is increasing in Massawa up to 440 per year in 1996. Despite these threats, the general status of Eritrean coral is good, and bleaching during the summers of 1997 to 1999, caused the some loss of foliaceous and branching Montipora corals in shallow areas (1m or less), but massive Porites and Platygyra were less affected. These reefs contain among the most resistant corals in the region, evidenced by rapid recovery from the bleaching. From: Meriwether Wilson (meriwether.wilson@glg.ed.ac.uk)

Jordan

The short coastline (27km) has fringing reefs along 50% the coast, with high coral diversity and associated fauna.

Saudi Arabia: Coral reefs fringe the entire Red Sea coastline and offshore islands, with several distinct areas of similar habitats and species composition: the Gulf of Aqaba in the north; the northern-central section from south of the Gulf of Aqaba to Jeddah; and the central-southern region from Jeddah to the Yemen border, including the Farasan Bank and Islands. The northern-central area has an almost continuous coral reef tract with many reef types: mainland and island fringing reefs; various forms of patch reef; coral pinnacles; and ribbon barrier reefs. Reefs fringe the mainland and often into the entrances and sides of sharms (lagoon-like inlets). Circular to elongate patch reefs are common in offshore waters less than 50m depth. Pinnacles (individual corals and coral 'bommies' surrounded by sand) occur in shallow waters (less than 10m), particularly in the Al-Wajh Bank and Tiran areas. Barrier reefs composed of platform and 'ribbon' reef structures occur further offshore on the 'continental' edge, where depths drop from 50m to more than 200m. The central-southern area is unique in having atoll-like or 'tower' reefs along the shelf edge and the outer Farasan Bank. Further south, the reefs are less well developed along the mainland coast because of high levels of fine sediments, however, complex reef structures developed further offshore on the Farasan Bank and islands.

Somalia

The coastline is divided into north and south sectors, separated by the Horn of Africa. There are flourishing coral reefs along the Gulf of Aden coastline near Raas Khansir, Raas Cuuda Siyara, and off El Girdi and west of Berbera.

Sudan

There are three coral habitat groups on the coast: barrier reefs; fringing reefs; and Sanganeb, an oceanic atoll towards the Egyptian border. Most of the coast is bordered by fringing reefs 1-3km wide which are separated by deep channels from a barrier reef 1-14km offshore, which then drops steeply to several hundred metres. The Sanganeb atoll is a unique reef structure rising abruptly from 800m depth to form an atoll that has been recognised as a regionally important conservation area such that it was proposed to UNESCO for World Heritage Status in the 1980s, but there has been no follow-up.

Yemen
Most of the coral reefs occur along the Red Sea coast and the Socotra archipelago, with some sites in the Gulf of Aden. Corals grow on the Red Sea coast as both coral reefs and coral communities on a variety of substrates. Coral reefs in the northern Gulf of Aden are limited by intense cold seasonal upwelling. Extensive coral reefs surround the Socotra archipelago.

STATUS OF CORAL REEF BENTHOS

There were many surveys and studies on coral reef communities between 1997 and 2000 in this region, including some that updated information collected previously, to allow comparisons over time at key locations. In some countries, these involved rapid assessments of coral reef habitats, with the most common methods being: belt transects for fish census, line intercept transects for substrate cover and composition, and timed swims for assessment of general reef condition and major disturbances. In Saudi Arabia and Yemen, satellite imagery was used to document the extent of reef habitats. Wide differences in the number of hard coral species were reported for the different countries, generally reflecting variance in the intensity and method of examination, rather than real differences. For example, the largest number of coral species have been reported in Saudi Arabia, where there has been a major effort to assess Red Sea reefs.

Coral cover on Red Sea reefs is generally high, in particular along the fringing reefs and offshore islands and barrier reefs. In the Gulf of Aden coral growth is limited by the seasonal upwelling of colder water, and reefs are patchily distributed, while reefs surrounding Socotra have an unusually high cover. Most areas of the Gulf of Aden and Socotra do not have true growing reefs, but some sites have live coral cover of over 50%, with very large colonies growing on rocky bases. Around Socotra, hard coral cover ranged from 1% to 75%, and there were some large patches (e.g. 1,000m^2) with about 100% cover. Overall, living hard coral cover averaged 20%, with highest cover (35%) on the Brothers (Samha, Darsa and including Sabunyah) Rocks, and on the north coasts (25%) as compared to 5% the south coasts. The following provides a synopsis of coral cover and health for each country based on recent surveys. (Results related to fish communities and links with respective fisheries are reviewed subsequently).

Djibouti
Two brief, but extensive, reef assessments were conducted in 1998, followed by a more comprehensive survey in 1999. These provided the first detailed information on the coral reefs, showing that there were 167 coral species dominated by *Acropora hemprichi, Echinophora fruticulosa* and *Porites nodifera*. Only 10% of the species were found at all sites, 40% were observed at several sites, and approximately 50% were found at only a few sites. *Acropora* spp. suffered high mortality in Khor Ambado and off Maskali. Percentage cover ranged from 5% (off the main tourism beach on Maskali) to 90% (at Hamra Island, Sept Freres), where the dominant *Acropora* formed coral gardens. However, the diversity was highest at Arta Plage, then Grande Isl. (in Sept Freres), and then Trois

Plages (Gulf of Tadjourah). At 26 sites coral cover was greater than 50%, and during swims along the reef edge, the percentage of living coral ranged from 5-70%, and exceeded 20% in all but 3 places.

Egypt

The reefs in southern Egypt are more diverse than those in the north, with nearly double the number of coral species. The distribution and development of reef-building corals is restricted in the Gulf of Suez by several factors, including temperature, sediment load, salinity and light penetration. During 1997 to 1999 three different, but coordinated, projects conducted surveys from approximately 130 reef sites between 1997 and 1999 from Hurghada to Shakateen (and more detailed studies to assess diving impacts at 11 sites near Hurghada). These have been summarised by respectively in reports by the: Egyptian Red Sea Coastal and Marine Resources Management Project; the Ecological Sustainable Tourism Project and the Coral Reef Biodiversity Project. Rapid Environmental Asssessments (REAs) have been made at 48 frequently visited dive sites as part of the Environmentally Sustainable Tourism Project. Collectively the results from the above surveys indicate there was 55% coral cover in non-sheltered areas, and 85% in sheltered areas. Live coral cover generally ranged from 11 to 35% on the reef flats, with the highest cover on reef walls (12-85%) and reef slopes (5-62%). Live coral cover was highly variable along the coast, with the highest being on reef walls and the leading edges of the reefs. A decline of 20-30% in coral cover has been recorded at most sites, and this corresponds with increases in the cover of recently dead coral, and crown-of-thorns starfish (COTS) outbreaks.

Jordan

Current research is focused on coral communities near the fertiliser industries but extends along the entire coastline to monitor biological and physical characteristics including currents, temperature, and nutrient levels. The Gulf of Aqaba is among the most diverse high latitude reefs in the world, with 158 species in 51 genera. The more tolerant and opportunistic *Stylophora pistillata* now dominates and replaces other corals, modifying the structure of the reefs. Recent surveys at 15 sites indicated the reefs were in good condition with over 90% hard coral cover, although it is unknown what proportion of this was alive.

Saudi Arabia

Coral reef habitats were assessed between 1997 and 1999 in the central-northern Red Sea, from to Haql in the Gulf of Aqaba, to Jeddah to produce detailed inventories for corals, fish, other benthos, algae, seagrasses, marine mammals, turtles, coastal vegetation and birds. These were combined with socioeconomic assessments of human use and detailed aerial photos and satellite image mapping to define key reef areas for conservation within MPAs. In the Farasan Islands Marine Protected Area (FIMPA), live coral, dead coral, and the coral predators (crown-of-thorns starfish and *Drupella* snails) were assessed in 1999. At least 260 species of hard corals have been identified from Red Sea waters, including 26 species not previously described and about 50 species as new records for the Red Sea. The predominant families were Acroporidae, Faviidae and Poritidae. Reefs also contained a diverse mix of soft corals, hydrozoan fire corals, gorgonians, corallimorphs and zoanthids. Hard coral species diversity ranged from 20-100 at different sites in the central to northern Red Sea, with a high degree of homogeneity in each coral community. Reefs with moderate to high species diversity and abundance were widely distributed, and these reefs will serve as good sources of larvae for damaged areas.

Living cover of reef-building corals ranged from 10% to 75%, while soft corals were up to 50% cover. Some reefs had more than 20% dead coral cover following coral bleaching or COTS predation. The highest cover of living corals occurred on reefs with relatively high wave energy and very clear water e.g. on the shallow reef slopes of exposed fringing, patch and barrier reefs. Most deeper reef slopes (below 10m), reefs in low wave energy environments and reefs in dirty water had lower living coral cover than shallow, more exposed parts, but there were some exceptions.

Somalia

Rapid assessment surveys in 1997 and 1999 on the north coast provided the majority of the current knowledge on the coral reefs, but detailed surveys are needed for the southern coast. At least 74 hard coral species, 11 soft coral species (Alcyonacean), and 2 fire coral species were found in 1999. Other more sparse reef organisms observed included lobsters (*Panulirus versicolor*), two species of anemones, giant clams (*Tridacna* spp.) along with sponges, ascidians, holothurians, echinoids, crinoids, molluscs and zoanthids. Living coral cover varied between 0-60%. The average coral cover on reefs affected by bleaching and COTS outbreaks was 2-5%. On unaffected reefs, which occur as a narrow fringing band on the outer perimeter of the reefs, coral cover ranged from 60-80%.

Sudan

Sudanese waters contain among the highest diversity of fishes and corals in the Red Sea, with reports of more than 200 species of scleractinian corals. Reef Check surveys in 1999 at Abu Hanish Jetty, Bashear Port and Arous to determine coral cover and assess reef health, built upon more detailed surveys in 1997 using 10m quadrat analysis and 20-minute timed swims of the coastal area from Port Sudan to Suakin. Over 80% of the coastal fringing reefs had high cover of thin turf algae (averaging 28.8%), with live coral cover ranging from 5 to 60%. Recently dead coral above 1% was noted at only 5 sites. Live hard coral cover at Abu Hashish Jetty ranged from 24% at 10m depth to 50% at 5m, while dead coral ranged from 2.5% at 10m depth to 0% at 5m. At Bashaer Oil Exporting Port, there was 38% live coral and 21% dead coral. Dead corals covered 51% of the substrate at Arous and no bleaching was observed below 4m. There were bleached corals on top of the fringing reef at 2m, with 14% cover of bleached corals.

Yemen

Survey methods have included satellite imagery (Socotra), rapid ecological and impact assessments (Red Sea, Gulf of Aden and Socotra), and more detailed biodiversity studies (Red Sea and Socotra) to start long-term monitoring programmes in the Red Sea and Socotra archipelago. Several major projects have assessed the living marine resources on much of the Yemen coastline, as well as the Gulf of Aden and Red Sea by the Arabian Seas Expedition. These show 176 species of stony corals, with the number at sites from 1 to 76. Almost 50% of the Red Sea sites had more than 40 species and 12% of the sites had over 50 species (similar to the central and northern Red Sea). At least 19 new distribution records for the southern Red Sea were identified, and more corals await identification. Diversity is lower along the Gulf of Aden coast area of Yemen, which may have about 100

MONITORING CORALS IN SOCOTRA (YEMEN)

The islands of Socotra are a little known part of the coral reef world in the Arabian Sea, north west Indian Ocean. Long renowned for their terrestrial biodiversity, the islands also support highly diverse coral and fish communities with unique biogeographic affinities to Arabia, east Africa, the Red Sea and greater Indo-Pacific. To assess present status and future trends in these communities, a long-term monitoring program was established in 11 monitoring sites scattered around the island group, as part of a GEF-UNDP Biodiversity Project and under the auspices of the Yemen Environment Protection Council. The GCRMN methods were modified for these sites which have little true reef development, and transects were set at just one depth range (3 – 5m or 8 – 11m), and aligned parallel to each other about 15m apart. Some sites were composed of large mono-specific stands of coral; others by more diverse communities of more than 50 hard coral species; and others were dominated by macro-algae with sparse corals. Coral cover ranged from less than 10% in algal areas exposed to seasonal cool upwelling to greater than 50% in the rich coral patches of the more-sheltered north coasts. These corals were affected by the 1998 global bleaching event, like other reef areas of the Indian Ocean. Several sites experienced major coral mortality, with loss of over half of total coral cover (approx. 25% decline in total area) and shifts in community structure, whereas others remained in a near-pristine state. Between 1999 and 2000, changes in these communities included coral death following flood run-off during intense rains of December 1999, and clear coral recruitment at sites badly impacted by bleaching in 1998. The coral recruits are growing rapidly, suggesting that here also some small juvenile corals survived the bleaching. At present, reef fishes are both diverse and abundant, with broad size distributions. Indeed, the waters above some of the small coral patches teem with fish – an impressive, but not high biomass. The reef fish are now being actively targeted by local fishermen, as national and international demand increases. The islands are also becoming an international tourism destination, with expanding infrastructure and development proceeding apace; and the recent Government proclamation of the Socotra Conservation and Development Zoning Plan, incorporating large multiple-use marine and terrestrial protected areas, is very timely. The islands should prove important as monitoring sites, being located in a key location for assessing effects of climate change and other impacts. The work to date has been undertaken by a joint international (Senckenberg Museum) and Yemen team. With completion of the GEF-UNDP Project, there is an urgent need for continued financial support for monitoring, management and protection of the area, if these communities are not to go the way of other unique marine ecosystems now being rapidly depleted. From: Lyndon DeVantier, Catherine Cheung, Malek AbdalAziz, Fuad Naseeb, Uwe Zajonz and Michael Apel coral species. There are about 240 hard coral species on the Socotra archipelago, making it one of the richest sites in the western Indian Ocean. The high diversity occurs because there is a mix of coral fauna from different part of Indian Ocean and Red Sea.

Cover of stony corals , dead corals, soft corals and algae is highly variable at sites in the Red Sea, Gulf of Aden and Socotra archipelago. In the northern Yemen Red Sea, there has been extensive coral mortality in the past 10 years with major reductions in living coral cover. Reefs of the southern Yemen Red Sea, and fringing offshore islands, were less disturbed, and had higher living coral cover. Semi-protected island reefs in the northern Yemen Red Sea had low average live coral cover (17%), high average dead coral cover (34%) and high macroalgae cover (20%). The northern and central coast and nearshore islands had very low live coral cover (3%), with very high recently dead coral (averaging 34%) and macroalgae cover (34%). Clear water reefs facing the open sea had the highest live coral cover (29%) and lowest dead coral cover (14%) in the Yemen Red Sea, along with the highest coral species diversity (46) and the largest coral colonies. Deep water pinnacles and submerged patch reefs once high coral cover (52%), but now there are similar levels of live (24%) and dead coral (28%). Exposed reefs with algal crests and mono-specific corals often had dead coral (branching and table *Acropora* colonies plus dead massive corals). Previously these had very high coral cover (averaging over 50%), but now it was mostly dead standing coral (44%). Southern fringing reefs also have more dead corals (23%) than live corals (15%), with cover of macroalgae (14%).

STATUS OF REEF COMMUNITIES AND FISHERIES

The surveys of 1997 - 2000 on coral reef communities also collected information on fish communities and biodiversity, through rapid habitat assessments, timed swims and belt transects. The following summarises artisanal and commercial fisheries for each country.

Djibouti
Fishes are relatively well distributed and are generally not over-exploited e.g. sharks were sighted at 4 places. Angelfish were observed in all surveys, with counts from 6 to 31, and frequently with 10 or more angelfish. Total butterflyfish counts ranged from 8 to 110, with *Gonochaetodon larvatus* the most abundant. Grouper abundance ranged from 0 to 56, and often with 20 or more. The most frequently observed species was *Cephalopholis hemistiktos*. Commercial fisheries are limited in Djibouti; however, subsistence fishing is important, with about 90 artisanal fishing boats (mostly small, 6-8m open boats with outboard engines and crews of 3 for day trips). There are 15 larger boats (10-14m with 5 man crews that can go out for 4 days). Thus it is mostly small scale, subsistence fishing with hook and line, and there are no fish processing plants. About 75% of the catch is landed at Boulaos between May, and September, with other landings at Escale, Tadjourah and Obock. Catches are composed of grouper (23%), Spanish mackerel (14%), red snappers (13%), antak (12%), blackspot snapper (10%), bonito (5%) and jacks (4%). Fisheries production increased from 200 metric tonnes in 1980 to 400mt in 1984 and 700mt in 1988. Production decreased dramatically between 1991 to 1994, due to political unrest.

Egypt
Fish surveys noted 261 species in 89 genera in the Egyptian Red Sea, with more on the southern reefs than those further north. Exposed reefs generally had more species than sheltered reefs, probably due to fewer divers and fishermen in these areas. Pomacentridae

(damselfishes) were the most abundant family (16-26 species across all sites), followed by 20 species of Labridae (wrasses). The most common damselfish was *Chromis dimidiata*, and the most common wrasse was *Labricus quadrilineatus*. Parrotfishes were least abundant with only 9 species; *Hipposcarus harid* and *Scarus ferrugineus* were the most common. Red Sea fisheries contribute approximately 11-14% of the total annual Egyptian fish production, with 44% of this from coral reefs. Large fishing boats generally fish in southern Red Sea waters, but land most of their catch in Suez. Fisheries are regulated by the General Authority for Fish Resources Development of the Ministry of Agriculture, which aims to increase fish catches to 70,000mt by 2017, but there is currently no active management of the Egyptian Red Sea fisheries. Over 7% of the national workforce are involved in fisheries, with 78% of marine fish landings occurring through Suez. There are 27 commercial reef fish species, but 5 constitute over 48% of the 22,000mt annual. The balance is made up of crustaceans, offshore pelagic fishes and demersal fishes (in equal proportions). Fish catches increased in 1993, and then decreased steadily.

Jordan
There are no recent data on reef fish biodiversity. Commercial and artisanal fishing is based in Aqaba with about 85 fishermen and 40 boats. Total catch in 1995 was 15mt, down from 103mt in 1993 and the maximum of 194mt in 1966. There are no cold storage facilities and catches are sold on landing.

Saudi Arabia
There are no recent data on coral reef fish fauna for the Red Sea, and past records vary greatly from 776 species in 1971, to 1,000 species in 1984, to 508 species in 1987, and 325 species in 1988. The differences reflect what definition of 'reef fish' was used. The fishery in Saudi Arabia was almost exclusively artisanal from small boats and larger Sambouks until 1981. Coral reef fisheries occur along the Red Sea, with most fishing boats in the south, many of which are prawn trawlers and pelagic fishing boats. Fishery statistics do not distinguish specific reef fisheries, or between Red Sea and Gulf fisheries. Reliable long-term catch and effort data are required to implement specific management for reef fish.

Somalia
Reef fish are diverse and abundant, with the presence of large schools as well as large fishes, which all indicate low exploitation. The reef fish community differed considerably from the eastern Arabian Peninsula to the north eastern Africa to the south, and the Red Sea to the west, particularly for the families Chaetodontidae, Acanthuridae and Balistidae. Fishing is limited and almost entirely artisanal, with scattered landing sites (Caluula, Xabo, Qandala, Laas Qoray, Berbera, Lughaye, and Saylac). These fisheries are essential for the livelihood of much of the coastal population. Somali fishermen target demersal stocks and some reef fish using basic fishing gear. Lobster are taken on nearshore reefs in the south east, and most commercial operations are through illegal foreign vessels (mainly from Yemen), that provide no statistics. The small northern Somali fishing industry is located at Berbera, Siyara and Karin using gill nets and hook and line methods from small canoes. Turtles are harvested opportunistically, both by harpooning at sea and capturing nesting turtles. There are permanent gill nets around the coral reefs at Siyara to catch sharks for the sharkfin export. The potential yield of small pelagic fish has been estimated at 70,000 to 100,000mt for the entire Somali coast.

Sudan

Based on surveys in Sudan, fish communities were healthy and abundant, with the prized humphead wrasse (*Cheilinius undulatus*) found in 3 of 25 sites. Butterflyfish and angelfish were observed at most sites and at least 15 sites had more than 10 angelfish. Triggerfish, which are often targets for fishermen, were only recorded at 7sites (maximum of 2 per site) and groupers were seen regularly, with 13 sites having more than 20 fish. Similarly, snappers, surgeonfish, including the endemic *Acanthurus sohal* and *Ctenochaetus striatus*, were abundant at most sites, and sharks were seen at 3 sites. Fisheries are minor in the national economy, but important for subsistence fishers. Neither commercial nor artisanal landings approach the estimated maximum sustainable yields, but there are no prospects for increasing the fishery because of insufficient refrigeration and transport. About 65 species are exploited, mostly fish, but also sharks, rays, prawns, lobsters, crabs, molluscs and sea cucumbers. There is also a trochus (*Trochus dentatus*) fishery. Over 80% of fish are caught with hook and line from an estimated 400 small fishing boats and about 300 slightly larger boats (9-10m with 4-5 crew). The Fisheries Administration of Sudan suggests that the maximum sustainable artisanal yield is around 10,000mt, with present annual production of 1,200mt. Peak landings occurred in 1984 and have gradually decreased by 30% since then. All the shallow water areas (mersas) along the Sudanese coast are potential spawning grounds.

Yemen

There are few published studies on fishes of the Gulf of Aden and the Arabian Sea coasts of Yemen, but they indicate that fish diversity is exceptionally high for the Arabian region. A preliminary study on Socotra in 1996 recorded 215 species of shore fish, 7 being first records for Arabian seas. Studies in the northern Gulf of Aden in 1998 showed 267 species, with 8 new records for Arabia. Fishing is a traditional profession along the entire coastline and on the islands, and annual catches vary between 90,000 and 95,000mt and more than 90% of the total fish catch is artisanal. Most landings come from trawling in the Red Sea and the pelagic fishery in the Gulf of Aden. Reef fisheries are undeveloped and predominantly subsistence in the Red Sea (from Midi, Khoba, Hodaida and Khaukha, and Mokha) and around Socotra. There are only minor reef fisheries in the Gulf of Aden. Large pelagic fish are caught including tuna, Spanish mackerel, sharks, jacks and marlins, but there has been gradual decline in pelagic catches since a peak in 1989, and bottom fish stocks have been declining sharply since 1987. Sharks are also fished, using trolling and surface longlining, with annual catches around 7,000mt. There is an artisanal fishery for lobsters (*Panulirus* spp.) in Hadhramut and Mahra and around Socotra. Reef fishing occurs along the entire coastline of Socotra, but there is only one processing plant for commercial catches, in Hadibo. Catch statistics are generally unreliable because sales are local for small size fish that are not weighed. There are also no accurate figures on the fishing effort.

THREATS TO CORAL REEF BIODIVERSITY

The level of anthropogenic threats to coral reefs throughout the region are much lower than other areas of the world. Most threats are shared by all countries due to the semi-enclosed nature of seas in this region, but are often more applicable or important to one country. Many are potential threats rather than existing ones e.g. coastal development in Somalia is virtually at a standstill, compared to Saudi Arabia, but with political stability, growth of coastal settlement will increase and result in sediment runoff, sewage pollution, and land-filling.

Habitat Destruction
Extensive coastal development, including dredging and filling, is destroying large tracts of coral reefs in Saudi Arabia (particularly around major urban centers along the Red Sea and in the Gulf), Egypt (around the Sinai and Hurghada), and parts of the Yemen coastline. Reef destruction also occurs in Egypt, Sudan and Djibouti through ship groundings, and from anchor and flipper damage by recreational divers. Urban, industrial and port development causes damage, because there is inadequate environmental planning, and few or no environmental assessments. Sedimentation invariably results from poor construction, dredging and land reclamation. There is a lack of management awareness, and enforcement of regulations, which often results in physical damage to coral reefs through ignorance or neglect.

Industrial Activities
Chronic industrial pollution has reduced water quality in Egypt, Jordan, Saudi Arabia, and major port areas in Sudan. This includes the discharge of untreated oily wastes from refineries in Egypt, Saudi Arabia and Sudan, and sewage and phosphate ore washing are principal caused of nutrient enrichment along the Egyptian and Jordanian coastlines. Rarely is sewage treated in the region, and most is discharged into the intertidal zones. Considerable solid wastes such as plastics and metal drums are dumped into the sea from urban areas and passing ships, particularly near the Suez Canal and passing trough the Straits of Bab al Mandab and major ports (Aden, Port Sudan, Suakin, Jeddah, Hurghada, Suez). Other major sources of pollution in Sudan are from a power station in inner Port Sudan harbour and carbon residues from a tyre manufacturer.

Oil and other Hydrocarbons
Threats comes from both exploration and transport; millions of tonnes of oil pass through the region. There have been more than 20 oil spills along the Egyptian Red Sea since 1982, which have smothered and poisoned corals and other organisms. Likewise many oil spills have affected the Saudi Arabian Gulf coast, and lesser portions of the Yemeni and Sudanese coasts. There is regular oil leakage from terminals and tankers in Port Sudan harbour and elsewhere, and also from ballast and bilge water discharges. Seismic blasts during oil exploration also threatens coral reefs. Virtually no ports have waste reception facilities and the problem will continue because of a lack of enforcement of existing regulations. There is inadequate control and monitoring of procedures, equipment and personnel and training, and the potential is always there for catastrophic oil spills, but there are no mechanisms to contain and clean such spills.

Maritime Transport
Major shipping routes run close to coral reefs, e.g. about 16,000 ships pass through the Strait of Bab al-Mandab each year, and 25,000 to 30,000 ships transit the Red Sea annually. Apart from ship-related pollution (e.g. discharges of garbage and oily wastes; bunkering activities), these ships often hit reefs and the reefs are regarded as navigation hazards, particularly near the ports of Djibouti, Jeddah, Port Sudan and Suakin, where ships pass through narrow, unmarked channels among large reef complexes. Sewage and discharges of solid waste pose additional threats. There are poor navigational control systems, and a lack of suitable moorings throughout the region.

Fisheries
Shark resources are depleting rapidly with rapid declines in shark-fin catches by local fishermen and foreign poaching vessels, particularly Yemen, Somalia, Djibouti and Sudan. Many fishermen operate without licences, catching shark by hook and line and nets, and damaging coral reefs. Large amounts of by-catch, including turtles, dolphins and finfish are killed. There is a lack of surveillance and enforcement of existing regulations, such as the unregulated use of spearguns in MPAs. Over-fishing of reef species is evident in Djibouti and parts of Yemen, with the removal of predators (snappers, triggerfish and pufferfish), possibly catalysing the crown-of-thorns starfish outbreaks that are causing major damage in Egypt and Yemen.

Destructive Fisheries
Unsustainable fishing practices include spearfishing, the use of fine mesh nets, and some dynamite (blast) fishing along the Egyptian coastline and other areas, but in general the incidence is much less than in Asia.

Recreational Scuba Diving
Some damage occurs around major tourist dive sites as anchor, trampling and flipper damage to fragile corals, particularly around the major tourist sites in Egypt at Ras Mohammed and Hurghada, in Sudan at Sanganeb and at Sept Freres and Moucha and Maskalia in Djibouti, where thousands of tourist divers visit each year. Large amounts of corals, molluscs and fish are collected for the curio and aquarium trades in Egypt, and was widespread in Saudi Arabia in the 70s and 80s, although this is somewhat curtailed today.

Sewage Pollution
Most sewage in the region is discharged untreated or partially treated, often directly onto coral reefs off major towns in Saudi Arabia (e.g. Jeddah, Jizan and Al-Wedj), Yemen, Sudan (Port Sudan and Suakin), and Djibouti. There virtually no sewage treatment plants in the region, existing plants lack regular maintenance, and coastal habitats are damaged because there are inadequate pollution control regulations, monitoring and enforcement. Algal booms have been reported on the coral of Sudan as a result of sewage discharges.

Natural Predators
There have been recent major outbreaks of the crown-of-thorns starfish (*Acanthaster planci*) and sea urchins (*Diadema* sp.) in most countries. Gastropod snails (*Coralliphyllia* sp. and *Drupella*) actively feed on *Porites* and branching *Acropora* at sites in Yemen, Saudi Arabia (southern Red Sea) and Jordan with partial mortality in coral colonies ranging from 10% to 70%. The large outbreak of *A. planci* (10,000 individuals) occurred around Gordon reef, near Tiran island, Egypt in 1998, and large numbers were also found at Khor Ambado in Djibouti. Recent extensive coral mortality on offshore Red Sea reefs in Yemen, resembled crown-of-thorns starfish damage. The urchins *Echinometra* and *Diadema* spp. occur in moderate to high abundance ($>10m^2$) at some sites, and are major contributors to bio-erosion along with grazing parrotfishes (Scaridae) and boring sponges. Bio-erosion was particularly noticeable in Yemen after the 1998 bleaching event.

Coral Bleaching

There was extensive recent coral mortality on many reefs, including those in the Arabian/Persian Gulf, the northern nearshore area of the Red Sea, in the southern Red Sea, the Socotra archipelago and north east Gulf of Aden. A number of Red Sea sites with healthy coral cover in the 1980s, experienced near total mortality. Bleaching around the Socotra islands and NE Gulf of Aden was patchy in 1998. At the worst affected sites in Yemen and Saudi Arabia, most species were injured and about half of the live coral cover was killed. In Djibouti over 30% of coral was killed, while no bleaching was reported in Jordan.

Coral Disease

These are apparently becoming more prevalent in the Red Sea, and include black-band and white-band disease, which may result from cumulative anthropogenic stresses such as high nutrient and sediment loads.

Desalination

There is extensive use of desalinated water to meet demands of the population and industry. There are at least 18 desalination plants along Saudi Arabian Red Sea coast which discharge warm brine and maintenance chemicals (chlorine and anti-scalants) directly near coral reefs. In Yemen, power stations at Mokha, Ras Katheeb and Hiswa (Aden) discharge saline high-temperature water which results in localised coral bleaching and mortality.

Floods

These are rare in this low rainfall region, however occasional heavy rainfall in Egypt, Yemen and northern Saudi Arabia, results in wadis delivering increased sediment loads and fresh water, mostly impacting reefs that fringe the mainland coasts.

MARINE PROTECTED AREAS (MPAS) AND LEVELS OF MANAGEMENT

Tourism has been the catalyst for the development and implementation of Marine Protected Areas (MPAs) in the region, with most of these in the northern Red Sea (e.g. Ras Mohammed and Hurghada areas) and Djibouti to the south. Recent awareness of the ecological importance of several sub-regions has resulted in areas being protected based on environmental qualities (such as Sanganeb atoll in Sudan, the Socotra archipelago in Yemen and the Jubail Wildlife Sanctuary in Saudi Arabia). The following provides a brief overview of MPAs in the region (additional details can be found in the 1998 Strategic Action Programme for the Red Sea and Gulf of Aden).

Djibouti

Two MPAs have been established for more than 10 years, and 2 more are proposed for protected status, including Sept Frères which is of regional importance.

Egypt

Ther are 4 MPAs that include coral reefs, and another two without reefs. Most of these are around the Sinai Peninsula at Ras Mohammed to support recreational scuba diving, but considerable anchor and flipper damage is evident. All of the offshore islands and mangroves are also legally protected but need better enforcement. Another 7 MPA areas of various scales have been proposed or suggested to the Government for protection though various projects e.g. GEF, USAID, etc.

Jordan
There are no existing MPAs, and the Aqaba Coral Reef Protected Area is the only area proposed for protected. This encompasses only 5% of the coastline.

Saudi Arabia
There is an extensive network of terrestrial protected areas, but the development and implementation of MPAs is less advanced. Many areas have been proposed and suggested, dating back to the mid- and late 1980s, but progress is slow. The Farasan islands in the far south were protected in 1996, and the Jubail Wildlife Sanctuary was developed shortly after the Gulf war, however, no other recent MPAs have been established. This is expected to change with the resurgence of PERSGA and its Strategic Action Plan, with up to 32 protected areas being proposed for the Red Sea alone.

Somalia
Three areas have been proposed for protection along the Gulf of Aden coast, however, only the Aibat, Saad ad-Din and Saba Wanak areas contain significant coral growth.

Sudan
The only MPA is the Sanganeb Marine National Park, established in 1990. This $12km^2$ atoll has highly diverse and complex coral reefs, but management is limited. Damage from recreation, such as anchor damage from tourist boats, and shipping is considered low.

Yemen
There is one protected area and 6 proposed MPAs, however the process to establish these is relatively new and will only proceed with funding and technical input from IUCN, the Global Environment Facility and PERSGA.

GOVERNMENT LEGISLATION, STRATEGIES AND POLICY ON REEF CONSERVATION

International Agreements: Not all States are Signatories to all the following Agreements, Protocols or Declarations, but these are the primary international agreements in force in the region, with the most pertinent for coral reef conservation and management marked with an *:

- The Protocol for Regional Cooperation for Combating Pollution by Oil and other Harmful Substances in Cases of Emergency (1982);
- The Convention for the Prevention of Pollution of the Sea by Oil (MARPOL)*;
- The Convention on International Trade in Endangered Species of Wild Fauna and Flora (CITES)*;
- The African Agreement for the Conservation of Nature and Natural Resources (Algiers 1988);
- The Bamako Convention on the Ban of the Import into Africa and the Control of Transboundary Movement and Management of Hazardous Wastes within Africa (1993);
- The Protocol Concerning Regional Cooperation in Combating Pollution by Oil and other Harmful Substances in Cases of Emergency (1984)*;
- The United Nations Convention on the Law of the Sea (1985);
- The Convention of the Prevention of Marine Pollution by Dumping Wastes and other Matter (London Convention) and its four annexes*;
- The Regional Convention for the Conservation of the Red Sea and the Gulf of Aden Environment (Jeddah Convention)*;
- The Convention for the Protection, Management and Development of the Marine and Coastal Environment of the Eastern African Region (1988)*;
- The Convention on the Conservation of Migratory Species of Wild Animals (1986);
- The Protocol Concerning Co-operation in Combating Marine Pollution in Cases of Emergency in the Eastern African Region (1988);
- The Convention on Biological Diversity (CBD)*.

Regional Agreements
One was signed by Yemen, Djibouti and Somalia to establish a sub-regional centre to combat oil pollution in the Gulf of Aden. Oil spill response facilities are stored at Djibouti. Yemen and Djibouti are currently negotiating a bilateral agreement regarding the use of this equipment. In 1986, Djibouti, and Somalia signed a bilateral fishing agreement.

National Legislation and Compliance
A number of Presidential decrees, Public Laws, Acts, Ordinances, Strategies and Regulations have been formulated and implemented for coral reef conservation, however, these are too numerous to list and there is considerable overlap in efforts to deal with oil and other forms of pollution, coastal development and tourism, including land-filling and dredging, sewage disposal and coral mining, through which coral reefs receive direct and indirect protection. More detailed listings of these instruments are provided in the complete country reports.

RECOMMENDATIONS TO IMPROVE CONSERVATION OF CORAL REEF RESOURCES

Summary of targeted national recommendations:

- Djibouti specifically needs improved legislation and enforcement, and a research and monitoring programme that supports coastal area management plans.
- Egypt needs to implement an integrated coastal area management plan and a review and upgrade of existing regulations to protect coral reefs that are coming under strong development pressures.
- In Jordan, pollution is limited and localised, and the main threats are oil spills and discharges, industrial discharges, municipal and ship-based sewage and solid waste, and the tourism sector.
- In Saudi Arabia, there is a need to establish MPAs and curb harmful development in urban areas. Threats originate primarily through residential and industrial development and maritime transport, including oil spills, landfilling, pollutant discharges, and effluents from desalination activities. The issues that remain unresolved or poorly addressed include enforcement of existing emission standards, industrial development, and integration of the public and private sectors.
- Improved conservation of coral reefs in Somalia will rely on increases in funding and personnel. Conservation is currently given a lower priority than the rebuilding of the nation and the eradication of poverty.
- Sudan has a weak legal framework for reef conservation and the absence of surveillance is resulting in damage to many reef areas.
- Yemen needs to develop a network of protected areas to conserve reef resources, as coastal development, the petroleum industry and maritime shipping pose significant risks in the form of untreated sewage, land filling, and hydrocarbon pollution.

Overview of recommendations applicable to all countries:

- Improvement of navigation conditions and waste handling facilities and navigation systems, including markers and updated charts are needed. The predominant threats to reefs in the region are from shipping, pollution from the petrochemical industries, industrial development, and coastal development that degrades of marine habitats through dredging and landfilling, and sewage discharge. There is a need for a more thorough network of navigation markers and a long-term maintenance programme for these. In addition, there is a need to establish sewage, sludge and oily waste holding and treatment facilities at the major regional ports (Suez, Port Sudan, Suakin, Djibouti, Aden, Jeddah, Jordan).
- There is a need to develop and implement coastal management programmes in each country, including the establishment of a system of marine protected areas. These are both needed in tandem to maximise ecologic and economic development and conservation goals. Underpinning legislation that curbs these environmentally degrading activities is also needed. Management plans should address landfilling, dredging and sedimentation in particular, listing sound environmental practices to control these activities.

- Better research facilities and improved capacity of local people in research and monitoring techniques are required, including reef assessment and monitoring of pollution sources and the development of databases and report writing capacity. These should include the translation of relevant training manuals into Arabic, and the development of field guides and other materials in various languages for the tourism sector. It should also address training of the tourism and public sectors to assist in data collection (such as through Reef Check).
- The development and establishment of monitoring network with a GCRMN Node would allow an upward flow of information to assist global efforts for reef conservation. The Node would channel information for each country to a regional level, which would make information more accessible at the global level.
- Finally, there is a need for more detailed reef studies in all countries except Saudi Arabia and Yemen, that go beyond rapid assessments. These studies should also address the status of coral reef fauna, and include a temporal scale to provide continued feedback on reef health. Coral reef management and monitoring should become a priority issue for environmental conservation at government levels but also at other levels through comprehensive public awareness programmes.

ACKNOWLEDGEMENTS AND SUPPORTING DOCUMENTS

This report was compiled from country reports prepared under the auspices of the Regional Organisation for the Conservation of the Environment of the Red Sea and Gulf of Aden (PERSGA). The regional report authors include: Nicolas Pilcher is based at the Institute of Biodiversity and Environmental Conservation, Universiti Malaysia Sarawak, Malaysia <nick@tualang.unimas.my>, and Abdullah Alsuhaibany works with PERSGA in Jeddah Saudi Arabia <abdullah.alsuhaibany@PERSGA.org>.

Each of the collaborators on each of those reports and all of the researchers who contributed to them are gratefully thanked. We would also like to thank Prof. Abdulaziz Abuzinada and Dr. Hany Tatwany at the National Commission for Wildlife Conservation and Development for giving access to material presented at the Regional Workshop on the Extent of Bleaching in the Arabian Region, held in February 2000. Finally we thank Fareed Krupp for critical reading of the manuscript. All assistance is gratefully acknowledged.

DeVantier, L. & N.J. Pilcher, 2000. Status of coral reefs in Saudi Arabia - 2000. PERSGA Technical Series Report, Jeddah. 45 pp.

Pilcher, N.J. & M. Abou Zaid, 2000. Status of coral reefs in Egypt - 2000. PERSGA Technical Series Report, Jeddah. 17 pp.

Pilcher, N.J. & S. Al-Moghrabi, 2000. Status of coral reefs in Jordan - 2000. PERSGA Technical Series Report, Jeddah. 13 pp.

Pilcher, N.J. & A. Alsuhaibany, 2000. Status of coral reefs in the PERSGA region - 2000. Technical Series Report. PERSGA, Jeddah. In Prep.

Pilcher, N.J. & L. DeVantier, 2000. Status of coral reefs in Yemen - 2000. PERSGA Technical Series Report, Jeddah. 47 pp.

Pilcher, N.J. & N. Djama, 2000. Status of coral reefs in Djibouti - 2000. PERSGA Technical Series Report, Jeddah. 29 pp.

Pilcher, N.J. & F. Krupp, 2000. Status of coral reefs in Somalia - 2000. PERSGA Technical Series Report, Jeddah. 21 pp.

Pilcher, N.J. & D. Nasr, 2000. Status of coral reefs in Sudan - 2000. PERSGA Technical Series Report, Jeddah. 28 pp.

World Bank, 1998. Strategic Action Programme for the Red Sea and Gulf of Aden. The World bank, Washington, DC: 89 pp.

3. Status of Coral Reefs in the Arabian/Persian Gulf and Arabian Sea Region (Middle East)

Nicolas J. Pilcher, Simon Wilson, Shaker H. Alhazeem and Mohammad Reza Shokri

Abstract

This report summarises the status of coral reefs in **Bahrain, Iran, Kuwait, Oman, Qatar, Saudi Arabia** and the **United Arab Emirates**, in which coral reef conservation is coordinated by the Regional Organisation for the Protection of the Marine Environment (ROPME). There are no reports of coral reefs of Iraq. Information was gathered from publications and reports provided by people from these countries and the key agencies operating in the region (some are listed at the end). In the north Arabian/Persian Gulf there are fringing hard corals along the Iranian side of the Gulf and around the most of the islands, from Saudi Arabia across to Iran. Most of the south is characterised by limestone or sandstone, which prevent significant reef accretion. Seawater temperatures and salinity show some of the widest fluctuations recorded anywhere in the world, and corals have adapted to survive in these especially harsh conditions. In the Arabian Sea, coral growth is limited to the southern shorelines, and corals grow beneath a thick algal canopy. There are areas in southern Oman where large corals exist, but these are not true coral reefs. Reefs also occur at Ras Madrakah, at Barr Al Hickman and in the shelter of Masirah Island, and around Ras al Hadd and the Daymaniyat islands.

The region contains complex and unique coral reefs with relatively low biological diversity, but with many endemic species. Large parts of the region are in a pristine state, but there are increasing environmental threats from habitat destruction, over-exploitation and pollution. Coral diversity is lower than in the Indian Ocean, but fish diversity is relatively high. Coral reefs are generally limited to a few areas due to extreme environmental conditions. Coral cover is generally low, with evidence of recent, widespread, coral mortality, as coral communities grow at the limits of tolerance to salinity, temperature, and sediment load. There were two major coral bleaching events, one in the summer of 1996 and another more severe case in the summer of 1998, which led to near-complete mortality of the reefs in Saudi Arabia, Bahrain, Qatar, and UAE. An average of around 50% mortality was experienced in Kuwait and Iran, and lower mortality was recorded in Oman. The major anthropogenic threats are land-filling and dredging of the reefs for coastal development, anchor damage from boats, discharges from industrial and desalination facilities, and solid waste disposal.

Throughout the region, there is a need to reduce boat anchor damage though the use of permanent mooring buoys. Dredging and land-filling need be curtailed and properly managed, and the risks of ship-based pollution should be countered through the provision of waste disposal facilities at major ports, monitoring on the high seas for pollution discharges, and through coordinated efforts to improve shipping navigation aids. Solid waste cleanup projects need to be organised to remove accumulated debris from the reefs, and public education campaigns would help increase understanding of the importance of coral reefs and their sensitivity to damage and pollution. There is also a need to develop and expand local capacity for monitoring and research on coral reefs, in connection with the designation of Marine Protected Areas as part of broader multi-sectoral integrated management plans. The major organisation assisting countries with coral reef conservation is the Regional Organisation for the Protection of the Marine Environment (ROPME), which refers to this region as the 'inner ROPME Sea Area', shortened to RSA.

INTRODUCTION

The Arabian/Persian Gulf (the Gulf) and Sea contain complex and unique tropical marine ecosystems, especially coral reefs, with relatively low biological diversity and many endemic species. The reefs are surrounded by some of the driest coastlines in the world, such that continental influences are limited. In addition these waters are also major shipping lanes due to petroleum industries, with a high-risk bottleneck at the narrow straits of Hormuz. While large parts of the region are still in a pristine state, environmental threats (notably from habitat destruction, over-exploitation and pollution) are increasing rapidly, requiring immediate action to protect the region's coastal and marine environment.

The Arabian Sea is subject to seasonal cold water upwellings due to the Indian Ocean monsoon system, which creates large temperature differences between seasons. This is reflected in the nature and distribution of coral communities. The Gulf is a semi-enclosed shallow continental sea measuring 1000km in length and varying in width from a maximum of 340 to 60km (at the Straits of Hormuz). These narrow straits restrict water exchange with the Arabian Sea, which means the waters become highly saline because of high evaporation and low inputs of fresh water. The Gulf is also subject to wide climatic fluctuations, with water temperatures ranging from 10 to 40°C and salinity from 28-60ppt. The coupling of seawater temperatures and salinity show some of the widest fluctuations recorded anywhere in the world, which would kill most reef-building corals elsewhere. Thus, corals growing in the Gulf have become adapted to survive the especially harsh conditions. This was thought to be the norm with temperature extremes until the major bleaching events in 1996 and 1998, which virtually obliterated all inshore reefs and badly depleted many of those offshore. The overall reef biodiversity of reefs in the Gulf is relatively impoverished compared to the Indian Ocean, with the exception of fish species. Further north there are fringing hard corals along the Iranian side of the Gulf and around the most of the islands, from Saudi Arabia across to Iran. The average depth is about 35m and maximum is 100m. The southern Gulf is characterised by shallow pre-Cretaceous limestone, sandstone or 'fasht' which is frequently insufficient to result in significant reef accretion.

Geographical Reef Coverage and Extent

Bahrain
Coral reefs in Bahrain are limited to a few areas. These include: Fasht Al Adhom, north Jabari, west Fasht Al Dibal, Khwar Fasht, Fasht Al Jarim, Samahij, and Abul Thama. Coral colonies are present in other areas such as Hayr Shutaya, but these are isolated colonies, not coral reefs.

Iran
The best-developed reefs on the Iranian coasts are found around Khark and Kharku islands in the far north and the southern islands from Lavan to Hormuz islands. Fringing reefs are the predominant reef structure on the Iranian coasts.

Kuwait
Coral reefs are largely restricted to the southern area and include a range of offshore platform and smaller patch reefs, and nearshore patch and fringing reef assemblages along the southern coastline. All are in shallow water (<15m) where the greatest growth and diversity is shallower than 10m. The best developed reefs are at Umm Al-Maradem, Qaro and Kubbar, which are typical coral 'cays'.

Oman
Coral growth is limited along the southern shorelines at Dhofhar, particularly on the islands of Al Hallaniyah, A'Sawda, Al Qibliah, where a low diversity of scattered corals grows beneath a tall and thick algal canopy. Bays on the mainland coast and on the Kuria Muria islands support coral communities, which grow on rocky substrate. There are areas in southern Oman where large corals exist, but these are not true coral reefs. Reefs occur at Ras Madrakah, at Barr Al Hickman and in the shelter of Masirah Island at about 20°N, and around Ras al Hadd and the Daymaniyat islands.

Qatar
Fringing reefs occur along the north and east coasts, but are limited by high sedimentation and periodic mortalities caused by freezing temperatures. Reefs on the west coast are sloping, non-growing domes below low tide mark, while true coral reefs occur further offshore. The Gulf of Salwa is too saline for coral development.

Saudi Arabia
The reefs are mostly small pinnacles or outcrops, and patch reefs between Ras Al-Mishab Saffaniyah and Abu Ali, and between Abu Ali and Ras Tanura, and as fringing reefs around the offshore islands of Karan, Kurayn, Jana and Jurayd.

United Arab Emirates
The southern Gulf coast is low-lying, swampy and rich in seagrasses. The water is shallow and unsuitable for good coral growth, although there are numerous patch reefs. The extensive areas of subtidal rocky platform along the Abu Dhabi coast were until recently dominated by corals and algae. Fringing reefs grow around many of the islands, which extend across to the coast of Qatar.

STATUS OF CORAL REEFS

Bahrain

The main coral reefs are limited to a few areas in extreme environmental conditions. Coral cover is generally low, with evidence of recent, widespread coral mortality. These coral communities are at the very limits of tolerance to salinity, temperature, and sediment load, but there are 28 species of coral in Bahrain. The reefs recently experienced two major coral bleaching events: summer of 1996; and more severely in the summer of 1998. These events resulted in the complete mortality of corals on Fasht Al Adhom, west Fasht Al Dibal, Khwar Fasht, north Jabari, Samahij, and Fasht Al Jarim. The only live coral reef surviving in Bahrain is on Abul Thama, a small raised area surrounded by 50m deep water about 72km north of the main island. In addition to coral bleaching in 1996, a large fish kill was reported in Ras Hayan lagoon.

Sea surface temperatures of 38°C were recorded in 1998 on Hayr Shutaya, about 30km north of Bahrain, and coral bleaching followed with morality between 85-90% of Fasht Al Ahdom, Jabari, Samahij, and Fasht Al Dibal. In 1996, corals were 100% bleached at Fasht Al Adhom, and 40% bleached (with 40% dead coral cover) at Jabari. At west Fasht Al Dibal there was 80% bleaching, and at Abul Thama corals were 33% bleached, with 33% dead coral cover. About 95% of the corals were dead on the Half Tanker wreck on Fasht Al Adhom by late-1996. On Abul Thama the hard corals on the seamount (which comes to 8m of the surface) were in better condition, because they are surrounded by 50m deep water. The cover of hard coral has remained relatively stable (25–35%), with a greater diversity of key invertebrate and fish species than Fasht Al Adhom.

The major direct anthropogenic factors impacting on corals in Bahrain are commercial trawl fishing, dredging and land reclamation, with little consideration given to maintaining the environment. Prior to 1998, there were about 300 artisanal shrimp trawlers and 6 larger fish trawlers, which often poached in shallow waters around the coral reefs. The reappearance of small soft corals and crinoids on Bahrain reefs is linked to the 1998 closure of the industrial fishery. There are major industrial plants for electricity generation, water desalination, oil refinery, aluminium refining, and for petrochemicals that discharge heated seawater, which increases temperatures during the summer months. Land reclamation on the northern and eastern coasts have increased the area by 11km^2 in less than 10 years, often covering valuable coastal resources, but not yet directly impacting coral reefs. However, there are proposals to reclaim part of major coral reef areas at Fasht Al Adhom. There are government regulations concerning land reclamation, but there is little enforcement and compliance with these regulations, and many projects are completed without government approval. Dredging is used to maintain navigation channels and collect sand for reclamation and construction. About 10 suction dredgers routinely operate in Bahrain waters, including specialised cutter and suction dredgers. During dredging operations, large amounts of silt flow directly onto corals from Muharraq dredging area, with about 182,000m^2 reef area lost between 1985 and 1992.

There was no regular monitoring of coral reefs in Bahrain until 1993, when a volunteer diving program started with assistance from the Department of Fisheries. These surveys show an almost total loss of live hard corals on Fasht Al Ahdom, with coral cover declining from 30-

40% to 0% between 1997 and 1999. In 1999, divers saw small coral colonies (1-3cm diameter), but these were too few to be detected by Reef Check surveys. Key fish species were also scarce at 4-8m deep Fasht Al Adhom sites.

Iran

There are fringing hard corals along some parts of the coast and around most of the islands. Very little information is available on their status, with the most recent information from Kish Island, 18km from the most southern point of the mainland in the Gulf (26°30'N; 53°54E'). Here there are sandy flats and mixed communities of live and dead corals, with the best live coral on north and northeastern margins. In shallower depths (8m) there is higher biodiversity of reef fishes, invertebrates and hard corals, compared to the deeper reef slope (13m). There was considerable destruction of branching corals during a storm in 1996, anchor damage is higher in deeper water (13m), and many corals in shallower waters are damaged by uncontrolled recreational activities.

There are at least 35 coral species around Hormuz Island, with branching *Acropora* dominating around islands in the Gulf, as well as some soft coral (*Sarcophyton*) and crown-of-thorns starfish (COTs) in the northern part of the Gulf (Tunb-e-Kuchak Island). At Kish Island, 19 coral species have been identified with Acroporidae, Favidae and Poritidae most frequently and Agariciidae and Dendrophyllidae are quite rare. Surveys showed that the upper slope at 8m depth had the best hard coral growth, with 22% cover, 21% dead cover coral, 44% sand and 13% for other invertebrates. Dead coral cover was 28% in deeper water, and 48% in shallow water. Reef Check surveys showed the highest diversity of fish in shallow water.

Coral bleaching was observed in mid-summer (July-August) of 1996 around Kish, Faroor and Hindurabi islands, mostly in massive (*Favia* sp.) and sub massive corals (*Porites* sp.), with about 15% of all hard coral colonies showing some evidence of bleaching. Local divers reported the recovery of these reefs in the following summer (1997). Storms, bleaching and extreme environmental conditions (especially large variations in temperature during midsummer) are the major natural factors threatening the coral reefs at Kish island. Anthropogenic impacts come from coastal construction as this has been declared the main Iranian 'free zone'. The ranking of impacts on reef corals around the Kish Islands are: over-exploitation of living resources; poor land use practices; pollution from land-based activities; extreme environmental conditions; pollution from maritime transport; tourism activities; tropical storms; coral bleaching; and destructive exploitation of living resources.

Kuwait

Corals in Kuwait are in the northern Gulf, hence the extreme northern limit of distribution. Coral patches occur from Kuwait City south to the border with Saudi Arabia. There are also a few coral-fringed islands, and some patch reefs on sea mounts (all south of the major oil terminal at Mina Al-Ahmadi). Maximum depth averages 15m, with greatest coral diversity above 10m. Coral diversity is limited on offshore reefs at Mudayrah by offshore currents and rough seas and the reefs are surrounded by 30m depth. Qaro in the south has the most diverse reefs, dominated by *Acropora* and *Porites*, with above 80% live cover at many sites. There is an extensive reef surrounding Um Al-Maradem enclosing a lagoon about 400m offshore. Nearshore reefs such as Qit'at uraifjan are limited by high sediment loads, although they survived the Gulf War oil spill and are dominated by massive *Porites*, *Montipora* and *Platygyra*.

At least 35 coral species are found on Kuwaiti reefs along with conspicuous sea urchins (*Echinometra mathaei* and *Diadema setosum*), which occur in dense populations (20-80 per m²) on many reefs. There are 124 fish species with the damselfish the most abundant, and hawksbill (*Eretmochelys imbricata*) and green turtles (*Chelonia mydas*) nest on the coral cays.

Threats to coral reefs include damage from fishing and anchors, flipper damage from divers, over-fishing, lost fishing gear, solid waste disposal, and the extreme environmental conditions. Most of the reefs are close to the shore and are major tourist and fishing areas. Anchors have destroyed large tracts of the reefs at all sites, and over-fishing has reduced populations of large predators, such that few fish greater than 20cm are seen. Oil pollution, however, has not caused massive mortalities to reefs, even though most were in the path of the massive oil spill during the Gulf War.

Oman

Seawater temperature is the major factor controlling the distribution of coral reefs in the northern Arabian Sea and Gulf of Oman. There are major temperature differences between the Gulf of Oman and the northern Arabian Sea during the summer southwest monsoon season, which show up in the nature and distribution of coral communities in Oman. Wind is the other factor influencing upwellings, which result in temperatures changing by as much as 8°C in 2 hours down to 10m depth. Such upwellings in the northern Arabian Sea result in seawater temperatures being around 19°C throughout summer.

There were 2 significant bleaching events during the past 10 years: in the summer of 1990 in the Gulf of Oman, seawater temperatures reached 39°C for several days; in May 1998 at Dhofhar, temperatures reached 30°C. Solar radiation during the monsoon (June-September 1998) was 170-180% higher than average because of low cloud density and thinner fog normally associated with the upwelling (all suggest a weak upwelling during that monsoon season). Between 75% and 95% of *Stylophora* colonies bleached and 50% of large *Porites* colonies were partially bleached in water less than 5m on the Marbat peninsula, Arabian Sea in May 1998. Other genera were also affected. Mortality was estimated to be around 5%, and temperatures were 29.5 and 31.5°C, but 50km to the east no bleaching was observed at temperatures between 25 and 25.5°C. No bleaching was observed or reported in the Muscat Area in the Gulf of Oman (23°37N; 58°35E; temperatures to 30.5°C).

Bleaching in Dhofar occurred in May immediately before the summer monsoon upwelling. After the upwelling in September 1999, only dead corals covered in algae were observed indicating almost 100% mortality from bleaching in very shallow (2-3m) waters. Corals in deeper water appeared unaffected with low levels of mortality. On the southwest tip of Masirah Island and southeast corner of the reef complex around Barr Al Hickman there was little or no mortality during 1998. No bleaching was seen in Hallaniyat Islands from February-April 1998 and corals around the islands of Al Hallaniyah, A'Sawda, and Al Qibliah were healthy in January 2000, with high coral cover and very large (5m diameter) tables of *Acropora clathrata* and *Porites* colonies greater than 4m in diameter.

Although seawater temperatures were raised (30-32°C) from mid-April until December 1998 in the Gulf of Oman, there was no bleaching (and no change in mid 2000 at the Daymaniyat Islands). However, low-density infestations of crown-of-thorns starfish were observed, causing significant damage in limited areas (100m x 50m) of the reef. Live coral cover at 53 sites around the Musandam peninsula was 10-20% at half of the sites, and 50-90% at over 25% of the sites. White band disease was seen in some *Acropora* and *Platygyra* at 5 sites, mostly in sheltered areas with high seawater temperatures.

Qatar

Fringing reefs occur along the north and east coasts, with a generally high coral cover but low species diversity (<20 species). Shallow (1-4m) reefs covering several hectares of *Acropora*, with *Porites* mounds east of Doha had high mortality in 1998, often at 100%. At other sites, there was about 10% live coral of *Porites* remaining, although reefs further offshore (up to 10m depth) had a higher coral cover. Results from October 1993 to March 1997 and later in March/April 1997 indicated large increase of coral bleaching in later surveys along the coast of Qatar (from Al Khor to Khor Al Oudeid). Other damage occurs from anchors of local fishing boats.

Saudi Arabia

Approximately 40% of the Gulf coast is highly developed with the presence of most of the oilfields. The reefs are mostly small pinnacles or outcrops, and patch reefs between Ras Al-Mishab Saffaniyah and Abu Ali, and between Abu Ali and Ras Tanura, and as fringing reefs around the offshore islands. Up to 50 species of coral and over 200 fish species have been reported around the offshore coral cays although diversity is lower on the inshore reefs where physical stresses are greater. The average live coral cover was 33%. However live coral cover dropped from 23% in 1994 to just 1% in 1999, on the reef slopes. At one site more than 99% of the colonies were dead with only small pockets of surviving coral tissue. Some reefs had more than 20% cover of dead coral following coral bleaching or COTS predation.

Threats to coral reefs come mostly from industrial development and maritime transport, including oil spills, landfilling, pollutant discharges, and effluents from desalination activities. Most acute damage to reefs is localised and restricted to offshore islands. The effects of the Gulf War oil spills on coral cover and community composition were minimal along the mainland coast and offshore islands. Fishing was almost exclusively artisanal from small boats and larger Sambouks until 1981. Fishery statistics cannot be used to distinguish specific reef fisheries, and reliable long-term catch and effort data are required to implement specific management for reef fish.

There is an extensive network of terrestrial protected areas, but the development and implementation of MPAs lags considerably behind other areas. Many areas have been proposed and suggested, dating back to the mid- and late 1980s, but progress is slow. The Jubail Wildlife Sanctuary was developed shortly after the Gulf war, however, no other recent MPAs have been established. Saudi Arabia has carried out a number of programmes and adopted a number of legal measures to conserve coral reefs, among them laws on pollution discharges and the establishment of protected areas. However, a number of issues remain unresolved or poorly addressed, such as enforcement of existing emission standards, industrial development which includes landfilling, and integration of the public and private sectors in reef conservation.

United Arab Emirates

Coastal waters support extensive areas of mostly mono-specific stands of coral, due to the extreme environmental conditions in the southwestern Gulf. Up to18 species of hard coral have been found on healthy coral reefs near Dalma Island off western Abu Dhabi. Here, the subtidal rocky outcrops and platforms were mainly covered with a veneer of corals in 1996, with large areas at depths of 1-4m covered by *Acropora* with an understorey of *Porites, Platygyra* and *Favia* spp. Greater depths were often dominated by large *Porites* colonies and an understorey of *Acropora*. *Porites*-dominated reefs were particularly well-developed on offshore reefs (such as Bu Tini shoals), where *Acropora* only survived in pockets protected by the *Porites* bommies.

There was extensive mortality of *Acropora* in Abu Dhabi during autumn 1996 and over the same period in 1998. There was more than 98% mortality of *Acropora* between August and September 1996 with seawater temperatures of 34°C or above for 10 weeks. During the summer of 1998 seawater temperatures remained at 34°C or above for 14 weeks, and most of the remaining *Acropora* colonies bleached and died. West of Abu Dhabi city, 60-80% of the other corals (*Porites, Platygyra* and *Favia*) were also bleached. An estimated 40-60% of the non-branching corals were bleached in 1998.

There were 77 fish species recorded in the Dubai area, and 29 species of corals, and now the Dubai Urban Area is protected following local order No. 2 of 1998. The genus *Acropora* suffered almost total mortality, where once it covered 51% of the coral substrate. Now the genera *Porites, Platygyra* and *Cyphastrea* have replaced *Acropora* as the dominant corals. The Jebel Ali reefs, which were previously one of the Gulf's richest ecosystems, lost most of the corals and other organisms in the summer of 1996. *Acropora* were the dominant coverage, and the coral cover was reduced by an average of 60% (ranging from 15-85%).

Dredging of Jebel Ali port and the hotel marina have affected offshore reefs through excessive sedimentation. Another threat to reefs is the change in coastal hydrodynamics due to the interference of the Jebel Ali port breakwater with inshore currents. Recreational use of the area is primarily by clients of the Jebel Ali Hotel and campers. Main activities are shore based: swimming, jet skiing, wind-surfing and surf fishing. A small-scale commercial fishery used to exist in the area, using mainly Ghargours (fish traps) placed in the coral area and targeting reef fish. This has practically ceased since the declaration of protected area status. A shore-based beach seine netting fishery still exists, but at low levels. There are also possible impacts on corals by dredging and infilling activities in certain parts of Abu Dhabi.

RECOMMENDATIONS TO IMPROVE CONSERVATION OF CORAL REEF RESOURCES

- There is a need to reduce diver and boat anchor damage. In all countries, mooring buoys should be installed at the major reef sites to prevent further anchor damage to corals. Additionally, spearfishing should be banned (and the ban enforced) to allow reef fish populations to recover.
- Solid waste cleanup projects need to be organised in each country to remove accumulated debris from the reefs, including plastics, metals, glass, and discarded fishing equipment.
- Public education campaigns are needed to increase understanding of the importance of coral reefs and their sensitivity to damage and pollution. These should highlight the use of moorings to prevent damage to corals, the need for restrictions on fishing on these reefs, the problems caused by littering and refuse, and the need for public and private participation in the management of coral reefs.
- There is a need to develop and expand local capacity to monitor and carry out research on coral reefs. This must also include the designation, where applicable, of competent authorities to manage and conserve coral reefs, and preparation of detailed management plans to promote the sustainable and wise use of reef resources.
- There is a need to designate additional Marine National Parks and Marine Protected Areas in the context of integrated management plans. Only within these protected areas will the countries be able to fully protect coral reefs.
- Coastal development, and in particular dredging and landfilling, should be curtailed and properly managed. The use of silt curtains should be mandatory in landfill operations, and only after stringent Environmental Impact Analyses. Given the limited extent of coral reefs, coral reef areas should never be approved for landfilling.
- Finally, there is a need to reduce further the risks of ship-based pollution through the provision of waste disposal facilities at major ports, monitoring on the high seas for pollution discharges, and through coordinated efforts to improve shipping navigation aids.

Acknowledgements and Supporting Documentation

Nicolas Pilcher is based at the Institute of Biodiversity and Environmental Conservation, Universiti Malaysia Sarawak, Malaysia <nick@tualang.unimas.my>. Simon Wilson is based in Oman <106422.2221@compuserve.com>, Shaker H. Alhazeem is at the Mariculture and Fisheries Department in Kuwait <shazeem@kisr.edu.kw>, and Mohammad Reza Shokri is based at the Iranian National Center for Oceanography <mrshok@hotmail.com>. The authors wish to thank the Regional Organisation for the Protection of the Marine Environment (ROPME) for assistance with writing this report.

References

Carpenter, K.E., P.L. Harrison, G. Hodgson, A.H. Alsaffar & S.H. Alhazeem, 1997. The corals and coral reef fishes of Kuwait. Kuwait institute of Scientific Research, Kuwait City. 166 pp.

IEC 1998. Study of Protected Areas for Nature Conservation in the Emirate of Dubai. Unpublished Report to Dubai Municipality. IX + 234 pp. International Environmental Consultants (IEC), Riyadh.

Pilcher, N.J., 2000. The status of coral reefs in the Arabian gulf and Arabian Sea region. Report to ROPME, Kuwait City.

Sheppard, C., A. Price, A. & C. Roberts, 1994. Marine ecology of the Arabian region. Academic Press, London. 359 pp.

4. STATUS OF CORAL REEFS IN EAST AFRICA: KENYA, MOZAMBIQUE, SOUTH AFRICA AND TANZANIA

DAVID OBURA, MOHAMMED SULEIMAN,
HELENA MOTTA AND MICHAEL SCHLEYER

ABSTRACT

The single largest threat to coral reefs ever documented in East Africa was the high temperature-related coral bleaching event caused by the El Niño Southern Oscillation of 1997-98. The degree of bleaching and mortality increased local water temperatures and reached northwards from South Africa (<1%) to Kenya (80% and greater in some areas). Recovery of affected reefs until early 2000 has been primarily through regrowth of the few surviving colonies. Since this time significant coral recruitment has been observed, raising hopes for recovery of a number of reefs, if other major threats don't transpire. Recovery of affected reefs has been variable, although in some cases faster for previously stressed reefs that had already lost sensitive coral species. Recovery is also likely to be influenced by the presence of other synergistic threats (natural and anthropogenic). Other threats to East African reefs continue since the last global report in 1998; these include: over-fishing, destructive fishing, pollution and, human settlement and development, mining and shipping industry activities. In the last two years, increased efforts in management and conservation of coral reefs and marine resources have advanced throughout all of the East African countries with improvements in institutional, technical and human resource capabilities being the most noteworthy. As elsewhere, East African countries are increasingly coordinating their activities through a number of informal and formal collaborations, leading to better integration and linkages of research and conservation projects related to coral reefs across regional and national levels.

INTRODUCTION

This regional review is focussed on coral reef status and related issues during the last 2-3 years. It builds on the GCRMN global report of 1998 and specifically summarises the coral reefs status for and across the 4 East African States. This review also draws upon biological, socioeconomic and institutional reviews commissioned for the WWF East Africa Marine Ecoregion programme and the annual national and regional reports prepared for the CORDIO programme.

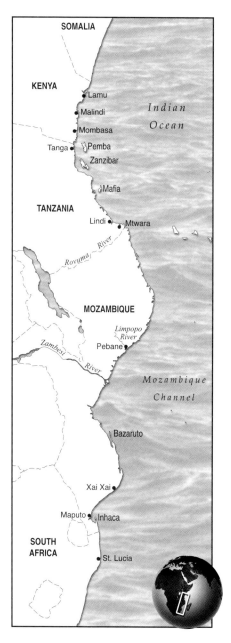

Geography, Environment and Pollution

Because of the north-south orientation of East Africa, there is a strong temperature gradient in the coastal waters of this region with a peak in the mid-latitudes of approximately 4-15°S (southern Kenya to northern Mozambique), and lower temperatures to the south and north. This geographic variation had a particularly strong influence on the health and status of reefs during 1998 when the region experienced the major El Niño climate event.

The East African coast is strongly influenced by river discharges and high nutrient levels, which affects coral reef distribution at both regional and local scales. Large expanses of the central Mozambique coast (800km) are devoid of reefs due to discharges from the large Limpopo and Zambezi rivers. Many fringing reefs have breaks opposite short coastal rivers that empty through deep mangrove creeks. The few measures of nutrient concentrations show high levels compared to other reefs systems, which are consistent with low levels of water clarity due to increased plankton and sediment loads. Information on coral reef status along the Indian Ocean coast of Somalia is poor, though patch reefs are known to extend to Mogadishu. It is likely that reefs do extend as far as the Horn of Africa, with expansion of many reef organisms onto the Socotra Archipelago (see Chapter 3).

South Africa

The coral reefs and coral communities of South Africa lie between 26-27°S. These are the most southerly reefs in the western Indian Ocean. They are generally deep (below 8m), offshore, have high energy banks, and normal temperatures ranging from 22-26°C to 29.5°C. The reefs grow on a narrow continental shelf only 2-7km wide, which is remote from major human populations. The principal use of the reefs has been for game fishing and diving.

Mozambique

The 2,700km coastline extends from 26°S to 12°S, encompassing a full spectrum of reef types from the high latitude reefs of South Africa to the fringing and island reef complexes along the Mozambique-Tanzanian border where the South Equatorial Current meets the

African coastline and splits north and south. The main reef system stretches for 770km from the Rovuma River in the north, to Pebane in the south (17°20'S). Smaller isolated reefs are dispersed along the 850km southern coast from Bazaruto Island to Ponta do Ouro (26°50'S). Mozambique's coastal population was estimated at just over 6.66 million people (42% of the total) in 1997. Artisanal and commercial fishing, and tourism are the dominant uses of coral reefs in Mozambique.

Tanzania
Two thirds of the 1000km coastline supports fringing and patch reefs on a narrow continental shelf. The main areas of reef growth are along the islands of Unguja (Zanzibar), Pemba and Mafia, and the mainland coast at Tanga, Pangani, Dar-es-Salaam Mkuranga, Lindi and Mtwara. Tanzania's coastal population was estimated at 8 million people in 2000, and is concentrated in the districts around the capital Dar es Salaam. Coral reef use is varied, with fishing and tourism supporting local and national economies.

Kenya
A continuous fringing reef dominates the southern 200km of Kenya's coast. The fringing reefs in the north are more patchy with influences from river discharges and influenced by colder water from seasonal upwelling of the Somali current system. Kenya's coastal population in 2000 has been estimated to reach 2 million people, concentrated around the main port of Mombasa. As with Tanzania, reef-based fishing and tourism are important components of the coastal economy.

PRINCIPAL THREATS AFFECTING CORAL REEFS: 1998-2000

El Niño
There was a strong south-north gradient in the extent and impacts of the El Niño bleaching event of 1998, which elevated normally warm local temperatures. Coral bleaching and mortality started in the south in late February to early March 1998, and finished in May in the north, directly corresponding to the 'movement' of the sun as it passed through the Inter-Tropical Convergence Zone. There were dramatic losses of coral cover (summarised below). After 1998, there were significant increases in fleshy, turf, calcareous and coralline algae growing on the newly dead coral surfaces on all reefs, which in turn was influenced by an abundance of herbivorous fish populations. The algae grew larger and more rapidly in areas where there were large reductions in herbivore fishes due to over-fishing. These coral to algal community shifts happened less in protected fish reserves. Non-reef building coelenterates (cnidarians), including anemones, corallimorphs and soft corals were also heavily damaged by bleaching and mortality on many reefs through East Africa. Changes in coral eating animals were minor, but there were some increases in herbivore populations, such as on Mafia Island.

- **South Africa:** The bleaching threshold for corals was exceeded for both hard and soft corals, but it was not excessive. Most corals on the reefs grow at 12m or deeper, which may have protected them from warmer temperatures in surface waters. Soft corals are more prevalent on shallow reef tops, with more conspicuous bleaching.

	1997/98	1999
KENYA		
Northern Kenya (>10m)		5.7+/-3.8 (3)
(< 3m)	13.2+/-2.2 (29)	5.1+/-1.9 (10)
Southern Kenya (> 10 m)	32.0+/-4.6 (6)	6.7+/-2.1 (1)
(< 3m)	30.1+/-3.0	11.4+/-2.6
Protected	39.6+/-2.9	11.4+/-1.4
Unprotected	20.6+/-0.7	11.4+/-2.2
OVERALL	26.3	8.1
TANZANIA		
Tanga	53.0+/-4.8 (4)	33.3+/-6.6 (4)
Pemba	53.7+/-12.8 (3)	12.3+/-3.7 (4)
Unguja	45.8+/-4.2 (5)	32.0+/-5.5 (5)
Kunduchi	43.0+/-6.0 (2)	35.0+/-0.0 (2)
Mafia	73.3+/-3.3 (3)	19.4+/-4.5 (5)
Songosongo	35.0+/-3.5 (4)	37.5+/-4.6 (4)
Mnazi Bay	60.0+/-0.0 (2)	20.0+/-0.0 (2)
OVERALL	52.0	27.1
MOZAMBIQUE		
Quirimbas		48.4+/-21.3 (2)
Mozambique I.		32.5+/-4.7 (2)
Bazaruto		69.5 (1)
Inhambane		13.8+/-1.9 (2)
Inhaca		50.0+/-10.5 (2)
OVERALL		42.8

These coral cover measures for East Africa taken before and after the 1998 El Niño bleaching and mortality event, clearly show the impact on the corals with high levels of mortality. Coral cover is in percent of the bottom (with the standard error and number of survey sites).

- **Mozambique:** There were low El Niño bleaching impacts in the south, although bleaching was seen at Iñhaca Island in early 1999. The most extensive impacts were on exposed reefs in the north, with up to 99% mortality on some patch reefs. Reefs in sheltered bays, which experience higher levels of nutrients and turbidity from land runoff, as well as variance in surface water temperatures were least affected.
- **Tanzania:** Bleaching and mortality levels were generally high but variable from Mozambique to the Kenyan border, with high bleaching (60-90%) at Tutia reef in Mafia Island Marine Park and Misali reef on the west coast of Pemba. There were some reefs Unguja Island, Zanzibar with low extent and impact of bleaching (10% or less).
- **Kenya:** These reefs were amongst the most severely damaged in the region, with levels of coral mortality between 50%-90%. Lagoon patch reefs and fore-reef slopes along the southern Kenya coast showed losses in coral cover of pre-impact levels of 30% going down to 5-11% for both protected and unprotected reefs. In northern Kenya, coral bleaching and mortality in shallow waters was as dramatic as in the southern reefs, however corals on reefs below 10 m depth suffered less mortality, although bleaching was often more than 50%.

El Niño Effects on Coral Diversity and Distributions

Stress from the El Niño event was large scale and intense, resulting in species-specific patterns of bleaching and mortality across the region. The fast growing genera *Acropora*, *Pocillopora*, *Stylophora* and *Seriatopora* showed up to 100% bleaching. Over 50% of species in these genera are still absent from many reefs 2 years after the event. High rates of bleaching and mortality also occurred in other genera such as *Galaxea*, *Echinopora*, and other minor acroporids. Low rates of mortality were observed in some corals even though they exhibited high levels of bleaching (e.g. *Fungia*, *Coscinaraea*, anemones). The majority of other coral species exhibited variable, and moderate to low bleaching and mortality levels, but because of their low abundance this did not contribute greatly to post-mortality abundance estimates, e.g. the faviids, acroporids in the genera *Montipora* and *Astreopora*, agariciids, poritids, siderastreids and most of the octocorals and zoanthids. In some adjacent colonies of the same species showed different levels of bleaching and mortality.

This 1998 bleaching event will have profound impacts on the structure and growth rates of these coral reefs for many years. The coral species that suffered the highest bleaching and mortality were typically fast growing, branching species, with high rates of reproduction (sexual or asexual) and competitive overgrowth. Many of the surviving species are slower growing massive, sub-massive and encrusting forms, which rarely dominate reef communities. Therefore any future El Niño events may have very different impacts on evolving reef community structure due to overall species changes in these coral communities. The El Niño event may be 'selecting' specific gene pools that are more resistant to temperature impacts.

Recovery Following El Niño

As noted above, there has been variable recovery of coral reefs that suffered high mortality. Coral recovery by regrowth of surviving colonies has been significant on some shallow lagoon reefs in Kenya (where coral cover has returned to the 1997 levels of approximately 15-20%). However, recruitment of larvae to all reef zones to replenish lost corals was low throughout 1999 according to sites surveyed. The first strong signs of major coral recruitment were first observed in November 1999, with increasing new coral numbers in

Country	Reef location	Populations of COTS	Impact of COTS on reefs
South Africa	Two Mile Reef, Maputaland	Spot outbreak of COTS to approx. 0.62m-2	Coral cover under long term decline where affected by COTS
Mozambique	Anchor Bay, Inhambane	3 year infestation to 1999, with recent feeding scars in 1999	Coral cover reduced to 2-5%, main survivors Pocillopora, Acropora
	Inner Two-Mile Reef, Bazaruto	Outbreak in 1995	80% mortality, with ongoing degradation
	Coral Garden, Bazaruto	COTS present in 1999	20% mortality, but of mixed causes
Tanzania	Changuu, Zanzibar	COTS at 0.8-1m-2	Coral cover reduced from 58% in 1996 to 25% in 1997
	Bawe, Zanzibar	New COTS aggregations in 1999	Mainly in Acropora thickets
Kenya	Shimoni	COTS reports from 70s, no new reports	No recent impacts

A summary of crown-of-thorns starfish (COTS) outbreaks in East Africa in the 1990s showing that, while these are a persistent problem, the threat is increased on reefs where coral cover has decreased meaning that recovery could be delayed.

2000. Some reefs that are showing significant recovery were already severely impacted by fishing and collecting, and occur in relatively marginal reef flat environments. This suggests that pre-exposure to temperature and other stresses may be a positive adaptation and result in faster recovery in some areas. However in others, damage to the reefs from overfishing and overgrowth by fleshy algae may have been delaying coral and reef recovery. Since these observations are based on reefs in early stages of succession no definitive conclusions can be made.

Crown-of-thorns Starfish (COTS)
These starfish (*Acanthaster planci*) have been a major threat to the reefs throughout Tanzania, Mozambique and South Africa during the past decade. Damage to deeper reefs in South Africa from COTS since 1990 appears to be shifting the balance towards soft corals, because the starfish target the hard corals.

Fishing
Excessive and destructive fishing in East Africa was the major anthropogenic problem for the reefs throughout the 1990s; only surpassed in impact by the El Niño bleaching event took which took over as the major threat in this region. Damaging fishing practices include: dynamite; pull-seine nets; poisons; over-exploitation of juveniles and small fish (direct targets and by-catch); and local intensive harvesting of specific resources (octopus, shellfish, and lobster). Sea cucumber populations have crashed following the introduction of export fisheries. Since the last status report of 1998, fishing pressures have not decreased in any country, yet there have been some specific improvements, notably:

- In Tanzania, increased public pressure by communities and NGOs has led to: a) effective reduction of dynamite fishing in key areas (Tanga, Mafia, Songosongo), and b) delayed implementation of a controversial aquaculture development project in the Rufiji Delta with its ultimate collapse in 2000;
- In Mozambique, there has been a ministerial ban on aquarium fish collection pending a study on the potential impacts and sustainable management of the trade; and
- There has been increasing efforts in all countries to involve and empower local users in monitoring and management e.g. in Tanga and Misali (Tanzania), Kiunga and Diani (Kenya), and Angoche (Mozambique).

Tourism
Coastal tourism in all the East African countries depends on healthy coral reefs as the primary attraction, however, increasing development of tourist facilities is also threatening the reefs. All countries are starting to promote 'ecologically friendly' and sustainable tourism developments, which include smaller, more exclusive hotels and resorts, diving and wilderness-oriented activities, as well as co-management of the reef resources between communities, developers and government. Some studies on tourism impacts on the physical, biological and socioeconomic status of coral reef areas have been conducted, however there are large gaps in the coverage of studies and in implementation of recommended actions.

Population, Pollution, Mining

Rising population and settlement issues throughout East Africa are the root causes behind all other anthropogenic threats to coral reefs. Related issues include: increasing sedimentation from rivers, coastal development and construction, shipping and harbour development, mining in the sea and shoreline, pollution and rubbish disposal, etc. There are major concerns about mining developments in the coastal areas and respective threats to coral reefs, particularly in Mozambique and Kenya. Both countries have been found rich deposits of titanium-containing sands in beach and inshore dunes. Mining of these sands pose potential risks to coral reefs and other marine habitats, including surface mining of the sands and associated oil spills during shipping and transport. Legally-required EIAs have been conducted and approval for mining is now dependent on these EIAs and feasibility studies.

MARINE PROTECTED AREAS

All countries in East Africa now have effective Marine Protected Areas, including new initiatives to establish more sites and improve existing networks of sites. National strategies and implementation methods vary among countries, ranging from a centrally-planned model in Kenya to the more diverse matrix of national and locally-implemented co-management models in Tanzania. There are now increasing efforts to coordinate MPA strategies between adjacent sites and across national boundaries, in light of shared MPA goals and oceanographic connections from the East Africa Coastal Current. The dominant MPA management issues in Kenya and Tanzania relate to control and access to resources by the local subsistence users. Historically, it has been hard to demonstrate to local users that a MPA management action which restricts fishing in some sites will be beneficial, or have 'spill over' to a larger scale or in the future because of the typically 3-5 year time horizon such refugia require. Poor participation and empowerment of user communities is the key limiting step in many existing protected areas in East Africa.

The widespread loss of reefs during the 1998 El Niño has established a sense of urgency to both increase the number of MPAs and improve respective management of old and new areas across East Africa. This urgency is now evident as more co-management initiatives between government, communities and NGOs.

Kenya

The MPA system is centrally managed by an independent parastatal, the Kenya Wildlife Service (KWS), as either parks (full protection) or reserves (traditional extraction allowed), and covers over 5% of the coastline. The well-established marine parks and reserves in Malindi/Watamu (park and reserve, established 1968), Mombasa (park and reserve, established 1989) and Kisite/Mpunguti (park and reserve, established 1978) are primarily oriented at conservation and tourism use, with significant monitoring and research in the last 15 years. In the Kiunga Marine Reserve, active management only started in the early 90s, and involves WWF as a primary partner in community participation and empowerment in the reserve. The Diani Marine Reserve was gazetted in parliament in 1995, but implementation was halted by local antagonism, predominantly from fishermen. This is now the focus of an Integrated Coastal Area Management project involving IUCN and the Coast Development Authority. IUCN is assisting KWS to develop models for stakeholder participation in the management of the Kisite-Mpunguti MPA.

Tanzania

There is only one national MPA, Mafia Island Marine Park, which is managed through the Tanzania Marine Parks and Reserves Board of Trustees, and has technical assistance from WWF. There are a number of local government/community-based MPA cooperatives at Tanga, Misali Island (Pemba), Maziwi Island and Menai Bay (Zanzibar). A new national MPA is under development at Mnazi Bay with assistance from Frontier-Tanzania. One private MPA exists, the Chumbe Island Coral Reef Park (Zanzibar), which is run by a local tourist operation. The total coastline length under protection has not been estimated, but is thought to be relatively low.

Mozambique

There are two MPAs: the Bazaruto National Park, under the Direcçao Nacional de Florestas e Fauna Bravia (DNFFB) with assistance from WWF, and the Iñhaca and Portuguese Islands Biological Reserve (Maputo), under the University of Edouardo Mondlane. The proportion of coastline covered is under 1%, although proposals for a number of new conservation areas exist, for example the Quirimbas Archipelago, Mozambique Island, Primeiros and Segundas Islands, and Ponto de Ouro.

South Africa

South Africa has one centrally managed MPA, the St. Lucia and Maputaland Marine Reserves, under the KwaZulu-Natal Conservation Service, and one user-management area, Aliwal Shoal in which managed zones cover all of the coral reef area.

GOVERNMENT LAWS AND POLICIES FOR CORAL REEFS

East African countries have a history of sectoral policies and legislation that relates to different ministries and government bodies (such as water resources, land, shipping, fisheries, forestry, etc.), which collectively have direct and indirect impacts on coral reefs, e.g. imprecise information and poor enforcement has generally resulted in reef degradation from human activities. Recently, policy and legislation fostering environmental impact assessments and integrated coastal zone management have attempted to bridge the gaps among the sectoral groups with varying degrees of success. Highlights include the establishment of the Ministry for the Coordination of Environmental Affairs (MICOA) in Mozambique and the revamping of the National Environment Management Council (NEMC) in Tanzania. Kenya passed an Environment Bill through Parliament in 2000 that should lead to similar multi-sectoral approaches to coastal environment issues. MICOA has drafted a National Coral Reef Management Plan for Mozambique.

REGIONAL CORAL REEF PROGRAMMES IN EAST AFRICA

During the past 2-3 years, regional cooperation is being increasingly supported at the strategic planning and institutional levels, especially with regard to areas of research, management, training, planning and policy. Coral reef management has been at the forefront of this trend and catalysed debates on marine and coastal issues, due to their recognised socioeconomic and biological importance. Leading global institutions active in the East Africa region are: UNEP-RCU, IOC/UNESCO, WWF, IUCN, with regional support from Sweden, the USA and the World Bank. Regional institutions and projects that are new or expanding significantly in the period 1998 to 2000 include:

- Western Indian Ocean Marine Science Association (WIOMSA): established in the last decade through a series of regional conventions led by the Arusha Conference in 1993, and supported by the SIDA East Africa Regional Marine Programme of Sweden. WIOMSA provides research grants for scientists in the region, supports scientific travel, training, and coordinates conferences and workshops;
- Secretariat for East Africa Coastal Area Management (SEACAM): established following a Ministerial Conference in the Seychelles in 1996 under the 'Arusha process'. SEACAM's mandate is to support coastal management in eastern Africa, through capacity building, coordinating seminars and workshops, training, publication and policy advice;
- Coral Reef Degradation in the Indian Ocean (CORDIO): established in 1999 following the regional reef devastation caused by the 1998 El Niño, and is involved with GCRMN in implementation of the East Africa node. CORDIO supports national coral reef monitoring teams, primary research on biological and socioeconomic aspects of reef impacts, reef rehabilitation trials, and investigation of alternative livelihood options for resource users affected by reef degradation; and
- WWF East African Marine Ecoregion: established in 1999 with reconnaissance surveys of biological, socioeconomic and institutional status of marine ecosystems in East Africa. The programme targets ecosystem and regional-level conservation planning and implementation, involving WWF protected area projects, partner organisations and policy-level interventions relevant to the regional scale.
- Coral Reef Conservation Project: established in 1987 conducts scientific studies on: human impacts on coral reefs: reef restoration; long-term monitoring of Kenyan reefs. It coordinates scientific studies on reefs, and trains of scientists in coral reef techniques.

GAPS IN MONITORING AND MANAGEMENT CAPACITY

While management results and actions for coral reefs and resources in East Africa varies greatly across countries in the region, there is however a vibrant and varied network of institutions and groups active in both coral reef management and research. A principal gap in reef health monitoring and management capacity, is the need for more effective communication and cooperation among the various groups with regard to sharing and collating information and experience. This is partially due to the small number of people with adequate training in this complex area of coral reef and coastal management and conservation. This capacity gap has been mentioned in all priority-setting workshops with donors and implementing agencies. The quality of coral reef monitoring methods is improving in all countries with increasing technical capability of staff, yet some monitoring programmes continue using outdated methods that are derived more by institutional history than data quality. Another recurring theme is the lack of connection between science and management, both from the outputs offered by scientists, and the use of scientific findings by managers. Greater involvement by managers and scientists together in the formulation and reporting of research and monitoring programmes is necessary. Another gap is in the integration of users and their knowledge base in management, particularly where there is a lack of information and expertise within the management team. Trans-boundary and cross-border issues in management of coral reefs are increasingly becoming apparent though there is currently little integration across national borders. Many of the threats are common to most countries, such as natural threats (El Niño) and artisanal fishing, or there is a spill-over of threats from one country to the next e.g. destructive fishing practises across the Tanzania-Kenya border and scuba diving tourism in South Africa moving into southern Mozambique. The most significant cross-border differences are in political and administrative areas, where technical resources, protected area and coastal management policy and practice may be completely different or even at odds. Improvements in these areas will require increased professionalism and commitment to objectives within East African institutions, as well as more rationalised donor and outside-agency support to coral reef and MPA initiatives in the region.

CORE CONCLUSIONS

- East African coral reefs have faced increasing threats throughout the last decades of the 20th century, culminating in the widespread devastation of approximately 30-50% of reefs during the 1998 El Niño.
- A large capacity of human resources and knowledge in research and management exists in East Africa to study the effects of this devastation on reefs and human populations dependent on them, and to formulate management plans to deal with degradation issues.
- All East African countries are improving legislative frameworks for implementing action. Nevertheless, more coordinated action is needed to better address the increasing likelihood of future threats on the scale of the recent El Niño.

RECOMMENDATIONS TO IMPROVE CORAL REEF CONSERVATION

Three primary recommendations are highlighted in this review, arising primarily from the massive impacts of the El Niño coral bleaching, and the regional scale of coral reef losses:

- National Monitoring Programmes: while there is improved capacity and rigour in coral reef and resource monitoring, more rational management planning remains a critical need. Increased standardisation of methods, data processing and archiving are all needed across research groups and countries. Longer term commitment of support, from local institutions to international donors, is needed to give the

necessary longevity to monitoring at national and regional levels. None of the countries have a functioning centralised database of coral reef monitoring data, though these nominally exist in some cases. Greater openness and collaboration among relevant institutions within countries (local, national and international), and financial support for appropriate staff to manage a national database are needed.
- Management Issues: An increased use of scientific information and monitoring resulting in management decision making is necessary to respond to dramatic changes in reef status in a timely manner. The establishment and support of forums to coordinate science-management integration at the scale of relevant issues would help break down these barriers. Co-management and empowerment of users and local populations is critical in instances where subsistence economies are reliant on reef resources, and where there is insufficient 'modern' information from science and management on critical issues.
- An improved framework for support of coral reef conservation is needed at national and regional levels, requiring rationalisation among sectoral interests within countries, political support and cooperation at the regional level, and donor commitment for regionally coordinated efforts. Support for region-wide initiatives, as well as localised cross-border issues, is needed.

ACKNOWLEDGEMENTS AND SUPPORTING DOCUMENTATION

David Obura (dobura@africaonline.co.ke) is the coordinator for CORDIO for Eastern Africa and works out of Mombasa, Kenya. Mohammed Suleiman (mohammed@zims.udsm.ac.tz) works for the Institute of Marine Science (IMS) in Zanziba. Helena Motta (coastal@tropical.co.mz) works for the Ministry for the Coordination of Environmental Affairs (MICOA) in Mozambique. Michael Schleyer (seaworld@dbn.lia.net) works at the Oceanographic Research Institute in South Africa.

NATIONAL AND OTHER REGIONAL REPORTS

Kemp, J. Hatton, J.C. and Sosovele, H. 2000. East African Marine Ecoregion Reconnaissance Reports, Vols 1-4. World Wide Fund for Nature (WWF), Dar es Salaam, Tanzania.

Motta, H., Rodrigues, M-J. and Schleyer, M. 2000. Coral Reef Monitoring and Management In Mozambique. Ministry for the Co-ordination of Environmental Affairs (MICOA), Maputo, Mozambique.

Muhando, C.A. and Mohammed, M.S. 2000. Status Of Coral Reefs Of Tanzania. Institute of Marine Sciences, P.O. Box 668 Zanzibar, Tanzania.

Obura, D., Uku, J.N., Wawiya, O.P., Mwachireya, S., Mdodo, R. 2000. Kenya, reef status and ecology. CORDIO-East Africa, P.O.BOX 10135, Mombasa, Kenya.

Schleyer, M.H. and Celliers, L. 2000. Status Report On South African Coral Reefs. Oceanographic Research Institute, P.O. Box 10712, Marine Parade 4056, Durban.

McClanahan, T. R., Sheppard, C.R., Obura, D.O. (eds) 2000. Coral reefs of the Indian Ocean: Their ecology and conservation. Oxford University Press, N.Y.

McClanahan, T. R., Muthiga, N.A., Mangi, S. 1999. The status of the coral reefs of Kenya's MPAs: 1987 to 1999. Coral Reef Conservation Project and Kenya Wildlife Service. P. O. Box 82144 Kenya.

ADDITIONAL NATIONAL INSTITUTIONS NOT MENTIONED IN THE TEXT FROM WHICH FURTHER SOURCE MATERIAL CAN BE OBTAINED

- Coast Environment Research Station (Moi University, Kenya)
- Coral Reef Conservation Project (Kenya)
- University of Edouardo Mondlane (Mozambique), Fisheries Research Institute (Mozambique),
- Frontier-Tanzania (Tanzania)

5. Status of Coral Reefs of the Southern Indian Ocean: the Indian Ocean Commission Node for Comoros, Madagascar, Mauritius, Reunion and Seychelles

Lionel Bigot, Loic Charpy, Jean Maharavo, Fouad Abdou Rabi, Naidoo Paupiah, Riaz Aumeeruddy, Christian Villedieu and Anne Lieutaud

Abstract

There were few baseline data on coral reef status across the 5 Indian Island nations of Comoros, Madagascar, Mauritius, Seychelles and France/La Reunion when the catastrophic 1998 El Niño associated coral bleaching and mortality hit. Just as the bleaching impacts were varied across this region, so are the condition of the reefs and how they are impacted by human activities. Reefs off the large island of Madagascar show distinct signs of human damage, with few reefs in good condition and even fewer protected. Likewise there is evidence of damage from the land to reefs on Mauritius and Reunion, but there are more efforts to conserve the reefs in protected areas; bleaching on these three countries was relatively minor. Prior to the 1998 bleaching, the reefs of Comoros and the Seychelles were in good to excellent condition except for some damage near centres of population and some over-fishing. But bleaching devastated these reefs with large scale mortalities leaving many reefs with less than 5% coral cover (down from levels over 50%). The 1998 bleaching event occurred just when the Regional Environmental Programme of the Indian Ocean Commission (REP-IOC) was establishing a coral reef monitoring programme in the Comoros, Madagascar, Mauritius, La Reunion and Seychelles, building on mandates already established by the IOC since its inception in 1982 for regional cooperation in economic, social and cultural fields. The REP-IOC programme aims to support national policies on Integrated Coastal Management (ICM) for long-term sustainable development of the region's coral reefs for future generations, including the establishment of permanent monitoring stations (of which there are now 44) in the 5 countries.

Introduction

The 5 IOC member states have large areas of coral reefs, therefore a unifying theme was the need to develop 'reef status monitoring' within a regional network. All countries are experiencing extremely strong human growth pressures that are resulting in reef deterioration and losses in reef resources that will have severe economic losses in fisheries, tourism, and shoreline protection, as well as losses in biodiversity heritage. Through REP-IOC, a regional reef network was fully endorsed by April 1998 to assist local, national or

Status of Coral Reefs of the World: 2000

regional level decision-makers. This regional monitoring programme is now able to alert reef managers in the event of problems, and is an integral tool for integrated coastal planning. Since then national networks have evolved, each with a National Focal Point, a National Committee and a Supervisory Body, all coordinated by a Regional Organizer. This network constitutes the regional Node for the Global Coral Reef Monitoring Network (GCRMN) for the 'South-west Indian Ocean Island States'. There are plans to establish more monitoring stations to increase the relevance of the results both in the region and within the international networks e.g. GCRMN, ICRI, GOOS etc.

STATUS OF CORAL REEFS

The following is a synopsis of the monitoring sites established, surveys conducted and findings for each country through the REP-IOC Programme:

Comoros

Two sectors in Grand Comore were monitored in 1998 and 1999: Mitsamiouli and Moroni (Comotel). After the installation of the 'Parc Marin de Moheli', a comprehensive assessment of reef status was required and 2 study sectors were selected: Itsamias in the Marine Park, which is a major marine turtle (*Chelonia mydas* and *Eretmochelys imbricata*) nesting site; and the small islands of Nioumachouoi with one transect site on the small island of Ouenefou, and another on Candzoni island, with reef flat and outer slope transects.

Mitsamiouli Site: This is on the north-west of Grande Comores, 42km from Moroni on 4km of coastline with rocky cliffs interspersed with white sand beaches where the major tourist sites are located (11°25' S; 43°18' E). The Mitsamiouli population of 4,500 has 3 extremely active associations for environmental protection, which have prohibited sand and coral mining, and fishing is a key economic activity. Permanent transects on the reef flat (8m depth) showed 46.9% live coral cover in 1998, and 44.0% in 1999, however, there was approximately 15% bleached and dead coral cover in 1999 that were covered with algae. Transects on the outer slope (19m depth) showed strongly degraded corals with cover of only 25% in 1999, whereas it had been 54% in 1998. This area was probably impacted by the EAL airline crash, leading to much coral death.

Itsamia Site: Here there are 376 people in over 80 families, including 39 fishing families, with agriculture as the main economic activity, and fishing is not well developed, although many people fish and glean reefs at low tides. There is an ecotourism resort as part of the Moheli Marine Park, with nesting turtles and seabirds, and an excellent coral reef (12°21'S; 43°52'E). The discontinuous reef at Itsamia has suffered huge anthropogenic damage, with the dominant branching and tabulate *Acropora* corals either broken or bleached, however, other hard corals (*Platygyra, Diploria, Favia* and *Porites*) are abundant. Estimated living coral cover is 40%, with 45% dead coral, and 10% algae. Most of the dead corals had been bleached and were covered with algae. Among the living corals, there was 50% *Porites*, 20% *Diploria*; 5% *Pavona*, 2% *Favites* and 19% of 4 *Acropora* species. At 10m depth below most human impacts, there was about 70% live coral cover mainly *Porites* (32.6%) and *Favites* (16.5%); dead coral covered 15%, algae 18.4% and rock 13%. Fish populations were fairly good, including parrotfish, grouper, butterflyfish, angelfish and damselfish, but most fish were small, indicating intensive fishing pressure.

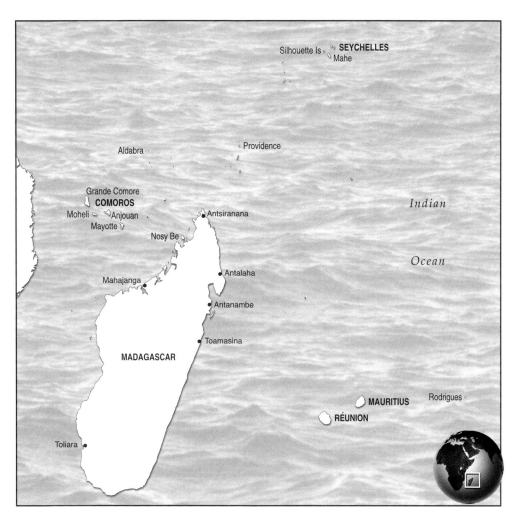

Nioumachouoi Site: The 5 small islands (Mea, Ouenefou, M'bouzi, Chandzi and Canzoni) and 3 large rocks (M'foro, M'bougo and Magnougni) were examined during exploratory snorkelling. Canzoni is remote and has been severely damaged by blast fishing and trampling by fishermen over the reef flat, and turtles have been poached. The outer slope of Candzoni is more protected, but there are still signs of dynamite use. Coral diversity is high with abundant *Acropora* and *Porites*, plus *Favites* and *Favia*. Living coral cover averaged 50% over large areas, with many giant clams (*Tridacna*) and other molluscs, and diverse fish populations. The reef flat transect at 5m depth over a well-preserved site had 55% live coral cover, mainly *Porites*, *Montipora* and *Pocillopora*, and 39% dead coral cover which was mostly acroporids that had died during the 1998 bleaching. The outer slope had 67% dead coral cover, mostly Acroporidae, which had been bleached. Live coral cover was no more than 10%, mainly *Porites*, *Fungia* and *Pocilopora*, and 11% algal cover. Evidence of high anthropogenic pressure was clear on this site (blast fishing, boat anchoring), explaining its poor condition.

The coral reef of Ouenefou is highly degraded and coral communities on the northern slope once started near the beaches and descended gently to the outer slope. *Acropora* corals have been completely destroyed and loose rubble covers the whole surface. Exploratory dives showed 25% living coral cover of *Favia, Favites*, plus *Pocillopora* and *Seriatopora*, with approximately 70% dead coral. The destruction was extensive and caused by dynamite smashing of the branching corals, but leaving some large soft corals, *Fungia* corals and giant clams. At another site on the outer slope, living coral cover was around 30%, with 25% algae and 60% dead coral. Fish diversity was low and all were very small, although there were many butterflyfishes. A permanent transect was established at 12m depth where there was 80% dead coral and rubble, with a few living *Porites, Pocillopora* and *Fungia*.

Madagascar
Reefs stretch over more than 1000km of the west coast from Toliara (Tulear) in the South to Antsiranana in the North with some breaks near Mahajanga. Reefs are only present between Antalaha and Toamasina on the east coast. Ten stations in 5 sites were established in the north: Tanikely and Dzamandjar in the northwest (Nosy-Be region); and Fouplointe, Nosy-Antafana and Antanambe on the east coast. Coral cover on the outer slopes was high: 50.9% at Dzamandjar; 68.3% at Tanikely; 85.7% at Nosy-Antafana; 83% at Antanambe; and lower at Foulpointe, 35%. There was evidence of deterioration on the reef flats with lower coral cover: 27% at Nosy-Antafana; 28.7% at Antanambe; 47% at Foulpointe; but higher on Nosy-Tanikely with 65.7%. The reef at Dzamandjar experiences higher anthropogenic pressures from industrial pollution (sugar refinery), intensive coral mining, domestic pollution from many hotels, no sewage treatment for the town of Dzamandjar, and overfishing. There was, however, relatively high coral cover of 42% in 1998 with many soft corals. Foulpointe reef is highly degraded, with large-scale collection of corals for sale to tourists and as building material. The reef flat is heavily sedimented with the remaining corals buried in sand, because the southern pass has been blocked and sediment build up has reduced the depth of the lagoon to 2m. Nosy-Antafana is in the Mananara-North Biosphere Reserve with water sports (spearfishing, diving, etc.) banned on the outer slope. Fishing for invertebrates, net and hook and line fishing appear to have few impacts on the outer slope with high coral cover. In 1998, cover was 46% on the reef flat, but it was lower (27%) in 1999, with coral regeneration evident around many dead coral colonies which had mostly died and were covered with algae lawns, although they remained fixed in place.

On the reef flat of Ifaty (South Toliara) in 1999, there was 25.3% coral cover (13.5% *Acropora* and 11.8% non-*Acropora*), and high cover of abiotic (sand, mud, rock) forms, (69.6%) mainly dead corals and rubble, and 4.7% algae. Monitoring in February 2000 showed an increase in algae cover (26.6%) due to the proliferation of macro-algae (14.8%), mainly standing algae such as *Turbinaria, Sargassum* and *Halimeda*. Also there was a small increase in live coral cover (28.8%). On the Ifaty outer slope, there were almost equal amounts of *Acropora* (20.1%) and non-*Acropora* (20.6%) in the 40.7% cover, and non-living cover of 52.5% in 1999. In 2000, there was a small increase in live coral (41.9%) with the appearance of many scattered juvenile colonies.

On the Great Barrier Reef of Toliara, exploratory snorkelling showed a relatively high living coral to dead coral ratio, and often 10% soft coral cover. Transect data showed the importance of abiotic forms (57.4%) compared with living corals (28.5%), which were mostly branching and digitate *Acropora* (22.3%).

Comparison of 1998 and 1999 monitoring data:

- Nosy Tanikely - coral cover increased on the reef flat from 46.9% in 1998 to 65.8% in 1999; and 56% to 68% on the outer slope. The increase is attributed to the increased level of protection initiated by tourist organisations;
- Dzamandjar - there was an increase of 5% in coral cover on the outer slope in 1999;
- Foulpointe - similar coral cover between the two years - 33.3% in 1998 and 35.0% in 1999, but an increase in rock and sand from 43.3% in 1998 to 54.3%, in 1999, reflects the increased sedimentation;
- Ifaty – on the reef flat, coral cover was 29% in 1998 and 25.3% in 1999, and rock and sand covered 63% against 69.6%. The reduction in cover reflects the continuing deterioration of the reef.

Mauritius

Two permanent monitoring sites were established in 1998: Trou Aux Biches (sheltered coast) near the major tourism developments and including a public beach; and Bambous Virieux (exposed coast). The reef flat at Trou Aux Biches is partially degraded by recreational activities, with coral cover at 2m depth being predominantly branching *Acropora*, showed little variation (45% in 1998, 41% in February 1999, 45% in July 1999). Algal coverage decreased to 10% and 5%, and dead coral cover rose from 15% to 29%. Fish populations were diverse with 6 major families. On the outer slope at 7-9m depth, coral cover (*Porites* dominating) was unchanged at 43% and 42% in February and July 1999 respectively, however, algal cover decreased considerably from 17% to 8% in February 1999, but increased to 10% in July 1999, which are seasonal changes.

The lagoon at Bambous Virieux is 5km wide, with an average depth of 2-3m. There is coral and sand mining and it is adjacent to an estuary draining a major agricultural area. Coral cover, predominantly *Pavona* and *Porites*, on the reef flat decreased from 56% in 1998 to 54% in March, 1999, and further to 30% in August 1999. Algae cover went from 32% in 1998 to 30% in March 1999, and 56% in August 1999, while dead coral increased from 15% to 29%. The second reef flat site also showed coral cover variations from 54% in 1998 to 48% in March 1999 and to 55% in August 1999, but algal cover showed a consistent decrease; 24% in 1998, to 17% in March 1999 and 11% in August 1999.

Comparison of 1998 and 1999 monitoring data:

- Trou aux Biches – there was no significant change in living coral cover between 1998 and 1999, however, algal cover was significantly reduced;
- Bambous Virieux - coral cover between July 1998 and July 1999 remained the same. However, a reduction was observed in March 1999. Algal cover rose in summer, but dropped back to the 1998 level in winter, indicating seasonal variations and the impacts of agricultural waste discharges.

La Reunion Island

Two different locations were chosen, each with 2 monitoring sites: St Gilles - La Saline with Toboggan, a healthy site, and Planch'Alizes, a disturbed site; and St Leu – Varangue a reference site, and Corne Nord, near the canal exit. Toboggan has thriving coral communities with healthy levels of new coral recruitment and a low level of disturbance. The outer reef-flat is characterised by alternating spurs and grooves towards the seaward edge. The reef-flat had living coral cover of 31% in 1998, including 22% branching (*Acropora formosa*) and digitate (*A. humilis*) and 9% other coral forms (foliose, encrusting). Algae cover (8%) was mainly turf algae in damselfish territories (*Stegastes nigricans*), and abiotic cover was 62%, mostly fine and coarse sediments. This apparently low level of living coral is normal for internal reef flats with many broad channels draining the reef flats. Live coral cover dropped to 22% in 1999, including 20% *A. formosa* and *A. humilis* and 2% other foliose and encrusting corals. Algae cover increased from 8% to 11% and abiotic cover was unchanged on 65%. The drop in coral cover in 1999 was attributed to extreme low tides that killed the tops of many colonies, which were then invaded by turf algae.

On the outer slope of Toboggan, living coral cover was 57% in 1998, including 26% digitate (*Acropora humilis, A. digitifera*) and submassive (*A. danae, A. robusta*) and 32% other coral forms, in particular encrusting corals. Algal coverage was 35%, including 31% calcareous algae, which are typical of wave impacted sites. The area had high coral species diversity like other turbulent outer slopes. In 1999, living coral cover was 38%, including 23% digitate and submassive *Acropora* and 15% other coral forms. Algae cover was 51%, including 39% calcareous algae. The drop from 57% to 38% was due to coral bleaching, particularly of *Pocillopora*, which started in April 98, but was much less than in other Indian Ocean countries.

The Planch'Alizes site is affected markedly by anthropogenic disturbances (coastal waste discharges, physical destruction, nutrient rich underground water discharges). Coral cover on the reef flat in 1998 was 18%, including 17% massive, submassive and foliose colonies. Algal cover was high (37%), mainly turf algae, soft macroalgae and calcareous algae on dead coral. Abiotic cover of 41% was mostly coarse sediment. Coral cover rose to 48% in 1999, nearly all massive, submassive and foliose forms. Algae coverage dropped to just 13% and abiotic cover was 32% of the total. The major changes are both seasonal and due to better protection because the area is now clearly marked. The outer slope at 12–13m depth amongst distinct spur and groove formations had living coral cover around 50%, with 20% algal cover in 1991, and in 1998, coral cover was 38%, including 24% digitate and submassive *Acropora*, 13% others, mainly encrusting forms, 7% soft corals, and 37% algae. In 1999, living coral cover was 42%, including 27% digitate and submassive *Acropora* and 15% other coral forms, in particular encrusting forms. Algae coverage was 39%, including 23% soft macroalgae and 16% calcareous algae. The variation between the 2 years reflects minor variations in communities and variations from sampling; the reef showed no real changes over this time.

The Varangue reference site is in the middle of St Leu Ville reef and has been a scientific study area for several years. Live coral cover on the reef flat was 51%, including 44% *Acropora* and 7.4% other foliose, massive and submassive corals in 1999. Algal cover was 18% mostly turf algae, and 31% abiotic cover. This relatively high cover indicates that this flat is free of excessive sediments and is well irrigated. On the outer slope, living coral cover was fairly high (53%) and included 23% *Acropora* and 30% other coral forms (mostly massive). The 35% algal coverage was mainly calcareous algae, indicating high construction capacity, and abiotic cover was very low (8%). This is a relatively healthy area with high cover of accreting corals and algae (88% accreting cover).

The Corne Nord site is an unusual zone on St Leu Ville reef as it is immediately adjacent to a reef pass, where much lagoon water drains out. The outer slope station has thriving corals, and is relatively remote from coastal disturbances. The reef flat site is close to the coast with dispersed coral patches, mostly *Porites*. Living coral cover in 1999 was 58%, including 54% *Acropora* and 4% other foliose and encrusting corals, 18% algae, 18% coverage of abiotic components, mostly coarse sediments. This area has the highest coral cover of all La Reunion reef flat sites. The outer slope had even higher coral cover (63%), with 54% *Acropora*. Both the reef flat and the outer slope have exceptionally diverse coral structures with abundant branching and massive *Acropora* communities. Algal cover was moderate (18%), and there was a low percentage of dead algal-covered corals (3.6%) which were a result of the quite minor 1998 coral bleaching.

Fish abundance on the reef flats was higher at Toboggan (246 fish $100m^{-2}$), then La Varangue (189 fish $100m^{-2}$), Planch'Alizes (117 fish $100m^{-2}$), finally Corne Nord Horn section of St-Leu reef (71 fish $100m^{-2}$ - due to intensive fishing). A large proportion of these were omnivores (maximum 52% at Corne Nord and minimum of 31% at Planch' Alizes), then next were herbivores (max. 22.5% at Planch' Alizes; min. 17% at La Varangue, with the exception of the Toboggan reef flat where just 4% of herbivores were observed), then benthic invertebrate grazers (max. 7% at Planch' Alizes; min. 3% at La Varangue). A key difference was in the population of planktivores, which were very abundant in Toboggan (46%) and absent at La Corne Nord, and between 18.5% and 23.5% at La Varangue and Planch' Alizes. A characteristic feature is the very low densities of top predators, which usually indicates line and spearfishing.

Relatively high densities of fish occur on the outer slopes of Corne Nord-St-Leu (189 fish $100m^{-2}$), and Planch' Alizes- La Saline (144 fish $100m^{-2}$). Lower abundances were observed at Trois-Chameaux-St-Gilles and in particular at La Varangue (51 fish $100m^{-2}$). Again, omnivores were prominent (between 12 and 22%), likewise the herbivores (between 12 and 22%) at Toboggan, Planch' Alizes, and La Varangue. At these 3 stations, planktivores varied from 30.5% (La Varangue) to 54% (Planch' Alizes). There was a high proportion of herbivores (59%) at the Corne Nord - St-Leu site near the pass. This indicates an imbalance in the fish populations because of the high concentration of spearfishers. Diurnal and nocturnal carnivores were relatively high (from 20% at La Varangue to 31.5% at Trois-Chameaux).

Seychelles
Three permanent monitoring sites were established: Trig Point in the Ste Anne Marine Park, East coast of Mah; Ternay Bay, West coast of Mahe, also in the marine park; and the Silhouette granite island 20 nautical miles north-west of Mahe. Trig Point is subjected to high sedimentation from the capital Victoria and nearby catchment areas. Furthermore, the area was dredged to build embankments on East Mahe between 1985 and 1992, but monitoring data are not available.

Ternay Bay is a more protected bay in the marine park, with reduced human impacts. Reef flat cover in 1999 was mainly abiotic forms e.g. dead coral, with no living coral on the transect. Turf algae (2.8%) were the only living material. This shallow zone (1m low tide) was severely impacted during the 1998 El Niño with around 95% coral mortality. In 2000, some living corals appeared (0.9%), along with zoanthids (*Palythoa* - 10%) and algae (12.3% - calcareous and macroalgae -*Turbinaria*). There were no new *Acropora* recruits, but the calcareous algae will constitute a good substrate for new coral settlement. On the outer slope, there was a low cover of living corals (2.1%), mostly massive forms and some soft corals. All the branching *Acropora* disappeared during the 1998 bleaching event. Some coral recovery was observed in 2000 with 0.5% branching *Acropora*, and a definite increase, although not quantified, of non-*Acropora* forms, mostly massive and soft corals. Algae were just 6.2% of cover, mostly calcareous algae (2.3%) and a few *Turbinaria* sp.

The Silhouette site was added in 2000. The reef flat is heavily impacted by the south-east monsoons and has always had low coral cover. The site at 2m depth has no *Acropora*, a 9% cover of non-*Acropora* (encrusting and massive corals), and some soft zoanthids (*Palythoa*). Most of the cover (over 50%) was dead coral rubble and sand. The outer slope (5-6m depth) had alive and dead massive corals (>15%) scattered over a sandy zone, with submassive corals (>15%), some encrusting corals, and 0.7% branching, very young *Acropora* which showed signs of fish grazing. This site appeared to escape most of the 1998 El Niño bleaching, possibly due to better water exchange with the cooler deeper waters.

COMPARISON OF 1997 – 2000 DATA

Coral cover on the Ternay Bay outer slope dropped from 54.6% in November 1997 to 3.5% in June 1998, 2.1% in January 1999 and 5.1% in March 2000. *Acropora* was more affected (29.1% in 1997, 0% in 1998 and 1999, and 0.5% in 2000) than non-*Acropora* massive corals (25.5% in 1997, 3.5% in 1998, 2.1% in 1999 and 4.6% in 2000). Soft corals appear to be fast growing, opportunistic species. As a reflection of these changes, abiotic forms went from 37% in 1997 to 92.3% in 1998, 92.8% in 1998 and 86.8% in 2000. These losses can be attributed to the 1998 bleaching phenomenon, as human activities are limited, and recovery is predicted for good coral cover, provided bleaching does not happen again within the next few years. A similar pattern was evident on the reef flat with a total loss of coral after the 1998 bleaching event; here recovery is predicted to be very slow.

MAIN PROBLEMS AFFECTING THE COASTAL ENVIRONMENT

Comoros

These islands experience both natural and anthropogenic disturbances, with the major impact being the global warming in 1997-98 which caused extensive coral bleaching and mortality in large areas, particularly to species of *Acropora*. The majority of reefs e.g. the small islands of Nioumachouioi, are subjected to strong swells from Monsoons and particularly from cyclones which destruct the more fragile corals, especially acroporids. However, this damage was always within natural recovery capacity and is now insignificant compared with anthropogenic degradations.

Blast or dynamite fishing is one of the major damaging factors, and is currently practised in Moheli, where the entire reef flat bordering these small islands is covered with newly-broken coral rubble. Reef walking by fishermen at low tide causes extensive damage to reef flat corals as they hunt for octopus and trap small fish. Intense over-fishing by over 4,500 fishermen (80% of the total) using traditional boats to fish in nearshore waters has been a continuing but growing problem for many years, and has caused major reef destruction. As in many countries, domestic wastes are discharged directly into the sea and result in the massive proliferation of algae (eutrophication) adjacent to the discharges. In addition, household refuse is dumped directly into the sea in the Comoros. The extraction of sand from beaches for building has caused coastal erosion and lead to imbalances in the ecosystems. Moreover, upstream deforestation and poor agricultural practices result in soil erosion and downstream smothering of corals.

Madagascar

The extraction of coral for building is a major problem in Madagascar, for example in the vicinity of Toliara continued, extensive removal of corals will result in large areas of the reef be totally destroyed within 2 years. There is an urgent need for legislation to limit or prohibit coral extraction. Although there are no data on subsistence fishing, recent studies in Toliara (a town with 140,000 inhabitants and high unemployment) show that family fishing from the shore, and gleaning on the reef flat is a common occupation, and sometimes provides a full-time livelihood. This type of fishing removes an estimated at 18mt $km^2 yr^{-1}$, higher than for dugout fishing (12mt $km^2 yr^{-1}$), and exceeds the regeneration capacity of the reef. Fish and octopus are over-exploited, and this has resulted in a shift towards less favoured species such as sea urchins. In order to reduce these fishing pressures on the reef flat, several Fish Aggregating Devices (FADs) have been anchored out at sea to permit catches of pelagic species. This is not an option for most areas of Madagascar, because deep waters are very close to the coast. There is an urgent need to decrease fishing pressures on reef resources and install more FADs for fishermen with dugouts and provide gill nets for selected fishermen.

Madagascar continues to be a minor tourist destination in the region, despite having exceptional potential with coral reefs and a beautiful coastline. Of the 3,040 hotel rooms available in 1995, only one third were on the coast and Nosy-Be, with approximately 200 rooms, is Madagascar's main coastal resort. The Government intends to make coastal tourism a key sector in future economic activity suggesting that visitor numbers of 75,000 in 1995, could triple in the next 5 years. This is however, a long way from the ecotourism policy encouraged by the ANGAP (National Agency for the Management of Protected Areas) in the adjacent Parks. It is highly likely that ecotourism will be limited to the vicinity of protected areas, while major hotel complexes built with international finance will be built along the rest of the coast. Tourism can therefore act as a catalyst for reef protection, as illustrated by a hotel manager near Toliara who asked local fishermen to decrease fishing on neighbouring tourist reefs, in exchange for the purchase of most of their production.

Mauritius

A million tons of coral sand are extracted annually from the lagoon to fill the high demand created by rapid economic growth. The value of this mining is about US$10 million and involves employment for 1,100 people, grouped in co-operatives. Sand is extracted from the shallows by hand and transported in dugout canoes. Mining is conducted in 4 zones: Mahebourg, Grande Riviere Sud-Est and Roches Noires/Poudre d'Or lagoons on Mauritius; and Rodrigues island. Although there is no legislation prohibiting the mining, the extractors must form co-operatives and are limited to 10 tons per day per license. New legislation will restrict extraction to fewer sites, with definitive cessation in 2001. About 2,700 jobs are generated by fishing activities on Mauritius and Rodrigues for a total catch of 6,800 tons per year. Amongst these fishermen, 786 have licenses and there are an indeterminate number of tuna fishermen and sport fishers. The main methods used are drag nets which are particularly destructive, but the practice has continued for over 30 years through licenses to fishing co-operatives. Legislation now limits the number of nets, their length and size and the Fishing Ministry has decided to ban these nets, which are not considered to be sufficiently selective. The incentives proposed are to provide each captain with 75,000 rupees (US$3,000) and 25,000 rupees (1,000) per sailor to turn in the licence and receive retraining, with the target of phasing out this fishing in 2007.

The use of toxic plant extracts and dynamite for fishing is prohibited and rare, and spearguns are banned due to frequent fish trap raiding by divers. No fish species are threatened but environmental pressures are very high due to fishing. Although there is no legal limit to the number of fishermen, numbers have remained rather stable in the last 10 years while the total population on both islands has risen. Chemical and bacterial pollution from industrial or domestic sources constitutes a serious public health problem and also damages the reefs. Residential and tourist developments are concentrated but lack adequate sewage treatment systems, therefore there is chronic water pollution around urban and tourist areas. Only the 3 main towns are connected to sewerage systems, which discharge into the sea near currents that carry the wastes offshore. A project at Montagne Jacquot is building a waste treatment plant capable of complete treatment.

Sugar cane farming is the main cause of agricultural pollution with cane farms occupying 88.5% of the 80,000 hectares of cultivated land with 17 sugar mills, which either discharge directly into the sea or into rivers. Rainwater runoff carries fertilisers, pesticides and large amounts of sediment into the lagoon, resulting in localised eutrophication and sediment damage to the corals. Laws from 1991 and 1993 require Environmental Impact Assessments (EIA) prior to all coastal installations, and these are carried out by private, national or international consultants for the Department of Marine Ecology at the Albion Fisheries Research Centre which checks them prior to the final decisions being made by the Department of the Environment. While EIA is effectively used as a management tool, its efficiency in the management of coastal zone activities can still be improved.

Tourism contributes 4% of GNP being the third largest sector in the economy and providing 10% of total employment (in 1995 there were 6,000 rooms and 51,300 jobs). Tourist arrivals increased by 31% in 2 years to 555,000 in 1997, and projected numbers are for 750,000 tourists in 2000. Most hotels are concentrated in a few areas, which results in serious sanitation problems. The 1991 'Environment Protection Act' requires all hotels over 75 rooms to be linked to sewage treatment, and be preceded by an EIA, however, many hotels were built prior to this and now do not comply with these standards. Hotel operators deal only with visual aspects (cleaning the adjacent beach and lagoon, and revegetation) and leave the State to deal with wastewater treatment. This is now an urgent issue to resolve before adverse publicity ruins tourist and public confidence in the clean image promoted by Mauritius. Currently there is a 30% tourist return rate, which is exceptional. Actions to improve the environmental quality of the reefs and lagoons will only occur if tourist operators are convinced that this is in their direct interest, and they should carefully consider tourist attitudes which are strongly towards environmental quality.

La Reunion Island
Pollution into the island lagoons is increasing in line with population growth, increased water consumption, and the proliferation of wastewater discharges. Urban developments on the coast, the sealing of lands for roads and other uses, and flood mitigation have lead to an increase in water runoff. Considerable efforts have been made since 1985 to improve urban sewage treatment, but problems still arise, including overflows from treatment works during peak tourist periods and heavy rains. This has resulted in bacterial pollution in the lagoons. Industrial wastes are generally discharged without treatment and are carried onto the reefs by ambient currents. Rainwater runoff carries fertilisers, heavy metals, hydrocarbons and sediments, particularly during cyclones, causing pollution to reef waters, particularly during the first rains of the rainy season. Eutrophication and adverse changes to coral communities have increased with increases in irrigation and the use of fertilisers and pesticides. La Reunion is a recent volcanic island and the reefs grow close to the shore, therefore, there is significant natural erosion of the land and coastline due to cyclones. This is now exacerbated by the high rate of development. Tourism is the major industry with over 300,000 visitors in 1995, and a target of 500,000 in the next few years, which makes La Reunion the second tourist destination, behind Mauritius. This concentration of tourists around limited reef resources is resulting in environmental damage, particularly through coral trampling and recreational fishing.

Fishing on the reefs was originally practised by a few professional fishing families, but it has increased greatly. Intensive fishing pressure on offshore stocks in the 1980s has undermined this economic activity, which was considered marginal despite having 1,600 to 3,200 fishermen, including 800 registered boats. In order to reduce pressures on reef fish stocks, FADs have been installed successfully and fishers encouraged to target deep sea pelagic stocks, and seek employment with the longline fishing fleet. Parallel to this shift in professional fishing towards pelagic stocks, there has been an increase in subsistence level fishing associated with a 40% rate of unemployment. These fishers are targeting shore and reef fish using low cost hook and line and net fishing. Regulations have been introduced, but poaching is prevalent and decreases in reef fish stocks are clearly apparent.

Seychelles
Intensive exploitation of reef resources is not a major problem in the Seychelles, because of the large areas of reef and oceanic waters for a relatively small population that favours pelagic fish over reef species. A reduction in biodiversity in the Saint Anne Marine Park, which is only a few kilometres off major industrial and domestic pollution sources from the capital, Victoria indicates that coastal pollution may be damaging the reefs. These are also impacted by increased sediment runoff from development in the adjacent catchment area. Relatively high concentrations of tourists in the bays of the main island result in acute pollution problems with some localised damage to nearshore reefs. Tourism, however, is the largest economic sector in the Seychelles, with 120,700 tourists in 1995 generating 70% of the foreign revenue. There are 4,430 rooms in 122 coastal resorts, including 19 large complexes.

Coral Bleaching and Mortality of 1998

Comoros

The coral reefs experienced extremely widespread coral bleaching in 1998. Approximately 50% of the corals were bleached at Moheli, which had been renowned for its pristine reefs. There was 40 to 50% coral bleaching at the Mitsamiouli monitoring site, with *Acropora* species being the worst affected. Socioeconomic impact assessments are being conducted under the CORDIO project, but already professional fishing has been severely hit, with reductions in benthic fish catches. There are no reliable data on the correlation between corals observed bleached and subsequent mortality, and many corals were partially bleached and recovered.

Madagascar

Coral bleaching affected many coral sites in Madagascar, few data were available at the time of this report, moreover the database software and the corresponding data were not functional in Madagascar. However, brief reports from the 1999 monitoring surveys indicate that the Nosy-Antafana Marine Park was affected directly, with a reduction in reef flat coral cover and increases in the cover of algae.

Bleaching in the Seychelles

The shallow coral reefs of the Seychelles granite islands suffered severe damage following the 1997/1998 mass coral bleaching event, and signs of recovery are slight. Live coral cover was reduced to less than 10% on most reefs around the inner islands, with high partial mortality of colonies. Dead standing coral is present on sheltered reefs, while exposed reefs have already been reduced to rubble. Zoanthids, anemones, and encrusting red and green calcareous algae have colonised shallow reef slopes and lagoons, and soft corals are growing on deeper reef slopes. Branching and tabular *Acropora* species and branching *Pocillopora* species have died on all reefs, with the only live corals left being massive species, (*Porites, Goniopora, Acanthastrea* and *Diploastrea*). The only areas with some coral cover either had low coral diversity or in shallow stressed environments such as the high turbidity Beau Vallon Bay and near the harbour on Mahe. Therefore most coral species have survived somewhere in the islands, but diversity on an individual reef has been severely reduced down to about 10 species. Recruitment of branching *Acropora* and *Pocillopora* corals is low, with 35% of sites showing no recruitment, and elsewhere some recruits (1-10cm diameter) were observed on limestone pavement, dead standing coral and rubble. These small colonies will be very vulnerable to predation from fish and urchins, and abrasion from mobile unconsolidated substrates during storms. Death and erosion of the reef edge has exposed many lagoons and shores to wave action, and there are signs of beach erosion on some islands such as La Digue. There is an urgent need to monitor recruitment and to protect live coral and recovering reefs, especially in those areas affected by activities on land, fishing and anchoring. Contributed by: John Turner, School of Ocean Sciences, University of Wales, Bangor, UK, e-mail J.turner @ bangor.ac.uk

Mauritius
Relatively minor coral bleaching was observed in mid-February 1998 in the lagoons and on outer slopes of Mauritius. Quantitative monitoring at 2 sites including Trou aux Biches from March-May 1998 showed degrees of bleaching at all sites, with less than 6% of corals totally bleached and 27% partially affected. Again *Acropora* species were the most affected. No bleaching was visible after 9 months at the monitored sites, with all corals apparently 'recovered' and the effects were much less than other IOC countries. Rapid surveys at 36 sites around the main island and inner islands showed that the reefs in Mauritius had suffered some bleaching in 1998, but most reefs were still healthy with the main signs of damage from boat and anchor damage and cyclones. The only large areas of dead standing coral were on the Barrier Reef off Mahebourg, and mean bleaching was less than 10% at all sites, often seen as partial bleaching of colonies. Mauritius probably escaped the mass bleaching event of 1998 because of cyclone Anacelle, which produced wet and cloudy weather in February 1998.

La Reunion Island
No widespread bleaching has occurred over the past few years, and the major El Niño event of 1998 affected less than 10% of the coral communities. Some bleaching was observed in 2000 on the west side of La Reunion, but this appears to be related to exposure during extreme low tides.

Seychelles
These reefs were amongst the worst affected in the Indian Ocean during the unprecedented events of 1998. There was extensive coral bleaching and mortality starting in February 1998 at the height of the southern summer and coinciding with the most intense El Niño event on record. Bleaching ceased in May 1998 after seawater temperatures had reached 34°C. It is estimated that 40-95% of corals in the Seychelles were bleached, with subsequent mortality of these varying from 50-95%, but the bleaching was not uniform across the archipelago. The southern islands (Aldabra, Providence, Alphonse) had only 40-50% bleaching, whereas it reached 95% mortality in the Ternay Bay Marine Park (Mahe). Coral regrowth and new larval recruitment has now started and small colonies are evident in various places, although the recruitment is still very weak reflecting the high mortality of adult corals. The largest pool for coral larvae appears to be from deeper zones unaffected by bleaching. If there are no further major bleaching events, these reefs should recover, although recovery will be slow.

LEGAL INSTRUMENTS, MONITORING AND MPAS

Legal Instruments - Creation of 'Coastal' or 'Reef' Laws
There are no specific laws for reef environments in the IOC countries, although most have legislation for coastal zones that could be applied to coral reefs. For example on La Reunion, which is a French territory, the coastal law of 1986 recognises that 'the coast is a geographic entity needing a specific planning, protection and development policy'. The decision to establish Schema d'Amenagement Regional (SAR or regional planning policy) on La Reunion predates the coastal law, and has probably reinforced the 'Schema de mise en valeur de la Mer' (SMVM or sea development policy) as the maritime arm of the SAR.

The national coastal law has helped consolidate the role of the 'Conservatoire du Littoral et des Rivages Lacustres'; a state-run administrative body dedicated to acquiring important natural coastal areas to protect them from development projects.

The environmental charter of 1990 in Madagascar has been the catalyst for a bill on coral and rocky reef state surveillance, which should be submitted to the Prime Minister shortly. If passed, Madagascar will be the first IOC country to have specific laws relating to reefs.

The IOC countries all have recent legislation on impact assessments, however, there are few examples of the laws being enforced. East African and West Indian Ocean countries have prepared a synthesis of legislation on impact assessments, and developed a code of good conduct for such assessments (SEACAM training seminar October 1998). Ecolabelling of tourist developments (ISO 14001 labels, Green Globe, etc.), which monitors resort operations, provides a powerful tool to encourage sustainable management of coastal resources, in parallel with impact assessments, which monitor the design-stage. The benefits are continued assessment of compliance following the pressures applied in design, however it is essential to provide information to the operators who are often unaware of environmental problems.

Environmental Monitoring
There are two major reef monitoring programmes in the region: the IOC programme summarised here; and the CORDIO programme. The latter (COral Reef Degradation in the Indian Ocean) was developed to assess the impacts of the 1998 exceptional bleaching phenomenon on the reefs and human communities of the wider Indian Ocean. Finance is provided by the World Bank, the Swedish Development Agency (SIDA) and other national agencies, and the WWF. CORDIO is now in its second operational phase, following the national assessments in 1999, and it is essential to ensure that there is coordination between the different monitoring programmes to avoid duplication and profit through synergies. Such coordination exists already in Eastern Africa and South Asia.

Marine Protected Areas

Comoros
Two marine park projects are proposed: the Coelacanth Marine Park, for which feasibility studies have been completed by the REP-IOC and sent to all stakeholders; and the Moheli Marine Park, which is currently being set up by the UNEP-GEF. These marine parks aim to preserve biodiversity, implement long-term marine resource management and develop ecotourism. The villages of Itsamia, Hamavouna, Nkangani, Ouanani, Ziroudani, Nioumachouoi, Ndondroni, Ouallah 2, Ouallah-Mirereni and Miringoni on Moheli island have been declared as 'Parc Marin de Moheli' National Park in line with the provisions of article 46 of the environment laws.

Madagascar
Only one marine park has been officially established, the northern Mananara Biosphere Reserve, but several are proposed in Masoala, near Toliana and on the island of Nosy Tany Kely, which has been considered as a marine reserve for many years but does not have legal status. The ANGAP (National Agency for the Management of Protected Areas) coordinates and manages marine protected areas, however, there are no exclusively marine

protected areas, and none are planned. All protected marine areas are within larger land-based parks, with the exception of the small island of Nosy Tany Kely. A bill originally proposed in 1997 has been passed to appoint ANGAP to manage all parks in Madagascar, under the supervision of the Department of the Environment.

Mauritius
The Department of Marine Parks was formed in 1995, and has established 2 MPAs with the main selection criterion being high coral diversity: Blue Bay with 320ha; and Balaclava with 482ha. A 10-year management plan was proposed and they were officially declared in October 1997. The constraints imposed in these park are minor: angling is the only allowable form of fishing; water-skiing and diving are authorised in limited areas only; and reef walking is prohibited.

La Reunion
There are no official marine protected areas in La Reunion, however, the long overdue 'Association Parc Marin' was created in 1997, following initiatives of local institutions and politicians. Before effective protection can be implemented, it will be necessary to create legal mechanisms and the current Association will evolve towards a management organisation run by an association of communes (with the probable title of Regional Natural Reserve). A major project is in progress to classify the southern and western reef zones as nature reserves, and status will facilitate the passing of legislation on permissible uses within zones created to conserve the various ecosystems.

Seychelles
Protected marine areas are the responsibility of the Seychelles Marine Parks Authority (MPA), which comes under the Environmental Division of the Department of the Environment and Transport. There are 3 objectives for these MPAs: to inform the general public; to monitor status of the reef ecosystem; and to introduce patrols and rangers to police the marine parks. Only the parks of Sainte Anne (location of MPA headquarters), Ternay Bay, Port Launay and Curieuse have effective management and benefit from the more or less permanent presence of MPA rangers. The other parks are visited regularly.

CONCLUSIONS

There was a serious lack of capacity and ongoing coral reef monitoring in most of these countries, and now the IOC regional reef network has gradually developed national monitoring networks and is constituted as the GCRMN and the ICRI Node for the Indian Ocean ('West Indian Ocean Island States'). This has proceeded over 2 years, through a series of regional workshops, forums and training sessions that allowed national delegates to exchange their experiences and request specific assistance. A network of 'national technical focal points', 'operational units' and 'supervisory bodies' has evolved with continual redefinition of the roles and functions.

As mentioned earlier, this network established and monitored 44 reef stations in the 5 countries in 1999. The increase in stations from 23 in 1998 reveals an increasing commitment by the IOC Member States to the recommendations in the 1998 first regional report of the REP-IOC. GEF (Global Environmental Facility) assistance in 2000 will allow existing monitoring programmes to be continued through provision of fundamental material support to the IOC countries. A methods guide 'Coral reef state monitoring in the south-west Indian Ocean' was developed from the GCRMN recommended methods and published in 1997 and is now available in both print and CD-ROM formats in English and French. The bleaching in 1998 caught the countries of the region off guard with only a few monitoring projects in place. If bleaching strikes again, the countries will be prepared for both rapid and long-term assessment of impacts and recovery.

Establishment of a regional database compatible for the countries of this region and the broader GCRMN global network is underway, utilizing AIMS Science Reef Monitoring Data Entry System which enables the storage and processing of monitoring data for benthic cover, incidence of bleaching, fish counts and separation into feeding groups, and is available in Microsoft Access 97. It is being successfully implemented in IOC countries via specific training courses.

RECOMMENDATIONS

To improve conservation in the region, there is a requirement to:

- improve capacity in coastal management (in parallel with monitoring capacity being developed) and ensure that countries have the funding to employ trained people in functional roles of coral reef resource management;
- establish many more and larger Marine Protected Areas with adequate planning, involvement of communities and effective enforcement so that these can serve as fisheries replenishment areas;
- develop better legal instruments for coastal conservation and ensure that inter-sectoral problems are minimised through the formation of high level coordination bodies in each country; and
- provide better education to all communities and other stakeholders in the value of coral reefs and in effective measures for their conservation.

The national monitoring network focal points have recommended the following to maintain long-term effectiveness of the network after 2000 through the regional network:

- facilitate the urgent appointment of a regional coordinating organisation by the IOC;
- request the regional coordinators develop an action plan, to facilitate continued reef monitoring, operation, workshops, reports;
- implement the GEF funding plan, to ensure the long-term functioning of the reef network and monitoring beyond 2000 and coordinate such with EU funds to ensure cooperation with relevant IOC programmes (ecotoxicology, coastal erosion, pollution, etc.);
- implement national 2000 monitoring programmes funded by the REP-IOC to enable continued monitoring and set up the database for decision-makers in coral reef management, and continue training, especially in method evaluation and data quality control;

- transfer equipment currently available to facilitate successful operation of the reef network and, specifically assist field teams undertake reef monitoring (data collection and acquisition);
- establish regular regional network events to link people and institutions for enhanced communication.

ACKNOWLEDGEMENTS

Lionel Bigot, Anne Lieutaud, Loic Charpy, Jean Maharavo, Fouad Abdou Rabi, Naidoo Paupiah, Riaz Aumeeruddy, Christian Villedieu served as primary authors. They thank all the other experts and network members who contributed to this document: Gilbert David, Martine Delmas-Ferre, Pascale Chabanet, Odile Naim, Marilene Moine-Picard, Said Ahamada, Edouard Mara and all the Albion Centre team. Contact: Lionel Bigot, ARVAM , 14 rue du Stade de l'est, 97490 Ste Clotilde, La Reunion, (lionelbigot.arvam@guetali.fr)

6. STATUS OF CORAL REEFS IN SOUTH ASIA: BANGLADESH, INDIA, MALDIVES AND SRI LANKA

ARJAN RAJASURIYA, HUSSEIN ZAHIR, E.V. MULEY, B.R. SUBRAMANIAN,
K. VENKATARAMAN, M.V.M. WAFAR, S.M. MUNJURUL HANNAN KHAN
AND EMMA WHITTINGHAM

ABSTRACT

The major coral reefs in South Asia surround the oceanic islands of Lakshadweep, Maldives, Chagos and the high islands of Andaman and Nicobar. Other extensive reefs are in the Gulf of Mannar region. There are also numerous fringing and patch reefs in India and Sri Lanka. In Bangladesh, the only coral reefs are around St. Martin's Island, and there are only scattered reef communities in Pakistan and little available information. Recent surveys indicate that recovery of corals bleached during high water temperatures associated with the 1998 El Niño event is poor. Natural and human disturbances, such as the crown-of-thorns starfish, coral mining, destructive and unmanaged resource harvesting, sedimentation and pollution continue to cause much damage to coral reefs in South Asia and reduce their capacity to recover from the 1998 bleaching event. Capacity for monitoring coral reefs has improved with donor assistance, however there is limited application of monitoring data due to a lack of management mechanisms. In the absence of proper management, the condition of marine protected areas in South Asia has degraded. Several new protected reef areas have recently been declared in the Maldives and another in the Andaman-Nicobar area. This report highlights the increasing population pressures on reef resources, lack of awareness and inadequate capacity for management. Regional and country-specific recommendations towards improved management, conservation and sustainable use of coral reefs in South Asia are identified.

INTRODUCTION

This status report focuses on the coral reefs of India, Maldives, and Sri Lanka, with additional information on the small coral resources of Bangladesh and Pakistan, and a brief status report on the vast Chagos Archipelago. The first summary report for this region was presented at the International Coral Reef Initiative, South Asia workshop held in the Maldives in December, 1995. This was updated at the International Tropical Marine Ecosystems Symposium in Townsville Australia in November 1998 and published in the first 1998 Status of Coral Reefs of the World. India, Maldives and Sri Lanka together form the 'South Asia Node' of the Global Coral Reef Monitoring Network (GCRMN), supported financially by the UK Department for International Development (DFID). This regional report addresses regional perspectives as well as summaries from country reports of India, Maldives and Sri Lanka.

The largest coral reef areas in South Asia include the atolls of Lakshadweep, Maldives, and Chagos. Corals also grow along the coast of the Indian subcontinent and around Sri Lanka, include extensive reefs around the Andaman and Nicobar Islands in the Bay of Bengal, the Gulf of Mannar and Gulf of Kutch on the mainland of India. Reefs in this region are strongly influenced by the southwest and northeast monsoons. There are no extensive coral reefs in Bangladesh except for small coral patches offshore around St. Martin's Island. In Pakistan, there are only small isolated coral colonies in highly turbid coastal conditions.

The South Asia region is characterised by large populations of very poor coastal people who depend on coral reef resources, particularly in India, and Sri Lanka. Coral reefs are important economic resources for India, Maldives, and Sri Lanka, but the perceived values varies amongst different sectors. In the Maldives, coral reefs are important primarily for tourism, followed by fisheries, coastal protection and the aquarium trade. In India, reefs are important mostly for fisheries and coastal protection, with strong potential for tourism developments in the Andaman and Nicobar Islands, and possibly on the Lakshadweep atolls in the future. The primary economic activity of reefs in Sri Lanka is fisheries, followed by coastal protection, ornamental fisheries, and tourism.

The coral reefs of India and Sri Lanka are being degraded rapidly by increasing human activities, particularly over-fishing, coral mining and the effects of sediment and nutrient pollution. In contrast, the remote reefs of the Maldives and Lakshadweep have been virtually unaffected. That was the case until the major climate related bleaching events during the first half of 1998, which destroyed many of the shallow water corals in the remote reefs of the Maldives and Chagos, as well as the Lakshadweep atoll reefs, and reefs around Sri Lanka and in the Gulf of Mannar of India. There are also reports of bleaching damage to coral reefs in the Gulf of Kutch and Andaman and Nicobar Islands, but these impacts were less severe than elsewhere. The greatest impact of bleaching was in shallow reef areas to about 10m depth. Branching and tabulate corals that dominated these reefs were the most affected, and appear to have been almost obliterated in many areas. Since the bleaching event, monitoring has been initiated in several reef areas to observe signs of recovery and new recruitment, as well as gathering sound baseline data to better understand reefs under highly stressful conditions.

The following provides an overview of each country in this node:

Bangladesh
The offshore island of St. Martin's is the only area with corals in Bangladesh and is heavily influenced by monsoons and frequent cyclones. There are no true coral reefs around this island, only coral aggregations in shallow waters along with seagrass beds, soft coral habitats and rocky habitats. There is heavy sedimentation from the combined discharge of the Ganges, Brahmaputra and Meghna rivers, which contribute about 6% of the world's total sediment input into the oceans.

Chagos
The Chagos Archipelago is the southern most group of atolls in the Laccadive-Chagos ridge and is located at the geographical centre of the Indian Ocean. Chagos is a British Territory and uninhabited except for the US military base on Diego Garcia. There are 6 major atolls, many

Status of Coral Reefs in South Asia

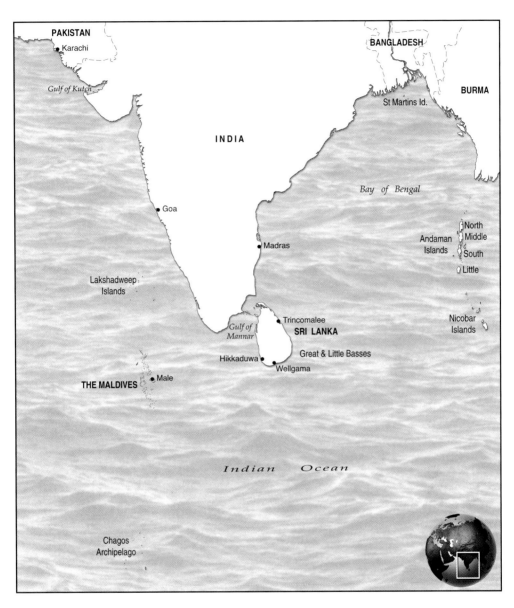

small islands, as well as many smaller atolls and submerged shoals. The central Great Chagos Bank has a large submerged reef area with 8 islands. This is the largest area of undisturbed reefs in the Indian Ocean, with probably the highest coral reef biodiversity in the region.

India
There are 4 major coral reef areas in India: Gulf of Mannar; Andaman and Nicobar Islands (1962km coastline); Lakshadweep Islands (132km coastline); and the Gulf of Kutch (Kachchh). There is also scattered coral growth on submerged banks along the east and west coasts of the mainland. Coral reefs are important economically for the livelihoods and social welfare of coastal communities providing up to 25% of the total fish catch.

Reefs in the **Gulf of Mannar** are found around a string of 21 islands, 8km off the southeast coast of India. The 3 island groups (Mandapam, Keelakari and Tuticorin) form the 'Pamban to Tuticorin barrier reef', which contains fringing, platform, patch and barrier reefs. Narrow fringing reefs surround the islands extending 100m from the shore. Patch reefs are also found and are typically 1-2km long, 50m wide and 2 to 9m deep. Reef flats are extensive on all islands. The total area includes approximately 65km^2 of reef flat and 14km^2 of algal growth. The major economic activities are fishing, coral mining for construction, harvesting of sacred chanks (*Turbinella pyrum*), sea cucumber, pipefishes, sea horses and seaweeds.

Of the 530 islands in the **Andaman and Nicobars,** only 38 are inhabited with 279,000 people as of 1991, but this is predicted to rise to 405,000 in 2001. The largest islands of North Andaman, Middle Andaman, Ritchie's Archipelago, South Andaman, Little Andaman, Baratang and Rutland Island are mountainous and forest covered, and are surrounded by some of the richest coral reefs in India.

The **Lakshadweep Islands** are true atolls at the northern end of the Laccadive-Chagos ridge, 225-450km west of the Kerala coast. There are 12 coral atolls with 36 islands and 5 submerged banks. Islands vary in size from 0.1km^2 to 4.8km^2 (total area 32km^2) and are surrounded by 4,200km^2 of lagoon, raised reefs and banks. The population on the 10 inhabited islands ranges from 100 on Bitra Island to 10,000 on Kavaratti. Offshore fishing is the most important activity, and reef fisheries are not economically important. Tourism is slowly developing, but provides little income for the local community.

There are 42 islands with fringing reefs in the southern part of the **Gulf of Kutch** along with extensive mangroves in the Indus River Delta. Corals survive through extreme environmental conditions such as high temperature, salinity changes, high-suspended particulate loads and extreme tides, as high as 12m.

Maldives

The Maldive archipelago is at the centre of the Laccadive-Chagos ridge, and is 864km long (north to south), 130km wide, and has a land area of approximately 300km^2. There are 1,190 coral islands, numerous sand cays and faroes within 23 atolls. The exclusive economic zone is approximately 90,000km^2. Islands of this archipelago have a maximum elevation of about 5m and a narrow fringing reef around each island, which slopes rapidly down to the seabed. The reefs are essential for shoreline protection. Until recently, the Maldives depended on the offshore tuna fisheries; however, tourism is now the mainstay of its economy.

Pakistan

Coral growth is inhibited by the high level of sedimentation, turbid conditions and limitations in the availability of suitable habitats for coral growth. The only known coral formations occur as small isolated patches growing on hard substrates. There is an almost total lack of published information on corals in Pakistan and no research programmes exist to monitor the corals and associated biota.

Sri Lanka

An estimated 2% of the 1,585km coastline has fringing reefs, with larger reef areas offshore in the Gulf of Mannar to the northwest and also along the east coast. Corals grow to varying extents on old limestone, sandstone and rocky reefs, with the location of reefs well known, but poorly mapped. Reefs are important for fisheries, coastal tourism, and preventing coastal erosion.

CORAL REEF STATUS AND BIODIVERSITY

Bangladesh
Coral communities extend to about 200m offshore of St. Martin's Island with maximum coral cover of 7.6% and colony density of $1.3m^{-2}$. These comprise 66 hard coral species, the most common are *Porites, Acropora, Favites, Goniopora, Cyphastrea* and *Goniastrea*. *Acropora* spp. are the target for coral harvesters, as well as *Favites* and *Goniastrea*. There are also many soft corals, sea fans, and sea whips. Other invertebrates are only represented by a few, with molluscs being the most abundant large invertebrates, however, these are declining due to unregulated harvesting. Reef fish diversity is low (86 species) with damselfish (Pomacentridae), surgeonfish (Acanthuridae) and parrotfish (Scaridae) being the most abundant. There are also 5 species of butterflyfish (Chaetodontidae) and one angelfish (*Pomacanthus annularis*). Predator species (groupers, snappers, and emperors) are heavily fished. There are no reports of coral bleaching from St. Martin's Island.

Chagos
These reefs have the highest species diversity of corals and molluscs in the Indian Ocean. There are no comprehensive studies on reef fish and other invertebrates, yet reef-building corals have been relatively well studied. The smaller islands support large colonies of sea birds, sea turtles and many species of small cetaceans. Formal protection for the Chagos reefs has been proposed. Corals around Chagos were seriously affected by the bleaching event in 1998. Before this bleaching event seaward reef status included: 50%-70% live coral cover, 10%-20% soft corals and approximately 10-20% bare substrate. Approximately 55% of the live coral cover on the outer reefs was lost after the bleaching, including most of the table corals. Soft corals, fire coral (*Millepora* sp.) and blue coral (*Heliopora caerulea*) were also heavily impacted during the 1998 event, while large *Porites* colonies on the outer slopes were only partially bleached, and calcareous algal ridges around the atolls were unaffected. The corals in the reef lagoons survived better than those on the outer reefs, potentially because they normally experience higher temperatures and have built up tolerance to temperature increases. (see Bleaching Chapter 2).

India
In the **Gulf of Mannar,** approximately 3,600 species have been recorded within the three main ecosystems (coral reefs, mangroves and sea grass beds) in the Gulf of Mannar. Biodiversity on the reefs include 117 hard coral species, with the most common corals being *Acropora, Montipora* and *Porites*. Other resources in the area include sacred sharks, pearl oysters, sea turtles, dugongs and dolphins. The main seaweeds are *Gracilaria, Gelidiella, Hypnea, Sarconema, Hydrodathrus, Caulerpa, Sargassum* and *Turbinaria*. Reef fish diversity and abundance has not been well documented. The bleaching event in 1998 destroyed most shallow water corals in the Gulf of Mannar, with live coral cover reduced by 60-80% and only about 25% of live corals remaining. The most affected species were the branching *Acropora* spp. and *Pocillopora* spp. All the *Montipora* spp. on Pullivasal Island (northeast Gulf of Mannar) were bleached during 1998 (although *Montipora aequituberculata* escaped bleaching in southern Sri Lanka). Massive corals are now

Country	Percentage destroyed prior to 1998	Percentage destroyed during 1998 bleaching	Potential for recovery
Bangladesh	20 – 30	None	Low
Chagos	1 – 2	70 – 90	Good
India – Gulf of Mannar	25 - 45	60 – 90	Medium – Low
India – Gulf of Kutch	15 – 25	50 - 70	Medium – Low
India – Lakshadweep	5 – 10	70 – 90	Good
India – Andaman & Nicobar Islands	2 – 5	15 – 25	Good
Maldives	2 – 5	70 – 90	Good – Medium
Sri Lanka	10 – 20	70 – 90	Medium – Low

These estimates were provided by National coral reef leaders in these regions and estimate the amount of reefs that were irreparably damaged prior to 1998, and those that suffered losses in the 1998 bleaching event, for which there should be good chances of recovery in South Asia.

dominant in all 3 island groups, with branching corals almost completely wiped out in the Tuticorin group, while only 1-2% survived in the other two island groups. Surveys carried out by the Zoological Survey of India one year after the bleaching observed patchy coral recruitment on the mainland coast.

In the **Andaman and Nicobar Islands,** there are 203 coral species, 120 species of algae, 70 species of sponge, 200 species of fish, 8 species of shark, and spiny lobsters on the islands. More than 1200 fish species have been recorded around the Andaman and Nicobar Islands, and recent random surveys have detected 571 species of reef fish. Dugongs, dolphins and sea turtles are also known from the islands. The bleaching event had less impact on the Andaman and Nicobar Islands as compared with other coral reef areas in India. Reports at the time indicated that 80% of live corals were destroyed, however, recent surveys at 5 different sites report an average of only 20% dead coral cover in shallow areas with 56% live coral cover and 11% coral rubble.

In the **Lakshadweep Islands** to date, only 95 hard coral species and 603 fish species (both reef dwelling and oceanic) have been recorded from Lakshadweep, but comprehensive biodiversity studies are lacking. Much of the living coral cover around Lakshadweep was destroyed in the 1998 bleaching event, with estimates ranging from 43% to 87% loss of live coral cover. Cover declined markedly to about 10% live coral in Kadmat Island, but there have been no apparent effects on fish populations. The status of many of the outer coral reefs is known only as anecdotal accounts. (see Bleaching Chapter 2).

In the **Gulf of Kutch** hard coral species diversity is low with only 37 species and a total absence of ramose growth forms. Reports of bleaching in 1998 vary considerably from about 70% of live coral loss, to much lower impacts. This highlights the need for permanent monitoring sites to provide adequate baseline data.

Maldives

Along with the Chagos Archipelago, the Maldives support the greatest diversity of corals and associated reef organisms in the region, with at least 209 species of stony corals. Coral reefs of the Maldives were in excellent condition prior to 1998, but were degraded heavily during the bleaching event. Surveys show that about 2% live coral remains on the reef tops at study sites in Haa Dhaal, North and South Male, Ari, Vaavu and Addu atolls. Observations by tourist divers and others indicate a similar status throughout the country, with approximately 20% loss of live coral cover compared to pre-bleaching observations. Recent surveys show that the bleaching has affected north and south Male atolls more than other areas of the country, however, encouraging levels of recruitment have been observed at all the sites, with many *Acropora* amongst the new recruits.

Pakistan
Detailed information on the biodiversity reef habitats in Pakistan is lacking, although a marine reference collection and resource centre was set up in the University of Karachi in 1969.

Sri Lanka
The most complete biodiversity information comes from Sri Lanka, with 183 hard coral species, 6 species of spiny lobsters, plus many other invertebrates, sea turtles and dolphins. Dugongs occur inshore of coral reef areas along the northwest Gulf of Mannar coast of Sri Lanka. The most economically important reef fish in Sri Lanka include: groupers, snappers, emperors, barracuda, jacks, sear and leatherskins, and fusiliers. Other reef fish important for the aquarium trade include: 35 species of butterfly fish, 6 species of large angelfish and pygmy angelfish, plus many others. The healthiest reefs in Sri Lanka were those remote from human settlements prior to 1998, with living coral cover estimates ranging from more than 80% cover on coral reefs at the Bar Reef Marine Sanctuary, to more than 50% cover at Weligama on the south coast, and about 35% in the Hikkaduwa Marine Sanctuary in the south.

In 1998, corals in most shallow reef areas of Sri Lanka were destroyed to depths of 3-5m including areas from the northwest around to the east coast, except near Trincomalee. Surveys at Pigeon Island near Trincomalee showed no bleaching in 1998. Many of the remaining shallow coral reefs lost virtually all coral cover, with the most affected species being branching and tabulate forms of *Acropora* spp, *Pocillopora* spp., *Millepora* spp. and corals in the family Faviidae. Several other species (*Montipora aequituberculata*, *Porites rus* and *Psammacora digitata*) were only marginally affected with most colonies showing no bleaching and the few affected colonies recovering within several months. Bleached corals were recorded at 42m depth off the east coast, but almost all bleached corals below 10-15m recovered after about 6 months. Surveys in 1999 and early 2000 showed that coral cover at Bar Reef Marine Sanctuary was near zero in shallow reef habitats (e.g. 3m depth). At Hikkaduwa Nature Reserve it was reduced to only 7%. At Weligama it was down to 28% and Rumassala reef had 20% live coral cover. Below 10m depth almost all corals had recovered completely, with no signs of bleaching. Coral species at different locations showed variable levels of impact and recovery from the bleaching event. Bleached *Acropora formosa* corals at Weligama recovered and remain healthy, whereas almost all other branching *Acropora* species were destroyed. At Rumassala, *Acropora formosa* did not survive, and other species such as *Acropora valida*, *A. aculeus*, *Echinopora lamellosa*, *Galaxea fascicularis* and *Montipora aequituberculata* have begun to recolonise damaged sections of the reef. Bleached soft corals (*Sarcophyton* spp., *Sinularia* spp., *Dendronephthya* spp., *Lobophyton* spp.) recovered quite rapidly and have begun to dominate some locations, such as Weligama. At many other locations, dead shallow coral areas have been overgrown by rapid growing algae and invertebrates (tunicates and corallimorphs). This overgrowth will inhibit the re-establishment of living corals at these locations. Some new coral recruits have been observed at many sites, but indications are that reef recovery will be very slow, although many genera have been observed to have recruited already (*Acropora, Platygyra, Stylocoeniella, Alveopora, Goniopora, Pocillopora, Galaxea*).

CORAL REEF FISHERIES

In general the reef fisheries of South Asia are poorly documented, with little information on the importance of these fisheries to local communities and economies. Reef fisheries are mostly for subsistence, therefore data are difficult to gather and do not appear in national fisheries statistics. Furthermore, the multi-species nature of reef fisheries makes it difficult to estimate a yield for each species and to predict sustainable levels of extraction. Similar problems exist for the marine ornamental fisheries, which are important in the Maldives and Sri Lanka.

Bangladesh

The main fishery in Bangladesh is offshore, with a small inshore fishery for croakers and snappers. These are caught using bottom-weighted gill nets and hook and line. Some snappers and medium sized groupers (approximately 40-50cm long) are caught near St. Martins Island, indicating that a relatively healthy population remains. Other fisheries include sea cucumber and molluscs. Unregulated harvesting of reef resources is contributing to a decline of reef biodiversity and abundance of resources.

India

Reef fisheries in India are not reflected in national fisheries statistics, as little significance is given to the fishery, however, they are important as a subsistence fishery for local people. The fishery includes: snappers, groupers, emperors, breams, barracuda, jacks, sprats, herrings and flying fish. There are also reef fisheries for sea horse, sea cucumber and sacred chanks.

In the **Gulf of Mannar,** specific information on reef fisheries is not available, but the annual catch of demersal fish, which includes reef fish, is about 45,000 metric tonnes per year. In the **Andaman and Nicobar Islands** fishing is mainly carried out around the Andaman Islands, with little around the Nicobar Islands. The main species targeted include sardines, anchovies, carangids, mackerel, mullets, perches, sharks and rays, catfish, pomfrets, silver bellies and catfish. Estimates of the fishery resource potential are highly variable and do not consider a maximum sustainable yield critical for proper resource management. In the period 1996-1997, the total fish catch was estimated at 26,550mt and the overall composition of the catch included; sardines (12-13%), perches (7-10%), carangids (6%) and mackerel (6-7%). There is also a sea cucumber fishery for which figures are not available. In the **Lakshadweep Islands** there is no organised commercial reef fishery for food or ornamental fishes, but there is a subsistence reef fishery. The local industry targets offshore pelagic fish, such as tuna, which require the harvesting of sprats (*Spratelloides* sp.) from reef lagoons for bait. No information is available on reef related fisheries in the **Gulf of Kutch.**

Maldives
The most important reef fisheries in the Maldives are for live bait for tuna fishing (including collection of silver sprat, blue sprat, fusiliers, cardinal fish, anchovies, damselfish and silversides) and the aquarium fishery. The aquarium fishery is relatively small, with around 100 species of reef fish collected, but the trade has expanded steadily in the last 20 years and there is now concern about over-exploitation. A quota system for harvesting and export is in place. A grouper fishery has also been developed in recent years exploiting some of the 40 species of grouper in the Maldives. The fishery initially began in the central section but has now spread to every atoll and supplies the local tourism market as well as export for the live fish restaurant trade. There is a shark fishery from some islands and 9 species of sea cucumbers are fished exclusively for export.

Pakistan
Information is not available on inshore or offshore reef fisheries in Pakistan.

Sri Lanka
There are 3 distinct components of the reef fisheries in Sri Lanka: 1) domestic trade in edible species; 2) small scale subsistence fisheries for village level consumption; and 3) fisheries primarily for export, such as spiny lobsters, sea cucumber, sacred chanks and ornamental fish. Reef fish catches and the trade are not clearly identified in fisheries statistics, but are included as 'rockfish' under the general heading of coastal fisheries. The National Aquatic Resources Research and Development Agency (NARA), the research arm of the Ministry of Fisheries and Aquatic Resources Development, monitors demersal fish catches at selected landing sites. Landing values for rockfish declined from 10,585mt in 1994 to 9,100mt in 1997 and increased slightly to 9,200mt in 1998. The reasons for these fluctuations are unknown, but may be related to limited access to some areas due to the ongoing military conflict. Most of the 250 fish species and 50 species of invertebrates used in the ornamental fishery are collected on reefs, which constituted about 40-50% of the US $6.6 million export in 1998. Other species harvested for export include sacred chanks (*Turbinella pyrum*; Turbinellidae), cowries, cones, murex and other shells. In 1998, 260,000kg of sea cucumber, 796,000kg of seashells and sacred chanks, and 11,400kg of molluscs were exported. Statistics for the spiny lobster catch cannot be isolated as they are now pooled with all crustaceans.

THREATS TO CORAL REEFS AND MANAGEMENT ISSUES

Bangladesh
The major threats to the coral habitats are high levels of sedimentation, cyclones, storm surges, freshwater and agricultural runoff, pollution from human settlements and the removal of coastal vegetation. There is also over-harvesting of corals, sea cucumbers and molluscs by excessive numbers of subsistence fishers. The removal of *Acropora* and other coral colonies for the curio trade is also a major threat to the reefs, such that *Acropora* are now rare. The main destructive fishing practice is using stones to weigh down the nets, which smash corals. There are no reports of blast fishing or the use of poisons. Large-scale removal of coral boulders and dredging of channels has caused considerable damage to the reefs, and a barrier wall built on the sea front has caused beach erosion. The removal of Pandanus trees for firewood has also caused much beach and dune erosion.

Chagos
Although there is no significant large-scale human damage to the reefs, there is illegal fishing around some reef areas, including the collection of sea cucumbers and shark. Occasionally poachers have been apprehended and vessels and equipment confiscated.

India
The **Gulf of Mannar** is one of the most heavily stressed coral reef regions in India, with impacts from destructive fishing, pollution and coral mining. Along the 140km coastline, there are 47 fishing villages with a combined population of approximately 50,000. There is severe over-exploitation of seaweeds, sacred chanks, pipefishes, sea horses and sea cucumber, and the extensive use of bamboo fish traps has seriously depleted fish stocks. Populations of pearl oysters, gorgonians and acorn worms (*Ptychodera flauva*) are also severely depleted due to over-harvesting. Approximately 1000 turtles are harvested annually and dugongs are also hunted. Increasing demand for grouper, snappers and emperors will put further pressures on these populations, and blast fishing has also been reported. Local fishermen complain that fish catches have declined both in the nearshore and offshore coral banks and islands. Sand mining, extraction of trochus shells, damage by crown-of-thorns starfish and sedimentation degrade corals in the area. About 250m^3 of coral is quarried per day from the Gulf of Mannar region. The islands in Tuticorin have been affected by industrial pollution and aquaculture. On the Keelakarai coast, sewage pollution has resulted in the overgrowth of corals by mats of green algae. Black and white band coral diseases have also been observed.

In the **Andaman and Nicobar Islands** deforestation has resulted in increased sediment flows to the nearshore reefs, however there are no quantitative assessments of the rate of sedimentation, nor the impacts on the reefs. Reports have indicated that corals have been killed by large quantities of sediment laden fresh water, as well as overgrowth by fleshy algae. Industrial pollution is also impacting the coral reefs around Port Blair. In **Lakshadweep Islands** crown-of-thorns starfish were first noticed at Agatti Island in 1977 and have spread to most islands and reefs causing loss of corals. Black and white band diseases and pink band disease have been observed in shallow coral areas, but bleaching has been the main cause of loss in reef biodiversity. There is some coral mining, dredging of navigational channels, unsustainable fishing practices, coastal development, and souvenir collection. Recently, blasting of corals to create navigational channels has been stopped, however, the construction of breakwaters on some islands is increasing coastal erosion. No sewage or oil pollution has been reported on these reefs. People have reported decreases in fish catches within the reef lagoons, which could be due to the loss of live corals after the bleaching event, or to increased harvesting due to population pressures (the population has tripled in the last 20 years). The methods used to catch live bait for tuna fishing cause damage to the reefs and reductions in live bait stocks have impacted on the local economy since the tuna fishery is the major industry in the islands. In the **Gulf of Kutch** major impacts on the reefs are associated with industrial development, ports and offshore moorings, pollution from large cities, and removal of mangroves. It is estimated that human activities have reduced the coral cover by more than 50% on most reefs here.

Maldives

Most of the reefs here are better protected than other reefs in South Asia; mainly because most are isolated from human activity. The main damage to reefs occurs around those islands that are heavily populated and where there is a high level of development. Impacts on the reefs include coral mining, pollution, dredging of channels for boats, coastal construction, increased reef fisheries and the crown-of-thorns starfish. Coral mining has lowered the reef flat around the capital of Male, such that concrete walls have been constructed to prevent shoreline erosion. Construction of wharves, groynes and breakwaters have resulted in increased erosion around some tourist resorts. However, the total reef area damaged due to human activities is relatively small compared to the extensive reefs present in the Maldives.

Pakistan

Although there are no extensive coral formations, marine life is affected by high levels of sedimentation, freshwater runoff, pollution from urban and industrial wastes, sewage discharge and dredging. Collection of marine organisms as souvenirs and for use in traditional medicines also has adverse impacts.

Sri Lanka

The major causes of reef degradation are: coral mining, sedimentation, destructive fishing practices (such as blast fishing), the use of bottom set nets, uncontrolled resource exploitation, including harvesting of ornamental fish and invertebrates, pollution from land based sources, and crown-of-thorns starfish. Site specific coral reef damage is associated with glass bottom boats, boat anchoring, destructive collecting techniques for ornamental species, urban pollution, coastal and harbour development, and high visitor pressure (causing damage through trampling and coral removal). Coastal erosion along the southwest and southern coasts has increased levels of sedimentation on the reefs. Rapid increases in the abundance of organisms such as tunicates, corallimorphs and algae (halimeda, caulerpa and filamentous algae) has smothered shallow reefs in certain locations, particularly following the bleaching events of 1998.

CLIMATE CHANGE AND IMPACTS

Predicted climate change impacts in South Asia include sea level rise and potential increases in the frequency and intensity of cyclones and storms, all of which will have adverse impacts on coastal areas. The most vulnerable areas are atolls and low-lying coastal regions, particularly deltas with mangrove swamps, wetlands, seagrass beds and sandy beaches that are prone to erosion, as well as highly developed coastal installations, such as harbours and ports. Cyclones, storms and heavy rainfall already have major impacts in the Bay of Bengal (Bangladesh and India). In other areas, coral reefs play a vital role in protecting shorelines and if their health deteriorates further, they may cease to protect shorelines from rising sea levels.

Seawater temperatures in the Indian Ocean have risen by $0.12°C$ per decade during the last 50 years. A continued increase will have significant impacts on coral reefs, ocean circulation, nutrients, primary production and fisheries. Increases in sea surface temperatures, and a cessation of the trade winds associated with the 1998 El Niño phenomena led to widespread coral bleaching and destruction of coral reefs in South Asia.

Only Bangladesh has information on recent sea level rise of 5.18mm per year in the Khulna region. If the predicted rises do occur, 50-70% of the mangroves will be adversely affected by 2050 due to inundation. In India there is a lack of awareness amongst the government and local population of the potential impact of bleaching on the reefs and the link to climate change. More efforts in India are needed to address climate change issues and to improve awareness. The low lying atolls of the Maldives, Lakshadweep, Chagos and river deltas areas (e.g. Bangladesh and Gulf of Kutch) are particularly vulnerable to sea level rise and potential climate change impacts. Coastal inundation, saline intrusion of fresh groundwater and coastal erosion are among the most serious impacts. The majority of all large urban centres in Sri Lanka are coastal, occupying 24% of the total land area and containing 32% of the population, and 67% of the industrial areas. Therefore, any sea level rise due to climate change will adversely affect the cities and agricultural and industrial lands due to saltwater intrusion. Currently there is significant coastal erosion along the western and southern coast of Sri Lanka and more erosion problems may be expected in the future.

STATUS AND GAPS OF MARINE PROTECTED AREAS (MPAS), MONITORING AND CAPACITY

Marine protected areas are not well managed across South Asia and there has been little improvement in the past 5 years, despite the recent declaration of Rani Jansi Marine National Park in the Andamans and 10 new protected sites in the Maldives. Moreover, these recent inclusions are placing increasing pressures on the already weak management systems, which lack motivation, trained personnel, equipment and funding. Implementation and enforcement of protected area management plans and regulations is generally poor or absent, and in some cases impossible because marine reserves typically lack physical boundary markers e.g. Hikkaduwa Nature Reserve, Sri Lanka. More importantly there are usually no alternative sources of income for those dependent on the resources from the protected areas. The problems with protected area and marine resource management are particularly serious in Sri Lanka, the Gulf of Mannar, and the Gulf of Kutch regions in India, and Bangladesh. MPA problems are less severe in the Maldives, much of the Andaman and Nicobar Islands, and the Lakshadweep Islands, primarily due to low population densities in these areas.

In spite of the above MPA scenarios, capacity to monitor reefs has improved in South Asia, particularly in India following training programmes conducted through the GCRMN South Asia Node, and more recently by the Indian Coral Reef Monitoring Network (ICRMN). Improvements are particularly evident as illustrated by the increasing level of information in national status reports, and a marked increase in data during the past 3 years. However, a capacity for reef management beyond monitoring has not improved and coral reefs continue to degrade in the region.

Bangladesh
St. Martin's Island has been identified for protection and management under the National Conservation Strategy of Bangladesh, but no analysis and identification of key sites has been carried out. There are no management plans and no trained staff to undertake conservation, therefore destructive human activities continue to degrade the reef resources. As there are no baseline data on St Martin's Island, surveys and taxonomic knowledge are needed,

particularly concerning fish and invertebrate diversity. There is no clear conservation policy and consultations with the local community on resource management have been limited. A review of the National Conservation Strategy showed that legal and institutional issues have largely been ignored

Chagos

The Corbett Action Plan for Protected Areas of the Indomalayan Realm identified the Chagos Archipelago as an area with marine conservation needs due to its extensive coral reef and unique terrestrial habitats. However, it is not known whether Chagos has been declared as a marine protected area.

India

There are 5 marine protected areas in India: Gulf of Mannar Biosphere Reserve (GoMBR - 10,500km^2); Gulf of Kutch Marine National Park (GoKMNP - 400km^2); Mahatma Gandhi Marine National Park (MGMNP) also known as the Wandoor Marine National Park in Andamans (282km^2); Great Nicobar Biosphere Reserve (GNBR - 885km^2); and Rani Jansi Marine National Park (RJMNP) in Ritchie's Archipelago in Andamans. A marine protected area (Perumal Marine Park) was proposed for Lakshadweep in 1996, but there is no evidence of its declaration. The management for these protected areas is weak, particularly those near the subcontinent where human impacts from resource use and urban and industrial development are high. Reefs in the Gulf of Kutch Marine Park have been neglected, with monitoring activities limited to occasional EIA studies associated with development activities and there are growing concerns that parts of the park may be rescinded for industrial development. Protected areas on Andaman and Nicobar Islands and in Lakshadweep are in better condition, but only because human impacts are less. These MPA areas are still vulnerable to impacts from the crown-of-thorns starfish and bleaching, which are largely beyond local management. There is a major need for training of conservation officers to manage the protected areas and funding for infrastructure development.

In general the existing marine protected areas noted above for India are not effectively managed and destructive activities continue. There is a need for studies on reef resource use, particularly harvesting of reef fish and the economic links between resource exploitation and coastal communities. Studies are also required to investigate the dynamics of the live bait fisheries. Surveys in the Lakshadweep, Andaman and Nicobar Islands are restricted to a few sites that are easily accessible. Fish censuses have not been included in monitoring programmes and a lack of trained divers and scuba gear prevents sampling in deeper areas.

No systematic monitoring or conservation strategy existed in India until recently, through initiatives mainly associated with the establishment of a GCRMN Node. There are few trained and skilled people for long-term monitoring, and there is little NGO involvement or community participation in reef management. There is also a lack of awareness of the value and importance of coral reefs among government agencies and local communities, although some academic institutions have conducted reef research. Thus, there is still a need for more coordinated approaches to coral reef management.

Coral Reef Monitoring Action Plans (CRMAPs) were prepared during the first phase of the GCRMN (1997-8) and have been launched by the ICRMN for all reef areas except the Gulf of

Kutch. Government support has been extended to implement the CRMAPs and train people to monitor the reefs, however, activities are still at an early stage and the capacity for monitoring and management is still poor.

Other significant international initiatives on Indian coral reefs include: UNDP-GEF Projects on the Gulf of Mannar and Andaman and Nicobar Islands; the Coral Reef Degradation in the Indian Ocean project (CORDIO); an Integrated Coastal Zone Management Training Project (ICZOMAT) funded by the UK Department for International Development (DFID); and an India-Australia Training and Capacity Building (IATCB) programme.

Maldives

There are 25 marine protected areas in the Maldives declared under the Environment Act of the Maldives and administered by the Ministry of Home Affairs, Housing and Environment (MHHE). All are important dive sites for tourists. These sites are generally small, without specific assessments of biodiversity, except for flagship species such as sharks, rays and groupers. The main threats to these MPAs are from overfishing, anchor damage, coral mining and diver damage. Other than sport diving, only a few activities are permitted, such as the collection of bait for the traditional pole and line tuna fishery. However, there are no restrictions on numbers of visitors to these sites and there is little management or protection, mainly due to a lack of management staff. Some islands of ecological importance as bird nesting sites are under consideration for protection. The Ministry of Fisheries, Agriculture and Marine Resources has recently proposed two large areas for complete protection to aid fish recruitment and stock replenishment. The Maldives Marine Protected Area Management Project plans to set up pilot demonstration MPAs with support from the Australian Government (AusAID). Approval has also been given for a UNDP/GEF 1 year projection preparation project for the conservation and sustainable use of biodiversity associated with coral reefs in the Maldives, to be implemented through the MHHE commencing 2000.

A major constraint in reef management in the Maldives is that monitoring information is not being used to support decision-making. Information is not effectively archived and reports are scattered in different government organisations. Inadequate financial and human resources also impede effective management. Greater collaboration is required between existing organisations and further training is required for research staff. Existing protected areas in Maldives are not actively managed, but are more or less protected by default because they are in tourism areas. For better conservation of marine resources, it is necessary to examine all resource uses, including reef fisheries, and develop a national programme for conservation.

Pakistan

There are no coral reef marine protected areas in Pakistan. The IUCN and the Pakistan Biodiversity Action Plan have identified areas for protection, but nothing has been declared and there is a large gap in capacity for conservation and management of marine resources, and a major lack of awareness at government levels about potential coral reef resources. Therefore the first need is to locate any offshore coral habitats, before considering building capacity.

Sri Lanka

There are two marine protected areas with coral reefs in Sri Lanka: the Hikkaduwa Nature Reserve (previously known as the Hikkaduwa Marine Sanctuary), declared in 1979; and the Bar Reef Marine Sanctuary, declared in 1992. These exist under the Fauna and Flora Protection Ordinance of the Department of Wild Life Conservation, which is technically responsible for their maintenance. A management plan was developed for the Hikkaduwa Marine Sanctuary in 1996, but has not been implemented satisfactorily. The coral reef zones declared in 1997 in the management plan have not been maintained and all lines and markers have been lost. Although the Department of Wild Life Conservation has an office in the sanctuary, no attempt has been made to reinstate sanctuary boundaries and there is no active control of damaging activities. There is also no management at the Bar Reef Marine Sanctuary. The coral bleaching in 1998 adversely affected shallow coral habitats in both sanctuaries and recovery is extremely slow.

There are plans to declare a Fisheries Management Area in the southeast, encompassing the Great and Little Basses reefs, that were identified for protection due to their unexploited reefs, unique setting and archaeological importance, including several ancient shipwrecks. Another 23 sites have been identified for Special Area Management planning in the Revised Coastal Zone Management Plan of 1997. The Bar Reef Marine Sanctuary and Hikkaduwa Nature Reserve are being considered for management under an Asian Development Bank (ADB) funded Coastal Zone Management Project.

Sri Lanka has a need for motivated manpower for conservation management as well as for infrastructure development, which needs to be combined with political motivation and support for resource users to find sustainable alternative employment. The sites identified in the Coral Reef Monitoring Action Plan for regular monitoring include the important coral reef areas in the country. The status of reef habitats at almost all of these sites is known from recent studies and as part of the reef monitoring programme of the National Aquatic Resources Research and Development Agency (NARA), the mandated government agency. The threats to and condition of reefs are relatively well known, however, this information is rarely used in management or to select suitable marine protected areas. There is little funding for conservation management and for infrastructure development for existing protected areas. The boundaries for existing MPAs are not marked as in the Hikkaduwa Nature Reserve, which was studied for many years under a Special Area Management project from 1990 to 1995 with financial and technical support from the USAID. Virtually none of the management procedures or activities recommended in this project still exist, there are no trained or motivated staff, the community has no commitment and the value in training and capacity building has been lost. These situations arise because there is a lack of political will for conservation of coastal resources; current security problems also provide a partial explanation.

In Sri Lanka, there are few socioeconomic data for management planning, e.g. there are no data on reef fisheries or associated economic returns. There have been some studies on the ornamental fisheries, but other resources are generally ignored and consequently there is poor understanding of the socioeconomic implications of not managing reef resources. There is little enforcement of regulations and laws relating to coastal resources, such that when blast fishermen and illegal coral miners are apprehended they are released with very low fines. Coral mining still occurs on the southern and eastern coasts, although the Department of Coast Conservation has removed the lime kilns from the coastal zone. Blast fishing has increased within the last two years in the south-western and southern coast and north of Trincomalee, despite the presence of security forces. A new licensing system for all fishing activities has not been implemented and few fishermen have licenses. The newly formed coastguard lacks boats or motivation for active surveillance and there is a shortage of people to monitor the use of fishing gear. The potential industries that could arise from bio-prospecting reef resources have not yet been recognised nor examined by the government.

Above is a summary of MPAs, their current status and the major limitations for effective conservation.

GOVERNMENT POLICIES, LAWS AND LEGISLATION

Bangladesh
The 'Bangladesh Wildlife Preservation Order, 1973' and the amended 'Bangladesh Wildlife Preservation Act 1974' provide government with the power to establish national parks, wildlife sanctuaries and game reserves. Provision for establishing marine reserves for flora and fauna was further strengthened by the enactment of the Marine Fisheries Ordinance, 1983. The relevant government agencies and semi-government organisations that need to coordinate this action are: Forest Department, Ministry of Environment and Forest; Directorate of Fisheries, Ministry of Fisheries and Livestock; Ministry of Aviation and Tourism; Department of Science and Technology, Ministry of Education; SPARRSO; Bangladesh Wildlife Advisory Board; and Environmental Pollution Control Board (EPCB).

India
The Federal government 'Coastal Regulation Zone Notification 1991' regulates onshore development activities which impact on coastal environments and strictly prohibits the collection and trade of corals. The 'Wildlife Protection Act 1972' provides protection for protected areas and some marine species. Efforts continue to bring corals under this act and to encourage more stringent enforcement of protection measures. Coral reef conservation is also included in the 'Environmental Protection Act 1986' and the 'National Conservation Strategy and Policy Statement on Environmental Development 1992'. The 'Action Plan of the Ministry of Environment and Forests' gives this ministry the mandate to conserve and manage coral reef resources and be the focal points for the Indian Coral Reef Monitoring Network and the International Coral Reef Initiative.

Maldives
There are two main pieces of legislation directly related to coral reef management and conservation: 'The Fisheries Law of the Maldives 1987' and 'The Environmental Protection and Preservation Act 1993'. Traditional management systems are still practiced in some atolls and these practices have been incorporated into fisheries law. Specific regulations associated with coral mining were introduced in 1992 and are currently under revision to be made applicable to the whole country. Many sectors of government are involved in managing coral reefs and reef resources, these include:

- Marine Research Centre of the Ministry of Fisheries, Agriculture and Marine Resources;
- The Environment Section of the Ministry of Home Affairs, Housing and Environment;
- Ministry of Construction and Public Works;
- Maldives Water and Sanitation Authority; and
- Ministry of Tourism.

Pakistan

There is potential legislation to conserve coral reefs and marine life (Biodiversity Action Plan 1997, Environmental Protection Ordinance 1983, the Wildlife Protection Ordinance 1972 and the Pakistan Environmental Protection Act 1995), but awareness by Government is required before these can be used.

Sri Lanka

There are many ministries and government departments that have responsibility for managing the marine waters around Sri Lanka, and those that have responsibility for coral reef management are:

- Ministry of Fisheries and Aquatic Resources Development (responsible for the development and management of fisheries, including licensing of fishermen, craft and gear);
- Ministry of Forestry and Environment (responsible for conservation of terrestrial as well as aquatic ecosystems);
- Department of Wildlife Conservation (conservation and management of terrestrial and marine protected areas and species);
- National Aquatic Resources Research and Development Agency (research, development and their coordination on all aquatic living and non- living resources);
- Coast Conservation Department (regulating development activities within the coastal zone and conservation and management of coastal resources. Responsible for implementing the Coastal Zone Management Plan and EIA process for development within the coastal zone);
- Central Environmental Authority (responsible for national environmental standards, coordination of environmental related matters, including the EIA process for development).

Protection for several marine organisms has been provided through the 'Fauna and Flora Protection Ordinance (amendment 1993)' of the Department of Wildlife Conservation and the 'Fisheries Act of 1996' of the Ministry of Fisheries and Aquatic Resources Development. New regulations were introduced in 1998 to the Fisheries Act to protect groupers from export, and they are listed under the restricted export category. Preliminary steps have been taken to establish a coastguard to control illegal fishing methods through the Ministry of Fisheries and Aquatic Resources Development.

CONCLUSIONS

- High water temperatures associated with the 1998 El Niño event caused widespread coral bleaching in the South Asia region and destroyed many of the shallow water corals (to 10m depth). Bleaching impacts were less severe in Gulf of Kutch and Andaman and Nicobar Islands. Surveys and observations since the 1998 bleaching indicate that recovery is slow, with patchy recruitment observed at many locations.
- Coral reefs continue to be degraded by human impacts associated with growing populations and coastal development and specifically related to uncontrolled resource exploitation, coral mining and the effects of sedimentation and pollution. Natural impacts also play a part in coral reef degradation with reefs threatened by crown-of-thorns starfish and impacts related to climate change, such as coral bleaching and cyclones.
- The capacity to monitor reef resources has improved with training activities undertaken by the GCRMN Node for India, Maldives and Sri Lanka. Advances in biophysical monitoring are particularly apparent in India, with continued training undertaken with support from ICRMN. However, the capacity for routine socioeconomic monitoring of reef resources is still lacking and there is limited application of monitoring information in management.
- Management of coral reef resources is lacking, such that coral reefs within designated marine protected areas continue to degrade. Weak management is linked to the absence of infrastructure or capacity for management, combined with a lack of funds and awareness.
- Support for further training in socioeconomic monitoring and for demonstration monitoring projects is being provided by the GCRMN Node and the CORDIO programme. National coral reef databases are also being developed through the GCRMN Node, to facilitate the application of socioeconomic and biophysical data for management.

REGIONAL RECOMMENDATIONS

- Strengthen socioeconomic monitoring of reef resources to provide information appropriate for coral reef management;
- Improve evaluation of reef fisheries and identify and develop viable alternative livelihoods for those dependent on threatened reef resources;
- Strengthen infrastructure and capacity for resource management, primarily targeting marine protected areas;
- Strengthen the capacity to develop and implement regulations relating to resource extraction;
- Create mechanisms to link monitoring information to management, through improved dialogue between government institutions and agencies;
- Undertake awareness raising activities to highlight the reef ecosystem and its interdependence with surrounding coastal ecosystems, threats to the reef and the options available for the future. Awareness should target all levels, including government ministers and departments, primary resource users, and schools, colleges and local groups; and

- Identify sustainable long-term funding mechanisms for protected area management and habitat conservation activities.

Bangladesh
- Develop and implement environmental, biological, socioeconomic and user monitoring programmes;
- Develop a clear government policy statement on the future conservation, management and protection objectives of marine resources for St. Martin's Island, which will also address the coordinated management of coastal lands; and
- Support the proposed management plan for St. Martin's Island and its coral resources with planning that involves all levels of government (i.e. an inter-governmental approach).

India
- Strengthen the role and powers of ICRMN to act as the central coordinating body for policy and programmes relating to coral reef resources, to provide better integration between government departments, institutions and local groups and to support the implementation of Management Action Plans;
- Strengthen capacity to monitor reefs and increase monitoring activities; and
- Provide training and awareness raising at all levels to better appreciate the concepts of conservation and sustainable use of coral reef resources.

Maldives
- Provide more training opportunities in marine resource conservation, management and assessment; and
- Enhance collaborative work between government research groups and local community, NGOs and international organisations.

Pakistan
- Develop capacity to locate and identify possible reef areas; and
- Develop capability in taxonomy of reef organisms.

Sri Lanka
- Improve monitoring of socioeconomic aspects of resource use;
- Develop alternative employment programmes for those engaged in fisheries activities that impact adversely on reef resources, such as the harvesting of sea cucumbers, ornamental fish, spiny lobsters and coral mining;
- Develop a practical plan for marine resource use and also develop capacity for resource management in the whole fishery sector;
- Carry out detailed studies to identify important reef resources such as sponges, and other organisms with a view to developing a bio-prospecting industry, which may have the potential to encourage improved conservation of resources in recognition of their economic value;
- Provide assistance for managing protected areas via financial support for protected area demarcation, patrolling capabilities and training of personnel; and
- Train the coastguard and purchase the necessary equipment for effective enforcement of resource use laws.

SUPPORTING DOCUMENTATION

Muley, E.V., Subramanian, B.R., Venkataraman, K and Wafar, M.V.M. 2000. Status of coral reefs of India. Paper presented to 9ICRS. (unpublished). EV Muley e-mail: muley@vsnl.com

Rajasuriya, A., Maniku, M.H., Subramanian, B.R. and Rubens, J. 1999. A review of the progress in implementation of management actions for the conservation and sustainable development of coral reef ecosystems in South Asia. Proceedings of the International Tropical Marine Ecosystems Symposium (ITMEMS), Townsville, Australia, November 1998. pp. 86 - 113.

Rajasuriya, A., Karunarathna, M.M.C., Vidanage, S. and Gunarathna, A.B.A.K. 2000. Status of coral reefs in Sri Lanka; community involvement and the use of data in management. National Aquatic Resources Research and Development Agency, Colombo, Sri Lanka. Paper presented to the 9ICRS (unpublished). A Rajasuriya e-mail: arjan@nara.ac.lk

Zahir, H and Saleem, M. 2000. Status of coral reefs in the Maldives. Marine Research Centre, Maldives. Paper presented to the 9ICRS. (unpublished). H Zahir e-mail: marine@fishagri.gov.mv

GCRMN South Asia Node

The South Asia node entered a second phase of activities in December 1999 following the successful completion of Phase 1 in March 1999. Support for the node comes from the Department for International Development (DFID) of the UK with GBP327,400 as an accountable grant to IOC-UNESCO over the next 2.5 years. The 3 core activities are:

1. support of demonstration site socioeconomic monitoring and the implementation of Coral Reef Monitoring Action Plans;
2. training in biophysical survey design and data analysis, and in socioeconomic monitoring; and
3. development of national and regional coral reef databases.

Biophysical monitoring capacity was expanded and strengthened during Phase 1 through training and pilot monitoring activities. Government and other project funds (e.g. CORDIO) now support continued biophysical monitoring, however, routine socioeconomic monitoring has not yet been established in the region, despite being recognised as critical for management. Socioeconomic monitoring is therefore a priority activity of Phase 2, with specific funds for monitoring at demonstration sites, where biophysical monitoring is also taking place. Socioeconomic monitoring will also be supported by on-site training in all aspects of resource use assessment. The Node is designing national coral reef databases for the GCRMN based on ReefBase and other successful databases (e.g. COREMAP), it is anticipated that all three countries will have functional national databases by July 2001.

Node Contacts and Representatives:

- India: the Department of Ocean Development (Dr B.R. Subramanian e-mail: brs@niot.ernet.in) and the Ministry of Forests and Environment (Dr E.V. Muley e.mail: muley@vsnl.com);
- Maldives: the Marine Research Centre of the Ministry of Fisheries, Agriculture and Marine Resources, (Hussein Zahir e.mail: marine@fishagri.gov.mv);
- Sri Lanka: National Aquatic Resources Research and Development Agency (Arjan Rajasuriya, e-mail: arjan@nara.ac.lk) ; the Regional Coordinator is Emma Whittingham, Address No.48 Vajira Road, Colombo 5, Sri Lanka; Tel: +941 505 649 / 858; Fax: +941 580 202; E-Mail: reefmonitor@eureka.lk

7. SOUTHEAST ASIAN REEFS - STATUS UPDATE: CAMBODIA, INDONESIA, MALAYSIA, PHILIPPINES, SINGAPORE, THAILAND AND VIETNAM

LOKE MING CHOU

ABSTRACT

Southeast Asian reefs continue to face development and exploitation pressures in spite of greater awareness of their ecological and economic importance. Reefs that were once considered remote have not escaped destruction from poison or blast fishing. Common threats from human activities are spreading throughout the region and there are no apparent signs of reversals or reductions in the trends of increasing reef degradation. Monitoring efforts on reef health status are expanding in tandem with increasing numbers of rehabilitation projects. Monitoring data are available for more and more reefs for which there was little previous information, and many countries have established national reef monitoring programmes. Reef Check surveys have increased steadily in the region. Monitoring by volunteers indicates little difference in reef condition between reefs in marine parks and non-protected areas. This confirms earlier observations that most marine protected areas are not meeting management objectives. Countering the patterns of losses

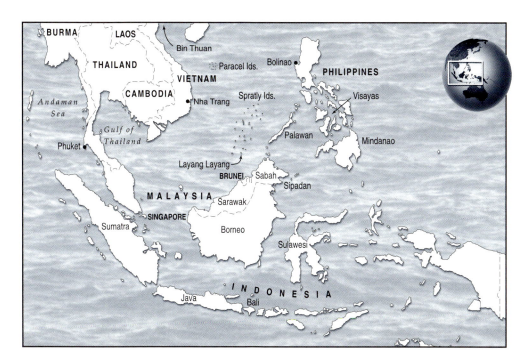

in reef integrity, there are isolated instances of management by local communities and coastal resorts. In addition to the impacts of human stresses, many Southeast Asian reefs were affected by severe bleaching in 1998 with recovery ranging from marginal to almost complete. Stronger, more effective and perhaps innovative management measures are necessary to facilitate the survival of Southeast Asian reefs in this new millennium.

INTRODUCTION

This report updates the information presented in the first edition in 1998, and focuses on changes since that time, which are unfortunately mostly negative. It also draws on recent assessments made by the World Resources Institute - Reefs at Risk project and another by UNEP on trans-boundary problems. The coral reefs of Southeast Asia are renowned for their exceptionally high biodiversity. Although the marine environment of Southeast Asia occupies 2.5% of the global ocean, it contains 27% of the world's coral reefs, including two large archipelagos which have 22% the global total; Indonesia has close to 18,000 islands, and the Philippines has over 7,000 islands. The extensive Sunda and Sahul continental shelves offer large shallow areas that favour reef development and all major reef formations (oceanic atolls, platform and barrier reefs) are represented, along with large areas of shallow fringing reefs on the coasts, which receive the full impacts from human activities from the land.

Economic growth and coastal population expansion have resulted in serious degradation of many reefs, and the trend continues in spite of increased awareness of the value of reefs to people. The coral reefs in the Malacca and Singapore Straits alone have a total assessed economic value of US$563 million for their roles and values as carbon sinks, tourist attractions, shoreline protectors, fishery resources and research potential. The problems of reef degradation are compounded by unabated and widespread destructive fishing practices, particularly poison and blast fishing. Unless sustainable management practices are implemented, the high biodiversity reefs of Southeast Asia will succumb further to economic development pressures in spite of quantified values. Various reef management models exist throughout the region, ranging from national government initiatives to local community involvement, and the successes of community-based management continues to spread throughout the region. Of the growing number of nationally legislated marine parks, less than 10% are adequately managed to meet protection objectives, and the situation does not seem to be improving sufficiently rapidly as Reef Check global analyses indicate little difference in the condition of reefs between protected and non-protected areas

STATUS OF CORAL REEF BENTHOS

This section updates information from the previous status reports in 1998, showing a continuation of decline in many of the reefs of the region from many similar causes. The national reports of natural disturbances and community structure of the reefs are available and will be published in the Proceedings of the 9th International Coral Reef Symposium in Bali.

Cambodia

The amount of information on Cambodian reefs is very limited, but recent efforts have assembled the most comprehensive information to date. However, there is still insufficient data on the status of coral reefs, their distribution and species composition. There are coral communities growing on rocky bottoms along much of the 435km coastline and some of the 52 islands, where there are often fringing reefs. Coral diversity is higher on the offshore reefs, whereas those inshore are in poor condition with low species diversity, dominated by massive corals.

At Koh Tang, an offshore island group in clearer water, there are 70 hard coral species including *Acropora* and *Montipora* which are absent on the inshore reefs, which in turn are dominated by *Porities*, mussids and faviids. *Acropora* spp. are now much less common on Cambodian reefs. Coral cover of up to 50% was found at Koh Rong Samlem in spite of the extensive bleaching impacts during 1998. This site also has few human impacts; therefore, these healthy reefs are not representative of the state of most coral reefs throughout Cambodia, which are more degraded. The reefs at Koh Tres, Poi Kompenhl and Koh Thas had live coral cover ranging from 21% to 70%. The black spiny sea urchin, *Diadema setosum*, is abundant at Koh Rong, Koh Tang and Koh Damlung.

There was bleaching in 1998 at reefs on Koh Rong Samlem, Koh Rong, Koh Tang, Koh Damlung and Koh Thas, but studies at Koh Rong Samlem suggest that recovery was strong. Crown-of-thorns starfish are not known on inshore reefs and their presence on the outer reefs is limited to small numbers. An outbreak was observed in February 1998 at a site on Koh Tong (20 adults within an area of 100m^2).

Indonesia

A comprehensive review of coral reef conditions in eastern and central Indonesia has just been completed by international and local experts for the Packard Foundation. This included considerable data from the Pusat Perelitian dan Pengembangan Oseanologi-LIPI reef monitoring database that covers 400 stations from 48 sites, as well as information from programme activities by local and international agencies. Indonesian reefs contain the richest species diversity of corals (450 species) and other reef-associated groups in the world. For example, a single reef can have more than 140 species of hard coral. About 40% of reefs reviewed are listed in the 'poor' category with live coral cover less than 25%, and only 29% considered as good to excellent (live coral cover above 50%), whereas previous reports suggest that most of Indonesian reefs were above this many years ago as there are few destructive storms and waves impacting on these reefs. Thus, there are clear indications of rapidly declining reef health throughout this area with the proportion of degraded reefs increasing from 10% to 50% over the past 50 years. The reefs of eastern Indonesia are in comparatively better condition, but are also declining quickly. There are many instances where blast fishing has reduced coral cover by 50-80%, and the widespread use of cyanide has resulted in large areas of dead coral. For example, reefs on Bali that had good coral cover of 73% with colonies 17-24cm in diameter on average in 1992, now have cover of 15% or less, with colonies averaging 2-3cm. Population outbreaks of crown-of-thorns starfish have been reported but appear scattered and not serious. The bleaching events of 1987-88 and 1992-93 did not appear to affect Indonesian reefs on a wide scale, however the 1998-99 bleaching event had a greater impact with many reefs affected.

Malaysia

There have been extensive surveys of the reefs of East Malaysia particularly in Sabah since 1996. The best reefs are the oceanic reefs at Sipadan (far east coast) and Layang Layang (southern Spratlys). Reef condition throughout East Malaysia varies widely, with an alarming amount of recently dead and shattered coral. Only 10% of reefs had dead coral cover of less than 10%, while dead coral cover of 10 to 20% was found for 70% of the reefs. At least 10% or reefs had approximately 40% dead coral cover, indicating recent losses. Evidence of destructive fishing practices was found on all reefs, except those under strict protection (e.g. Sipadan and Semporna Islands), and there are reports of more than 4 bomb blasts per hour in many offshore reef areas. In large areas of the Tunku Abdul Rahman Park, live coral cover has dropped due to a combination of storms, blast fishing and pollution from a mean of 30% in 1994, to less than 5% now. Some of the individual reef statistics from this park are illuminating, with only one reef, Manukan, remaining stable.

Live coral cover changes	1987	1991	1994	1999
Merangis Reef	40.5	41.5	34.5	14.2
Sapi Reef	47.0	30.5	37.5	4.1
Staghorn Patch Reef	30.0	36.5	33.0	1.6
Manukan Reef	30.0	36.5	38.5	35.0
Sulug Reef	-	32.5	19.0	2.9
Mamutik Reef	-	18.0	19.5	12.3
Tanjong Wokong Reef	-	-	31.0	5.4

This summarises changes in coral cover on reefs in the Tunku Abdul Rahman Park, Sabah, Malaysia over more than 10 years, with unfortunately a decline at most sites.

There are few reefs in Sarawak because of high levels of sediment runoff from several large rivers. On the Talang Talang islands, there are occasional patches of coral cover of about 30%, but most do not exceed 20%. There was large scale bleaching in 1998 with close to 40% of coral colonies in shallow water (10m) and 25% in deeper water 10-20m bleached. By late 1999, colonies either recovered or were overgrown by zoanthids or soft corals. Bleaching in Tunku Abdul Rahman Park was up to 30-40%, but 6 months later, all colonies had recovered except for very few that had died.

Philippines

There are over 400 hard coral (scleractinian) species known from the Philippines, of which 12 are endemic. Data from the 1990s show a decline in reef condition, with reefs in the Visayas area most at risk. An analysis of more than 600 data sets showed that 'excellent' reefs (live hard and soft coral cover above 75%) has reduced from 5.3% to 4.3% since the late 1970s. If hard corals alone are considered, only 1.9% of the reefs can be called 'excellent', with average hard coral cover on all reefs at 32.3%, whereas it used to be much higher. *Acropora* covered 8.1% of Philippines reefs, and the decline is thought to be due primarily to human impacts, particularly blast fishing, as well as infestations of coral eating crown-of-thorns starfish and drupellids. The 1998 bleaching started at Batangas in June 1998 and other reefs were affected in an almost clockwise sequence around the Philippines. Most reefs in northern Luzon, Palawan, most of the Visayas, northern and eastern Mindanao were affected. The most severe impact occurred at Bolinao, Pangasinan where 80% of the corals were bleached. Most vulnerable were *Acropora* and pocilloporids and even *Porites*, faviids, fungiids, caryophilids and hydrocorals were seen to bleach.

Singapore
Continued monitoring in 1998 and 1999 of permanent locations at 5 reefs showed no reversal in the declining trend of live coral cover. Most reefs have lost up to 65% of live coral cover since 1986. The best reef at Pulau Satumu, furthest from the mainland, also showed a reduction in live coral cover of 37% over the past 13 years. The 1998 bleaching affected all reefs at a scale never previously experienced. About 90% of all corals bleached of which 25% failed to recover. The prognosis for these reefs is not good, when coupled with national strategies to expand the island into the ocean and increase shipping activities.

Thailand
A comprehensive reef survey programme, using primarily manta tows, was conducted between 1995 and 1998 at 251 sites in the Gulf of Thailand and 169 sites in the Andaman Sea. All sites were fringing reefs, with most less than 1km^2, within the total reef area of Thailand estimated at 153km^2, distributed equally between both seas. The condition of reefs in the Gulf of Thailand were: 16.4% excellent, 29% good, 30.8% fair, 23.8% poor. In the Andaman Sea, 4.6% were excellent, 12% good, 33.6% fair, and 49.8% poor. Reef condition in the Gulf of Thailand has worsened compared to the late 1980s, while that of the Andaman Sea remained comparable or improved slightly. About 80 permanent study sites at the Andaman Sea side of Thailand have been established during long-term monitoring programmes, which began in 1981 by the Phuket Marine Biological Center.

The 1998 bleaching event affected reefs in the Gulf e.g. at the Sichang Islands, 40-50% of live coral cover was affected with no sign of recovery. Few *Acropora* survived while *Porites* and faviids were slightly affected, and there are no signs of coral recovery or coral recruitment. The bleaching killed 60-70% of live coral at Sattahip with *Acropora* being most affected. The reefs at Rayong Chantaburi were similarly affected, however, there was strong coral recruitment at both sites. In most cases, the loss of live coral cover did not appear to affect reef fish communities.

Vietnam
Coral reefs are the richest marine habitats in Vietnam and extend along the 3,260km coastline and on more than 3,000 inshore and offshore islands. Reefs in the north support fewer species and are mainly fringing reefs, whereas in the south, there are also platform reefs. Of the 300 coral species, 277 species (72 genera) are found on reefs in the south, while reefs in the north have 165 species in 52 genera. Surveys of 142 sites from 15 of the 28 reef areas between 1994 and 1997 show that only 1.4% have live coral cover above 75%. 'Poor' reefs with less than 25% coral cover occurred at 37.3% of the sites. Of the remaining sites, 48.6% had coral cover between 25 and 50%, and 31% between 50 and 75%. There is a distinct correlation between healthier reefs and remoteness from human population centres with the best coral cover on offshore islands or remote coastal locations.

Typhoon 'Linda' caused extensive damage to the best protected reefs at Con Dao islands in November 1997. Bleaching also affected the reefs of Con Dao islands, north Binh Thuan province and Nha Trang bay during the summer of 1998. Monitoring in 1999 showed very slow

> ## THE SPRATLYS AND PARACEL ARCHIPELAGOS
>
> These islands sit across major shipping lanes in the South China Sea, hence are of great strategic importance. Ownership of them is currently disputed with Brunei, China, Malaysia, Philippines, Taiwan and Vietnam having claims on all or part of the Spratlys; and China, and Vietnam claiming the Paracels. These are coral reef islands that sit on seamounts rising from oceanic depths in clear waters remote from continental influences, except for fishers that come out using dynamite and cyanide to catch reef fish, and the construction of military installations. The reefs are thought to be significant as possible sources of larvae for nearby reefs and are important in completing reef connectivity throughout the South China Sea. Proposals that they should be declared a 'global' marine protected area are not supported by all of the claimant states. Malaysia, Vietnam and the Philippines have conducted some joint studies on some of these reefs in recent years, however, it is essential that more joint efforts be initiated so that the full significance of these reefs can be understood.

recovery of the Con Dao reefs from the double impacts of typhoon and bleaching. The reefs at north Binh Thuan however recovered well due to the June-September annual upwelling.

STATUS OF CORAL REEF FISHES

Southeast Asian reefs contain a rich diversity of reef fish and other reef species, including many endemic species. The region's reefs have also been described as an important source of larval recruits to reefs in adjacent regions. The joint project between ASEAN and Australia reported 787 reef fish species, including 41 species of butterflyfish at study sites in Indonesia, Malaysia, Philippines, Singapore and Thailand. However, a recent analysis in the Philippines revealed 915 species from 63 families, including 48 species of butterflyfish. This clearly indicates that with more intensive studies, many more species will be found.

Indonesia has the richest reef fish diversity in the world, particularly on the eastern reefs, e.g. 123 pomacentrid species, and 83 species of angelfishes and butterflyfishes, far exceeding other areas of the world. The offshore reefs and less disturbed reefs have higher species diversity and abundance, e.g. in Vietnam there is higher reef fish diversity in the south and central regions, but offshore reefs have higher fish abundance. The reefs of the Gulf of Tonkin have low species richness. The oceanic reefs of Sipadan and Layang Layang (Sabah), have higher numbers of butterflyfish and groupers and other species compared to inshore sites. In Cambodia, fish diversity is low, particularly at inshore reefs, characterised by few fish, mainly damselfish and butterflyfish. Some reefs such as Poi Tamoung about 50m from shore with better corals support greater diversity, including angelfish, butterflyfish, rabbitfish and parrotfish. Thus there appears to be a clear correlation between healthy corals, larger fish diversity, abundance and lower levels of human interference at these sites.

Throughout the region, reef fish diversity and abundance are threatened by reef degradation, destructive fishing and over-fishing. Cambodian reefs are affected by over-fishing of large commercial fishes and invertebrates. Only 11% of the Philippine reef sites had high standing stocks of fish above 35mt km^{-2}. Species richness was poor for 35% of the sites (27-47 species 1000m^{-2}) and high for 25% of the sites (>75 species 1000m^{-2}). Fish abundance was high at 31% of the sites (>2268 indiv. 1000m^{-2}) and poor at 24% of the sites (<676 indiv. 1000m^{-2}). High species diversity was observed on the reefs of Palawan, Tubbataha and Turtle Islands – all remote and mostly protected sites. Fish abundance was high at Cagayan, Calauag Bay, Palawan, Bais Bay, Surigao del Norte and Tubbataha reefs. The Bolinao reefs had poor fish abundance.

ANTHROPOGENIC THREATS TO CORAL REEF BIODIVERSITY

Over-exploitation of fishes was identified by 6 countries (Cambodia, Malaysia, Indonesia, Philippines, Thailand, Vietnam) as the most serious cause of reef degradation in an assessment of trans-boundary problems with the help of UNEP. Next were destructive fishing methods, and sedimentation with equal weighting, and then pollution associated with coastal development. This over-exploitation has led to the collapse of many local fisheries e.g. sea cucumbers are rare on most reefs in Cambodia, Thailand and Vietnam, and many Indonesian reefs face similar fates. Target species are being totally removed in response to external market demand. The consequences of destructive fishing methods (blast and cyanide fishing and the 'muro ami' method of bashing the corals to drive fish into nets) are well documented and in widespread use throughout the region, including remote reefs and those within protected areas.

In Cambodia, dynamite fishing is the most pressing problem and has extensively damaged many reefs off the coastal towns of Kep, Sihanoukville and some of the reefs in the Gulf of Thailand. Reef damage at other areas such as the Koh Sdach group is not as serious, although blast fishing is reported. Dynamite fishing has had moderate to serious impacts on the reefs around Sihanoukville, with many areas reduced to rubble. On Koh Rong, over-

	Over-exploitation	Destructive Fishing	Sedimentation	Pollution
Cambodia	x	x		
Malaysia	x	x	x	x
Indonesia	x	x	x	
Philippines	x	x	x	x
Thailand	x		x	x
Singapore	x		x	
Vietnam	x	x	x	x

Countries in the region, in association with UNEP, assessed the critical anthropogenic threats to coral reefs in their areas. In all cases, over-exploitation topped the list.

fishing has depleted stocks of large fish of commercial value (e.g. humphead wrasse, grouper, sweetlips), sea cucumbers and giant clams. Cyanide fishing is a relatively new threat targeting fish such as groupers and snappers for grow-out in cages at Tumnup Rolork and Stoeng Hav, although the Fisheries Department is discouraging this practice. Coral collecting was a large threat between 1995 and 1997 when foreign companies collected and exported container-loads of corals, but the business has been suspended by the Ministry of Environment and the confiscation of corals on sale in Sihanoukville has reduced this trade. Selective coral collecting for the marine ornamental trade is regarded as a cause for reduced *Acropora* abundance on many reefs.

Over-fishing and use of destructive methods (poison, blast and fine mesh nets) has resulted in widespread destruction of reefs in Quang Ninh, Nghe An, Quang Binh, Thua Thien-Hue, Quang nam, Da Nang, Quang Ngai and Khanh Hoa provinces, Vietnam. Cyanide fishing for groupers for the live fish trade is widespread in the northern and central parts of Vietnam, and even near the Con Dao National Park. Sea cucumbers are heavily exploited, and collection of ornamental fish for the aquarium trade is a rising problem as poisons are used to stun fish. Lobsters have declined from some reefs e.g. at Cu Lau Cham, and pearl shell (*Pinctada martensii* and *Lutraria philippinarium*) have disappeared from many reefs in the north. Exploitation of species listed as endangered in Vietnam's Red Book still occurs (e.g. four lobster species, two abalone species, the squid *Loligo formosana*, and three clupeid species). Non-resident marine fishers from China and Hong Kong compete with local fishermen in offshore waters. Marine tourism is fuelling the demand for souvenirs, including endangered and protected species such as turtles. *Acropora* is becoming rare in places such as Nha Trang, the centre of coral trade in Vietnam, and other invertebrates are also harvested as curios.

Similarly bleak pictures of over-fishing and destructive fishing apply throughout East Malaysia, Philippines and Indonesia. Blast fishing and the use of cyanide for the live fish and aquarium trade has destroyed many reefs in East Malaysia, often with losses of 80% of original coral cover. There have been reports of decreased blast fishing in most parts of the Philippines. However this practice is still common in the Palawan group of islands, Sulu Archipelago and western Mindanao, and coral recovery from blasting in these areas remains slow. Two other fishing practices can potentially deplete marine resources; spearfishing using scuba or hookah, and drift net fishing. Blast fishing however is not practiced in West Malaysia possibly due to strong enforcement, and also in Singapore due to all reefs being close to major shipping activities.

A special problem has occurred in Indonesia since the economic downturn since the late 1990s. Many people who were employed in industries have returned to exploiting coral reef resources as industries closed. Moreover, the collapse in local economies has meant that population control programmes have stopped and the population of over 200 million is likely to expand further. Also funding for government conservation measures has decreased, with parallel increases in deforestation, including mangrove forests, such that runoff of sediments and nutrients are impacting heavily on coastal reefs.

Other anthropogenic impacts in the region include coastal development and marine pollution with primary sources from industry, agriculture, shrimp pond effluent, unsustainable logging, and domestic waste; all of which increase sedimentation, nutrients and other pollutants in the water. Heavy sedimentation has degraded Vietnam's reefs at Cat Ba Islands and Ha Long Bay.

CURRENT AND POTENTIAL CLIMATE CHANGE IMPACTS

The 1998 bleaching of corals was widespread and on a scale and intensity not previously encountered in the region. This has highlighted the potential for climate change to impact on coral reefs. Slow rates of sea level rise will provide the necessary environmental conditions for reefs to optimise structure and orientation. However, erosion factors caused by increased rainfall may reverse this upward growth. Sea surface temperature changes will trigger further bleaching events, which are expected to be more severe than the 1998 bleaching. The rapid recovery of reefs from the 1998 bleaching at north Binh Thuan in Vietnam was attributed to the annual upwelling, which brought cold waters to the surface. Reefs elsewhere in Vietnam recovered at a slower rate, implying that reefs near major upwelling areas may suffer less from bleaching events.

The collapse of reef systems can be translated directly to loss of goods and services that they generate. In terms of carbon sequestration alone, using the avoided future cost of climate change of $20/tonne of carbon, the estimated carbon storage value of coral reefs is US$240/ha/year. Thus, just the reefs of the Malacca Straits have annual values of US$237,000 for the coral reef area of 1,317.5ha on the Malaysian side, US$93,863,000 for 521,462ha of Indonesian reefs and US$51,700 for the 287ha of Singapore reefs.

Sea-level rises will have a magnified impact in this region as 70% of the population live in the coastal area. In December 1999 abnormally high tides and waves caused extensive flooding and erosion of coastal areas in many parts of the region. High tides resulted in floating fish farms being pulled from their anchoring points, coastal resorts being flooded, rivers overflowing their banks, and disruption of basic infrastructure services throughout the area. Submergence of coastlines considered well above high tide level resulted in excess nutrient runoff and macroalgal blooms in some coastal lagoons in Singapore. There was complete flooding of the Turtle Islands by up to 50cm for 5 hours resulting in significant erosion of shoreline and loss of an inland turtle hatchery.

CURRENT MPAs AND MONITORING AND CONSERVATION MANAGEMENT CAPACITY

A variety of reef management systems exist in the region, with an increasing number of Marine Protected Areas (MPAs) being declared throughout Southeast Asia. Unfortunately many occur only on paper with only limited government commitments to staffing and operational funding, thus few MPAs are effectively managed. Of the 109 MPAs established by ASEAN countries as of 1994, 65% contained coral reef resources. Many of these MPAs were former terrestrial reserves with jurisdiction only recently extended to include marine resources. Vietnam has only 3 protected areas (Cat Ba, Con Dao and Halong bay) that include marine components and reefs are not adequately represented in the country's protected areas.

Frequently when there are management-oriented staff in MPAs, they lack adequate training and skills and are not provided with logistic resources. In addition, many problems arise due to conflicting responsibilities for management of the resources within the MPAs because responsibilities are often based on different sectoral interests, e.g. exploitation agencies attempt to maximise fisheries and act against environmental bodies seeking conservation. Frequently agencies managing different aspects of the coastal area have poor communication between them, resulting in uncoordinated efforts which undermine conservation and protection of reef resources. Increased political commitment is necessary to address the deficiency seen in these MPAs and respective links with broader coastal management strategies.

Community-based management systems are apparently having more widespread success and different models have been applied to suit local situations. Successes in the Philippines, Thailand and Indonesia show that it is effective for small areas where local communities are directly involved in management. Community-based management provides users with a better sense of propriety and greater motivation to manage the very resources that they themselves are dependent on. Such communities are effective in controlling destructive activities caused by other users as well as themselves. For larger areas, the co-management system approach is more effective where management is shared between government agencies, local communities and non-governmental organisations.

Many Southeast Asian countries are signatories to international conventions and agreements to maintain natural ecosystems. Of relevance to coral reefs is the Convention on International Trade in Endangered Species (CITES), which has been effective in slowing the export of live and dead corals and other reef organisms from the region. Another international convention addressing the global loss of plant and animal species is the Convention on Biological Diversity which the majority of the states have signed and ratified. Informal regional agreements such as the 1989 Langkawi Declaration on the Environment, the 1990 Baguio Resolution on Coastal Resource Management, the 1990 Kuala Lumpur Accord on Environment and Development, and the 1994 Bangkok Resolutions on Living Coastal Resources, all attest to increased regional awareness of the need to manage coastal resources including coral reefs for sustainable use.

GOVERNMENT POLICIES, LAWS AND LEGISLATION

Some countries have numerous policies and laws relating directly to coral reef conservation, while others do not. For example, in Cambodia and Singapore the only laws pertinent to coral reefs relate to fishing and not directly to conservation. More targeted national policies or conservation strategies for coral reefs are required to ensure that reefs are given the deserved protection, including clear goals for the conservation and sustainability of reef resources. Laws for many marine parks were extensions of laws establishing terrestrial parks and did not contain relevant or applicable provisions to cover the marine environment and respective ecological and economic management differences. Such policies and laws for reef conservation need to be drafted within the context of integrated coastal management to be effective and ensure that development and conservation concerns are reconciled in the interest of sustainable development. Indonesia did not establish a systematic coastal and marine management regime before 1999 even though over 12 national Ministries have coastal and marine management responsibilities. To address this need for better co-ordination and effective management, the Indonesian government created a new Ministry of Marine Exploration and Fisheries. Regardless of adequate administrative and legal framework, problems will arise from lack of political will, corruption, lack of resources, and lack of understanding of the role of coral reefs. These are issues that are best addressed through the adoption of an integrated coastal management strategy.

GAPS IN CURRENT MONITORING AND CONSERVATION CAPACITY

Coral reef monitoring has expanded throughout the region during the 1990s with reef surveys increasingly being used for management assessment. However, it is clear that monitoring capacity varies widely between the countries based on those participating in an ASEAN-Australian cooperation project (Indonesia, Malaysia, Philippines, Singapore, Thailand), even though most of these countries have institutions with trained staff capable of monitoring reefs and are jointly using accepted methodology. The use of common survey methods permits comparisons and regional analysis of trends in reef health, and monitoring is conducted fairly routinely. The capacity in Brunei Darussalam, Cambodia, Myanmar and Vietnam is developing as well, albeit at different levels, e.g. some countries require not only trained personnel, while others need equipment and facilities. For countries with regular monitoring programmes, the sites selected are sometimes not well distributed. In some cases there are clusters of intensively studied sites, while large areas of reefs remain un-monitored. Thus compiling an adequate picture of reef status throughout the region is not possible. The addition of Reef Check surveys involving volunteer divers has supplemented national programmes.

THE ASEAN-AUSTRALIA LIVING COASTAL RESOURCES PROJECT

The effects of a collaborative project between 5 ASEAN countries (Indonesia, Malaysia, Philippines, Singapore, Thailand) and Australia between 1984 and 1994 are still having an influence in the region. These countries have considerable capacity to monitor coral reefs and all continue monitoring at many of the sites that were established in 1986 or earlier. The emphasis, under funding from Australian AusAID, was on developing capacity to research and monitor coral reefs, mangrove forests, seagrass beds and the fisheries over soft bottoms. A major task was to develop common methodologies and this resulted in a manual now in global use. A common reef survey method was proposed, tested and developed to monitor reefs over spatial and temporal scales throughout the region. Using this Line-Intercept Technique, participating countries implemented reef monitoring programmes and gathered information on the condition of reefs and how they changed over time. Use of common methods facilitated comparisons across the region. While the project enhanced reef monitoring capacity in these countries and enabled reef condition to be quantified, the sites selected did not provide an even distribution throughout the region. Most reef sites surveyed were within convenient reach of participating agencies, resulting in thorough information of some reefs and none of others. English, S., Wilkinson, C. and Baker, V. 1997, Survey Manual for Tropical Marine Resources. Australian Institute of Marine Science, Townsville, 390 pp.

CONCLUSIONS AND RECOMMENDATIONS

Southeast Asian reefs continue to face increasing threats from economic and population growth, and these pressures are overriding the increasing numbers of conservation initiatives at national, local and community levels. The benefits and significance of reef systems are now better known and there is greater awareness of the importance of managing and protecting the resources; however, reef degradation continues to outpace reef protection, as the former is driven by short-term economic development and market pressures.

Management capacity is clearly insufficient in a region heavily reliant on natural resource exploitation and where large segments of the population derive a subsistence living from direct exploitation of these resources. During the Asian economic crisis since 1997, economic recovery took priority and slowed down most reef management activities. However, the Blongko Marine Sanctuary in North Sulawesi, which is fully managed by the local community, demonstrated that effective reef resource management ensured the continued generation of goods and services to the community and provided the locals with adequate financial security throughout the crisis. All levels of management capacity need strengthening, whether at community, local government or national government levels. Management at community level can be relatively straightforward and simpler to implement provided conflicts amongst groups or individuals do not override the larger goals. At local government and national government levels, reef conservation should be implemented within the context of integrated coastal management approaches. Many large marine parks not only have problems with surveillance and enforcement deficiencies, but also the lack of recognition of local community needs. Both integrated coastal management and community-based management approaches can be combined to provide greater and more effective management efficiency as they involve the local government and local communities.

This region illustrates a wide diversity of reef management experiences. Dive resorts for example, appear to perform a more active role in conserving the core resources, which attract their visitors. Many 'house reefs' are well-protected and some resorts provide help (e.g. fast boats and fuel) to local agencies to improve surveillance. Systematic and innovative approaches should be considered to compile all types of reef management models operating within the region and to analyse what works best for certain situations. There are numerous case studies, but the lessons from failures or successes have not been fully synthesised and applied.

A more coordinated approach to coral reef monitoring should be established and a regional mechanism will permit more effective sharing of information at national and regional levels. This will enable the region as a whole to adopt better strategic approaches to protect reefs. Areas that have little or no monitoring, but are thought to have a strategic role in larval supply routes, need to be identified and targeted for investigation. Public education will remain an important process to reef management. It has to be promoted, expanded and implemented with greater urgency.

Acknowledgements and Supporting Documentation

I must record my full appreciation to all colleagues for responding to my request for information and/or for making their national reports available. The report would not have been possible without their full cooperation. They are: Annadel Cabanban, Ian Dutton, Edgardo Gomez, David Hopley, Hugh Kirkman, Al Licuanan, Jeffrey Low, Vipoosit Manthachitra, Nicolas Pilcher, Karenne Tun, Vo Si Tuan, Matt Wheeler, Clive Wilkinson.

Loke Ming Chou is a Professor of Biological Sciences in the National University of Singapore. (dbsclm@nus.edu.sg)

Alcala AC (2000) Blast fishing in the Philippines, with notes on two destructive fishing activities. Workshop on the Status of Philippine Reefs, 24 Jan. 2000, Marine Science Institute, University of the Philippines

Dahuri R, Dutton IM (2000) Integrated coastal and marine management enters a new era in Indonesia. Integrated Coastal Zone Management 1: 11-16.

MPP-EAS (1999) Total economic valuation: Coastal and marine resources in the Straits of Malacca. MPP-EAS Technical Report No. 24/PEMSEA Technical Report No. 2. Global Environment Facility/United Nations Development Programme/International Maritime Organisation Regional Programme for the Prevention and Management of Marine Pollution in the East Asian Seas (MPP-EAS)/Partnerships in Environmental Management for the Seas of East Asia (PEMSEA), Quezon City, Philippines. 52pp.

Talaue-McManus L (2000) Transboundary Diagnostic Analysis for the South China Sea.

EAS/RCU Technical Report Series No 14. UNEP Bangkok, Thailand. 106pp.

Vo ST (2000) Preliminary report on the status of coral reefs in Vietnam: 2000 (thuysinh@dng.vnn.vn)

The following reports were prepared for the Packard Foundation and copies may be available from the authors:

Hopley, D. and Suharsono (2000). The Status of Coral Reefs of Eastern Indonesia. Pp. 114 (David.Hopley@ultra.net.au)

Licuanan, W.Y, Gomez ED (2000) Philippine coral reefs, reef fishes, and associated fisheries: Status and recommendations to improve their management. Pp. 44. (licuanan@msi01.cs.upd.edu.ph)

Pilcher, N, Cabanban, A. (2000) The status of coral reefs in Sabah, Labuan and Sarawak, east Malaysia. Pp. 62 (nick@tualang.unimas.my)

8. Status of Coral Reefs of East and North Asia: China, Japan and Taiwan

Shuichi Fujiwara, Takuro Shibuno, K. Mito, Tatsuo Nakai, Yasunori Sasaki, Dai Chang-feng and Chen Gang

Abstract

These reefs sit at the northern edge of the Southeast Asian centre of biodiversity and as such share many of the species and characteristics of reefs to the south. Coral bleaching and mortality seriously damaged the reefs of southern Japan and Taiwan, and possibly of China, from July to September, 1998, when seasonal winds and currents slowed during the La Niña climate change. Bleaching stopped in late September when the first typhoon of the season came. There were many reports of coral losses of 30-60% and some as high as 80-90%, with some localised extinctions of prominent corals reported. The Japanese government has established an international coral reef centre on Ishigaki Island (the southern islands of Okinawa) to facilitate coral reef conservation in the region and assist the GCRMN with monitoring. All countries have improved policies concerning coral reefs, but more attention is needed for designating and managing MPAs and building capacity as reefs in this region come under extreme pressures from over-fishing, as well as high levels of sediment and nutrient pollution arising from activities on land.

Introduction

The coral reefs of Taiwan and Japan are closely linked by the northward flowing Kuroshio current which brings warm water and coral reef larvae from reefs in the South China Sea and the Philippines. This creates ideal connections for corals and other species to move northward to the Japanese main islands, and the warmer conditions make these some of the most northern coral reefs in the world. The Chinese reefs, mainly on Hainan Island are at the northern end of the South China Sea, therefore they have links with reefs of Vietnam and the Spratly Archipelago to the south. These reefs are also adjacent to the largest concentration of people on earth, such that they have provided resources of food and other materials for thousands of years. However, the pressures of these populations on the reefs has been increasingly resulting in considerable damage to the resources, which was exacerbated by heavy coral bleaching and mortality during the massive climate change events of 1998.

China

The coral reefs of Hainan are on the northern margin of Indo-Pacific coral reef centres of high biodiversity, and presumably there are also large numbers of species here, but there is lack of taxonomic capacity to confirm this. Studies are required to assess the possible important role that these reefs play in the global reef system. The major coral reefs areas

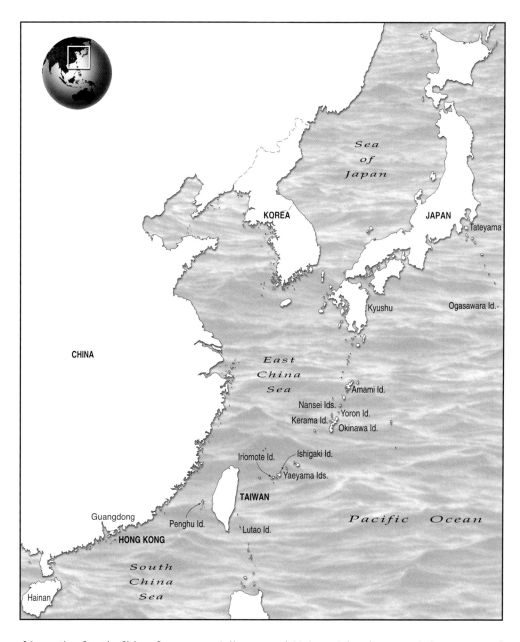

fringe the South China Sea, especially around Hainan Island, around the coasts of Guangdong and Guangxi Autonomous Region, and also around Hong Kong. These last reefs are the only ones for which there is significant descriptive material. China also has strong interests in the Paracel and Spratly islands (Xisha and Nansha Islands; See Box in the SE Asian report on Spratlys)

Japan

There are two major currents that influence coral growth on the Japanese Islands, but in two different ways. The Kuroshio current from the south warms the waters around these high latitude islands, bringing larvae and juveniles of tropical species. The Oyashio cold current from the north blocks the Kuroshio around the latitude of Tokyo, which corresponds to the limits of coral growth in this region. The Kuroshio maintains warm water temperatures above16°C in winter around the southern half of the country, and warms the Ryukyu Islands to a minimum of 20°C. Corals are therefore found on the coast of Japan up to Tateyama, near Tokyo Bay, at latitude 35°N, even though the water temperature can drop as low as 12°C in winter, however the maximum extent of true reef growth is around 30°N.

There are three types of reefs in Japan: fringing; barrier; and one raised atoll (Daito Island). There are mostly fringing reefs in the Nansei Islands area, whereas around the Ogasawara Islands small fringing reefs predominate, called 'apron reefs'. The total reef area in the Nansei Islands is about 100,000ha, with 41% of these in the Yaeyama Islands, 29% around Okinawa and nearby islands, and 19% in the Amami Islands. There are 460ha in the Ogasawara Islands, just south of the mainland (Muko, Chichi and Haha Islands). There are about 400 hard coral species, with the highest number at Yaeyama (Iriomote Island, Sekisei Lagoon, and Shiraho of Ishigaki Island), and decreasing towards the north. The Yaeyama Islands are particularly important because they have high species diversity, with many rare or endemic species (e.g. *Leptoseris amitoriensis*), and serve as 'stepping stones' for coral larval dispersal from the Philippines to the Ryukyu Islands, and then further north.

Taiwan
There are extensive fringing reefs and some platform reefs around the main island and some of the smaller islands. The species diversity is relatively high, with about 300 hard coral species, 70 soft coral and gorgonian species, and 1200 fish species. The reefs are under the impact of typhoons and have not been invaded by the crown-of-thorns starfish.

STATUS OF CORAL REEF BENTHOS

China
Hainan Province is the main coral reef region of China, however, little is known of the recent status of coral reefs of the province, except for Sanya in the southeast where there are most of the fringing reefs. Surveys of the reefs inside the Ya Long Wan Marine Park show that the corals are in a healthy condition, but outside the park they have been badly damaged by sedimentation, dynamite and poison fishing. In all locations, most edible reef organisms occur in very low numbers indicating serious over-fishing and harvesting. These reefs occur within the tropical and equatorial monsoon belt, characterised by high temperatures, high rainfall, monsoon currents and waves, distinct dry and wet seasons, and occasional tropical cyclones. The reef corals share close similarities with the islands and coastlines of the South China Sea, and through them to the centres of reef coral diversity to the south in the Philippines and Indonesia.

Japan
An aerial survey by the Environment Agency of Japan (EAJ) from 1990-1992 revealed that in approximately 34,200ha of coral communities in the Nansei Islands, there was generally low coral cover as follows: 61.3% of communities with coral cover less than 5%; 30.6% of communities with coral cover between 5% and 50%; 8.2% of communities with coral cover over 50%. In the moats (shallow lagoons) of the Okinawa main islands, 90.8% of the

coral communities had less than 5% cover. On Ishigaki Island moats (Yaeyama Islands), coral communities had less than 5% coral cover, whereas cover was much higher in some communities on the east and northwest side. Sekisei lagoon between Ishigaki and Iriomote Islands, has over 13,000ha of coral communities with 9.9% having over 50% cover, mostly in the north of the lagoon.

The reef margins in the Nansei Islands (surveyed with 1,300km of manta tows) had 52% coral cover ranging from as low as 5%, with the margins higher percentages than the moats. On the main island of Okinawa, 67% of the reef margins had coral cover less than 5%, whereas in the nearby Kerama Islands, most coral communities had high coral cover. There are many seagrass beds in the Ryukyu Islands, estimated at 6,900ha, and smaller ones around Amami Islands (130ha). The largest area of seagrass beds was in Yaeyama Islands (approx. 60% of the Ryukyu Islands) and on the east coast of Iriomote Island, seagrass beds covered 1,400ha.

The most severe coral bleaching and mortality event ever observed occurred around southern Japan during the summer of 1998, where 97 municipalities reported bleaching from: the Ryukyu Islands to Hachijo Island south of Tokyo; to the Amami Islands; southern Kyushu; Shikoku; and southern Honshu. No bleaching was observed in western Shikoku, Miyake Island or Ogasawara Islands in the Kuroshio region. Reported bleaching was 40-60% in Nansei Island; under 20%, in Koshikijima Island western Kyushu, eastern Shikoku and Kushimoto; 30-40% on Hachijo Island (most severe was 80-90% on Hachijo Island). Mortality was 70-90% on south Yoron Island and 30-60 % on north Yoron Island. Surveys before and after the bleaching in the Sekisei Lagoon, showed that 40% of *Acropora* died after bleaching and 8% of all coral cover was lost. In Ishigaki Island, coral cover showed a 62% decrease after bleaching, with the most damaged corals being *Acropora*. Whereas mortality in *Heliopora coerulea, Porites cylindrica* and branching *Montipora* was not significant.

Taiwan
Due to coastal development and pollution, there has been substantial degradation to some reefs during the past 10 years, with living coral cover being reduced from approximately 50% to 30% and some species of reef fishes, gastropods, and crustaceans becoming locally extinct. Destructive fishing practices (especially dynamite), sedimentation from construction and dredging, coral collection, sewage pollution, aquarium fish collecting, and unregulated tourist activities have caused this damage.

Reefs in the Penghu Islands (west of Taiwan on the Tropic of Cancer) have been extensively damaged by dynamite fishing, trawling, and sedimentation. The Lutao and Lanyu reefs (southeast of Taiwan) are being damaged by destructive fishing practices and intensive tourist activities. Reef fish populations are very low because of aquarium fish collecting and spearfishing. In 1998, extensive coral bleaching occurred on the reefs in southern Taiwan, Penghu Islands, Lutao, and Lanyu, with approximately 30% to 50% of coral colonies bleached. Surveys in 1999 and 2000 revealed that about 20% of coral colonies had died during this bleaching event.

STATUS OF CORAL REEF FISHES AND FISHERIES

The effects of the severe coral bleaching of 1998 on the fish community at Urasoko Bay and Ishigaki Island (Ryukyus Islands) differed between the two habitats. On the outer reef, most of the *Acropora* spp. corals had died by late September 1998 and were coated with filamentous algae by late October 1998. There were fewer coral-polyp feeders (butterflyfish etc.) but more herbivores, particularly two acanthurids (*Ctenochaetus striatus*, *Acanthurus nigrofuscus*). The numbers of individual fish per transect increased, but the species diversity decreased after the coral bleaching. In the moat where the bottom is mostly covered by coral rubble and sand with several large *Porites* microatolls, there was little effect of coral bleaching, and surgeonfish populations did not change, but the number of individuals and species per transect, and species diversity increased after the bleaching.

Although fisheries statistics are collected in all prefectures of Japan, only Okinawa Prefecture has fisheries statistics for coral reef fishes. The total catch of coral reef fishes in 1993 was 6,066mt, whereas in 1998 it dropped 21% to 4,792mt, with the catch by family: 895mt Lethrinidae (including coastal Lutjanidae); 807mt Scaridae; 50mt Caesionidae; 412mt Serranidae; 308mt Carangidae; and 271mt Siganidae. The catch of Carangidae has decreased considerably between 1993 and 1998, whereas the Scaridae catch was stable. Fishing effort appears to be decreasing, for example: the record of only 328 gillnet fishers in 1998 is a drop of 12% in the past 5 years. The 177 longline fishers has decreased by 24%, however with hook and line fishers, there was a 7% increase to 1,168 between these years.

ANTHROPOGENIC THREATS TO CORAL REEF BIODIVERSITY

Sedimentation from Land Development

The most critical environmental issue in the Ryukyu Islands is terrestrial runoff of red clay soils, which is affecting the coral reefs of Amami-Oshima Id, Tokuno-Shima Island, and Okinoerabu Island in Kagoshima Prefecture and Okinawa Island, Miyako Islands, Ishigaki Island and others in Okinawa Prefecture. The sediment arises from poor land development and use of farmlands, road building and forestry. Activities within US military sites are also a factor. On Ogasawara Islands, run-off has been caused by the wild goats stripping vegetation. Attempts have been made to resolve these problems, and the Okinawa Prefecture established the 'Red-Silt and Other Soil Particles Outflow Prevention Ordinance' for large developments; however, runoff still continues from smaller constructing sites and others completed before the Ordinance.

Construction of an airport adjacent to the Shiraho coral reef of Ishigaki Island will probably release large amounts of sediment and damage the reef that attracts to tourists to the island. Workshops on watershed and environment management have been held in northern Okinawa and Ishigaki with the support of administrators, farmers, fishermen, and the public in order to reduce red clay pollution. Suggestions to solve the problem include: reducing the modification of rivers; preventing developments from altering natural groundwater flows; and controlling domestic drainage and wastes. Similar reports are of damage to coral reefs around Hainan Island.

Development of Coastal Areas Near Coral Reefs
Construction of the airport on Ishigaki had raised public awareness about the need for coral reef conservation and the impacts of development. However, large-scale land reclamation has been performed on the coral reefs near Naha airport (Okinawa) and port and harbour development throughout the islands with considerable damage to coral reefs. Ironically land reclamation is planned on coral reefs to attract more tourists e.g. on Yoron Island (Kagoshima prefecture) in the 1990s. A particular centre of interest is construction of a military airport by the US Marine Corps on Henoko, Okinawa, near an area inhabited by endangered dugongs.

Over-use of Coral Reefs
There are over 4 million tourists to the Ryukyu Islands every year and the marine leisure industry is a major component of the local economy. However, insufficient rules and management do not ensure that coral reefs conservation is sustained, even though it is essential to the survival of the industry. Tourists on Miyako and Ishigaki Islands continue to trample over live corals on the reef flat and there is no education or regulation to prevent this damage. The increasing use of jet skies and underwater walking with helmets is creating new threats. Scuba diving is particularly popular and high concentrations of divers at key sites result in considerable damage to the corals. Diving instructors are aware of the problems and educating tourism-oriented divers about the importance of coral reef conservation. For example, diving locations are rotated to prevent overuse of some diving spots. Nevertheless, there remains further need for regulation of leisure activities in the Ryukyu Islands, and increased coordination between the tourism and fishery stakeholders. Some cooperation in Yomitan-Son (Okinawa) and other areas has started between leisure service providers and fishers to increase local tourism while preserving the reefs.

Collection of living resources such as reef fish, shells, octopus, cuttlefish, crabs and algae is important as livelihoods for people from the Ryukyu Islands; yet, these resources have decreased in quantity and quality through over-fishing and environmental degradation. The Okinawa Prefecture Government operates some fishery projects for sustainable fishery such as release of juveniles, establishing artificial reefs and fisheries management. These stories are also applicable for China and Taiwan.

Conservation Efforts
Although various measures have been introduced to conserve the coral reefs, these are insufficient, especially the lack of coordinated conservation of adjacent land and the sea areas. Laws and plans for land use stipulate that the reefs be conserved, but these are only general recommendations. Similarly, conservation of the natural environment is virtually ignored outside of the protected areas, and development always has priority over conservation. To improve conservation of coral reefs, it is essential to introduce the concept of sustainable use of nature, and apply more regulations and zoning of the land. This will require better environmental education for the whole community, especially local residents. The Ryukyu Islands are densely populated and reefs are heavily exploited, however, there were once traditional rules about sustainable use of the resources. These rules should be re-introduced into the culture in order to conserve the coral reefs of the Ryukyu Islands.

CURRENT AND POTENTIAL CLIMATE CHANGE IMPACTS

The Japan Marine Science and Technology Center (JAMSTEC) carried out extensive surveys of the 1998 bleaching in Akajima Island and the main island of Okinawa, and also collected meteorological and oceanographic data which were closely correlated with the bleaching. A simulation model was developed to estimate how variations in meteorological and oceanographic conditions affect the recovery rate of bleached coral reefs. Remote sensing from aircraft was also used to assess the impact of the 1998 bleaching, after attempts using satellite images were ineffective. The meteorological parameters of air temperature, cloud cover, solar irradiation, wind, rainfall, water temperatures, Kuroshio current strength, salinity, tidal level etc. were measured seeking correlations with the bleaching data. The analysis showed that variations in water temperatures in the Equatorial areas were reflected in Okinawa some months later and large areas with high surface water temperature (29°C or higher) occurred in the summer season of 1998. The conclusion was that global oceanographic variations were the driving forces of the bleaching of coral reefs, but there was no clear link that meteorological conditions could initiate the bleaching, and further studies were recommended. The other conclusion was that recovery rates will be different among species, and strongly dependent on water temperatures and turbidity.

MPAs, MONITORING AND MANAGEMENT CAPACITY

China
Research and monitoring capacity is generally low and coral reefs have had little previous prominence in policy making. In late 2000, the first Reef Check-GCRMN training will be held in Hainan, supported by UNEP. Participants from three coastal provinces with coral reefs: Hainan, Guangdong and Guangxi Autonomous Region, will work on training materials translated into Chinese by the Chinese State Oceanic Administration. Earlier in 2000, Reef Check in Hong Kong assembled 16 teams including government, academia, and the private sector to survey reefs throughout Hong Kong; probably the most concentrated volunteer monitoring programme in the world.

Japan
The establishment of Natural Parks under the Natural Parks Law is an effective means for environmental conservation in Japan. There are 6 Natural Parks with coral reefs in the Amami, Ryukyu and Ogasawara Islands, with a land area of 45,854ha, approximately 11% of total area of the coral reef islands. There are also 23 Marine Park Zones in National and Quasi-National Parks in these coral reef islands (total areas 1,615ha). The total area of these Marine Park Zones is approximately 1.7% of the coral reef area of Japan.

- Ogasawara National Park, managed by the Environment Agency of Japan (EAJ). There are 7 Marine Park Zones in Chichi and Haha Islands with open ocean characteristics of high transparency, high species richness of fishes and abundant *Tridacna maxima* (giant clams). These reefs are small and appear to grow slowly, with some interesting limestone structures.
- Iriomote National Park, managed by EAJ, including 3 Marine Park Zones in Sekisei Lagoon, and 1 in the moat around a coral reef island. The moat area has widely

distributed thickets of branching *Acropora*, whereas the other Marine Park Zones are dominated by table *Acropora*. The Marine Parks Center of Japan has a coral reef research station on one of the islands.
- Amami Islands Quasi-national Park, managed by the Kagoshima Prefectural Government. This park has 9 Marine Park Zones, including 5 on Amami-Oshima Island, 1 on Tokunoshima Island and 3 on Yoron Island. Each is located in the moat or channel. Yoron Island has a wide moat with much branching *Acropora*. Some ancient raised reefs are also present.
- Okinawa Coast Quasi-national Park, managed by the Okinawa Prefectural Government. There are 3 Marine Park Zones: 1 on the main island of Okinawa with a wide moat and an underwater observatory; and 2 on the Kerama Islands, where the reef flats are poorly developed. However, the reef margins are covered with table *Acropora* and have high coral cover, making their undersea scenery excellent.
- Marine Special Zone in Sakiyamawan Nature Conservation Area, by the EAJ; Marine Special Zone (Nature Conservation Areas under the Nature Conservation Law) on Sakiyama Inlet. This is a coral reef on the western coast of Iriomote Island and is 128 hectares of coral reef without any land. It includes a large colony of *Galaxea fascicularis* and seagrass beds, mainly *Enhalus acoroides*.
- International Coral Reef Research and Monitoring Center, constructed and managed by the Okinawa Regional Office for EAJ Japan established for coral reef monitoring for South East Asia at the ICRI regional meeting in 1997 at Okinawa. In 2000, the Environment Agency opened the International Coral Reef Research and Monitoring Center at Ishigaki Island, to coordinate research for better coral reef conservation.

The Environment Agency of Japan established the International Coral Reef Research and Monitoring Center as an East Asian Seas node for the GCRMN at Ishigaki Island, Okinawa in May 2000. The Center started monitoring coral reefs in the area, developed a database on coral reef information; and formed a community framework for conservation. Some monitoring of coral communities around Japan has been carried out by the Environment Agency as the National Survey on the Natural Environment since 1978 with data collected on distribution, area, dominant genera and including growth forms.

There are problems with the protected areas with coral reefs, particularly as the total protected area is 1943.5ha in 24 units. This is a small area for conservation compared to the total area of Japan and the amount should be re-examined to assess the appropriateness of existing areas and seek the designation of more areas. The most important issue to resolve in the near future is to minimise disturbance to reefs from land based activities, especially sediment and pollution runoff, and over-fishing.

Taiwan
Encouragingly, the Taiwanese Government and scientists are promoting conservation and sustainable use of the coral reefs by establishing marine protected areas (MPAs). Currently, a proposal for implementation of MPAs is under discussion within government.

GOVERNMENT POLICIES, LAWS AND LEGISLATION

The Hainan marine authorities of China strictly enforced the law against collection of corals for conversion to lime in 1999, and all lime kilns (more than 90) were shut down and destroyed. Current protection and conservation activities for the coral reefs in Hainan Province are proving to be satisfactory, with some reversals of previous years of degradation and neglect.

Environmental administration of Japan is outlined in the Basic Environment Plan of 1994, with 4 goals:

- 'sustainable cycles of materials arising from human and natural activities';
- 'harmonious coexistence between people and nature';
- 'participation of all stakeholders in sharing benefits and undertaking conservation'; and
- 'international activities'.

Harmonious coexistence is invoked for conservation of coastal wildlife such as coral reefs through the direct action of declaring Marine Parks and Reserves. Improved education and interpretation of nature will be achieved with improved participation of people. Furthermore, Japan seeks to assist and cooperate in environmental governance in the Asia and Pacific areas for the global environment and participate in international activities.

The National Strategy on Biological Diversity in 1995 derived from the Convention on Biological Diversity with distinct policies on conservation of biodiversity and sustainable use of natural resources. The following regulations pertain for coral reef conservation in Japan:

- Natural Parks Law (designation and management of Marine Park Zones in Natural Parks);
- Nature Conservation Law (designation and management of Nature Conservation Areas and Wilderness Areas);
- Fisheries Resources Protection Law (protection of fishery resources and designation and management of fishery protected areas);
- Red-Silt and Other Soil Particles Outflow Prevention Ordinance (prevention of runoff of soil in Okinawa prefecture).

CONCLUSIONS AND RECOMMENDATIONS

Many coral communities in the Ryukyu Islands were degraded by outbreaks of the crown-of-thorns starfish in 1970s and terrestrial run-off after that outbreak. Coral communities, particularly in moats, have declined due to the increased sedimentation that accumulates readily because of closed topography. To prevent further damage to the reefs from land, integrated coastal management should be launched as soon as possible to improve performance in management, understanding and community participation. A coordination mechanism involving all stakeholders concerned with coral reefs should be formed with a priority action to develop active conservation measures and programmes. The International Coral Reef Research and Monitoring Center is expected to play a role as a base for such activities as well as promoting monitoring of the reefs.

Monitoring capacity and the implementation of regular monitoring programmes on coral reefs of the region are well below that for other regions of the world, therefore urgent attention should be given to improving capacity through training in appropriate monitoring and management techniques. Effective conservation of coral reef resources will eventually rely on enhanced commitment amongst all levels of government and with the participation of local stakeholders. Education and involvement are low in the region and should be enhanced. Regulations for the use and conservation of coral reef resources are also inadequate and enforcement is generally poor of existing laws; these are areas for urgent attention.

AUTHORS AND CONTACTS

Shuichi Fujiwara is from the Marine Parks Center of Japan; Takuro Shibuno and K. Mito are at the Ishigaki Tropical Station, Seikai National Fisheries Research Institute; Tatsuo Nakai is from the Nature Conservation Society of Japan, Yasunori Sasaki is at the Japan Marine Science and Technology Center and Dai Chang-feng is from the Institute of Oceanography, National Taiwan University, and Chen Gang is at the Hainan Marine and Fishery Department (gchen@han.cei.gov.cn). For further information contact: Shuichi Fujiwara or Tadashi Kimura e-mail: marpark@blue.ocn.ne.jp

9. Status of Coral Reefs of Australasia: Australia and Papua New Guinea

Thomas Maniwavie, Hugh Sweatman, Paul Marshall, Phil Munday and Vagi Rei

Abstract

Australia's coral reefs are well described and monitored, and are generally in good condition. These reefs have exceptionally high biodiversity, favoured by the massive size and diversity of habitats. This biodiversity is, in general, well studied. They are well protected from the relatively low level of human pressures resulting from a small population that is not dependent on reefs for subsistence. An extensive system of marine protected areas is being implemented, the best known of these is the Great Barrier Reef Marine Park (which is also a World Heritage Area). This is the largest marine protected area in the world and serves as a model for the establishment of many other similar multi-user areas. The monitoring programmes on the Great Barrier Reef (GBR) are also probably the largest and most extensive in the world and are used as models for other projects. These are amongst the best studied coral reefs in the world with very high capacity in all areas of coral reef science, management and education. Large numbers of crown-of-thorns starfish have damaged some regions of the GBR in the past, although recovery is good in most areas. A damaging outbreak is again threatening. Coral bleaching seriously affected a small part of

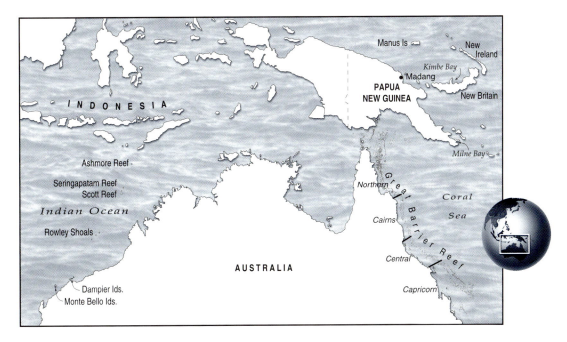

the inner GBR in 1998 with relatively low levels of mortality generally confined to shallower areas (<6m), whereas there was extensive coral mortality on the offshore reefs on the NW Shelf off Western Australia at the same time.

By contrast, the coral reefs of PNG have been relatively poorly described, and have much lower levels of formal management, although there is still strong traditional management. The resources include large areas of reefs with exceptionally high biodiversity about which there is virtually no written information. Human pressures are generally low, but some are concentrated around centres of population and now adjacent to major deforestation activities. Fishing pressures are steadily growing, but do not appear to have had large impacts as yet. Attempts to assess the threats and anthropogenic impacts to coral reefs in PNG are severely limited by a lack of data.

INTRODUCTION

Australia's extensive coral reefs are concentrated on the NE coast, with lesser development on the NW coast and limited reef development along the north coast. The major reef structures form the Great Barrier Reef (GBR), a province of nearly 3,000 fringing reefs, submerged reefs, platform and barrier reefs spanning the continental shelf of the Pacific coast of Queensland and merging to the north with the reefs of the Torres Strait and Papua New Guinea. The area is covered by the multi-use GBR Marine Park (339,750km^2) within the GBR World Heritage Area. There are also scattered and remote reefs further east in the Coral Sea and a number of subtropical reefs (or coral communities growing on rocks) to the south including the Gneering Shoals, Solitary Is., Elizabeth and Middleton Reefs and Lord Howe Is.

On the West coast, there are considerable areas of coastal fringing reefs, notably more than 4,000km^2 at Ningaloo Reef and the Dampier and Monte Bello archipelagos and along the Kimberley coast. There are some significant reefs offshore on the NW Shelf: Ashmore Reef, Scott Reef and the Rowley Shoals. There are some coral reefs and areas of coral growing on rocky reefs in the Abrolhos to the South.

All these reefs occur off low to very low populated coastlines with no subsistence fishing pressures. Moreover the economy is strong and the Australian people have a particularly strong conservation ethic and demand that coral reef resources be given high priority for conservation. This conservation is supported by extremely strong capacity in all aspects of coral reef science, assessment, management and education, as well as strong legislation and capable enforcement.

Papua New Guinea (PNG) encompasses the eastern half of the large island of New Guinea, the islands of New Britain, New Ireland and Bougainville and many smaller islands. The country contains a diversity of cultural groups, with over 800 languages and a wide range of land tenure customs and systems of natural resource use. The coastline is over 10,000km and the EEZ is several million square kilometres. Coral reefs occur along most of the mainland and island coasts with many offshore reefs; thus there are all major reef types, however, their distribution is poorly mapped and the only estimate of total reef area is probably an underestimate at 40,000km^2. The southern reefs of Papua New Guinea are a continuation of the GBR, whereas the reefs to the north and around the islands have strong affinities with reefs of the Solomon Islands to the east and Indonesia to the west. Coral reefs are an integral part of the subsistence economy in most coastal regions where people rely on them for coastal protection, food, medicines and cultural properties. They are also essential for export fisheries and a growing tourism industry. The national economy is poorly developed although the islands are rich in resources and foreign aid plays a major role in this economy.

This paper builds on the previous GCRMN publication in 1998 and new material is based on reports on the status of the GBR by the Australian Institute of Marine Science, which coordinates probably the largest coral reef monitoring programme in the world on the GBR and Western Australia (Box P. 155), and the Great Barrier Reef Marine Park Authority (GBRMPA – Box P. 153). The PNG component builds on the 2000 report on the status of coral reefs in Papua New Guinea prepared by the GCRMN for the Packard Foundation.

STATUS OF CORALS, OTHER BENTHOS AND FISHES

PNG
Recent surveys and anecdotal accounts indicate that most reefs in PNG are in very good condition. Reefs surveyed recently had relatively high coral cover and little evidence of damage from human activity. Over 40% coral cover is common, but this varies with location, reef type, depth and other variables. It also appears that total cover of algae above 20% is not uncommon on apparently healthy reefs in PNG. Subsistence and artisinal fishing is the predominant human activity on PNG reefs. In general, reef fish harvests are thought to be below sustainable levels, however, there is evidence of over-fishing around Port Moresby and other large coastal centres. There is also good evidence of substantial overfishing of invertebrates such as sea cucumbers, trochus, green snail and clams in many locations. The pressures on reef resources in PNG will almost certainly increase as the population continues to grow, especially in large coastal towns.

Australia
While there is considerable information on the status of Australian reefs, there are still gaps for some significant parts. Much of the Great Barrier Reef has been studied extensively and the area is the subject of several monitoring programmes, the largest by the Australian Institute of Marine Science (AIMS). Reef organisms are monitored at Ningaloo, Scott Reef and Rowley Shoals in the West and in the Solitary and Lord Howe islands in the southeast, however, the far northern reefs of the GBR, most of northern Australia and the coastline of northern Western Australia are little known.

The status of the GBR is summarised using the four sections in the original declaration of the GBR Marine Park in 1975. Each section contains hundreds of coral reefs, so comments on reef condition are based on observations of a small proportion of the reefs. Reefs outside the GBR Marine Park in the East and the reefs of Western Australia are then discussed. There are also large areas of sparse coral communities, but with low coral cover in northern Australia around Darwin. A major protected area is the Cobourg Peninsula Marine Park and Sanctuary, managed by the Northern Territory Parks and Wildlife Commission.

EASTERN AUSTRALIA

GBR Far Northern Section
These reefs are fairly remote, so there is relatively little regular monitoring; 9 reefs were surveyed by manta tow in 1999. They had average reef-wide coral cover of about 26% (range 12 - 40%) which is slightly higher than the long term average for the GBR. Crown-of-thorns starfish (COTS) were present at very low densities on most reefs.

GBR Cairns Section
This section includes some of the most accessible reefs on the GBR. Much of the adjacent hinterland receives high rainfall, causing reefs closer to shore to be regularly subjected to low salinities, high sedimentation and potentially, enriched nutrient levels. In 1999, 49 reefs were surveyed by manta tow and had mean reef-wide hard coral cover of 21% (range 3-40%); a low value caused by extensive COTS outbreaks on mid-shelf reefs. Coral bleaching affected inner-shelf reefs in this section in 1998, which caused some mortality. Video surveys of sites on the northeast sides of 19 reefs showed a similar picture, with outer shelf reefs in the north of the Section having some of the highest coral cover on the GBR (53-61%).

GBR Central Section
This section also includes the accessible reefs for tourists and fishers in the Whitsunday Islands. Manta tows of 33 reefs found reef-wide coral cover to be 23% (range 3-51%). This is because inshore reefs in the north of the section suffered some mortality from the bleaching in 1998. Intensive surveys of NE facing sites on 18 reefs estimated hard coral cover at 32%. Outbreaks of COTS are occurring in the North of the Section. This pattern of outbreaks appearing first in the north of the Cairns Section, followed by a wave of outbreaks moving south was observed in the 1980s. It is presumably driven by southerly drift of larvae with the East Australian Current.

GBR Mackay-Capricorn Section
This most southerly section of the GBR contains some remote reefs in the Pompey Group and the Swain Reefs as well as the Capricorn-Bunkers which are some of the best-known because of research stations on Heron and One Tree Islands. There was high average coral cover on all parts of 12 reefs of 33% (range 16-49%), with consistently high reef-wide coral cover on outer-shelf reefs in the Pompey Group and the Capricorn-Bunkers (above 36%). Intensive surveys of NE facing sites on 11 reefs reflect this, with mean hard coral cover estimated at 41%. There are chronic outbreaks of COTS on some Swain reefs, which do not follow the episodic pattern of outbreaks in northern sections.

Elizabeth and Middleton Reefs
These are large atolls about 120km north of Lord Howe Island and well south of the GBR. Surveys in the 1980s and 1990s showed there was more diversity and more tropical species than on Lord Howe. Crown-of-thorns starfish did some damage throughout the 1980s, but the full impacts are not known. Human impacts are negligible, but storms can cause substantial changes.

Interpretation of Values for Coral Cover

This report refers frequently to percent coral cover. The average reef-wide cover of hard coral on GBR reefs was estimated at 23% in 2000, with values ranging from 3-51%. More precise estimates from underwater video records on a range of NE facing slopes show relatively higher cover of hard coral with a mean value of 28% and range of 4-68%. These relatively low values reflect fairly frequent disturbances such as cyclones and crown-of-thorns starfish outbreaks during the last few decades. Much of the change in coral cover on the GBR is due to growth and destruction of table *Acropora* spp. (e.g. *A. cytherea* and *A. hyacinthus*). These table corals live in areas of high water movement and grow rapidly to cover a large area, but they become increasingly susceptible to storm damage as they grow. On reefs close to the equator, where cyclones are rare, such corals cover much of the reef and average values for coral cover will be higher (up to 80% or more in places). Thus different disturbance regimes will lead to different norms for the extent of coral cover. Therefore, it is essential to know the environmental regime of an area to interpret coral cover. An estimate of 30% on the GBR is often an indication that the reefs are in excellent health as storms and COTS tend to keep coral populations down. Whereas in other areas, cover of 30% would be viewed with alarm, as reefs in parts of the Red Sea, the Maldives, Indonesia, and parts of the Caribbean often have levels way over 70%, due to a lack of tropical storms or they are in protected waterways like the Red Sea.

Lord Howe Island

This high volcanic island (14.6km^2) is 603km off the east coast of Australia and one of the most southern coral fringing reefs in the world (31°40'S). The reefs have low species diversity, but good coral cover on the rocky slopes in passes and lagoons. The island and surrounding seas were declared a World Heritage Area in 1982, a Marine Park was declared in June 2000, and a reef-monitoring program is being implemented. Two previous surveys found significant differences in the composition of coral species between 1978 and 1993, suggesting some turnover of rare species. There was an increase in crown-of-thorns starfish numbers in the early 1990s and some coral bleaching in the late 1990s, but impacts have been minimal and the condition of the reefs is good. Fish, plants and invertebrates have a mix of tropical and temperate species and a number of endemic species. There is a management plan to protect the World Heritage values while allowing recreational fishing, and fishing to supply locals and tourists.

There is coral growth on rocky reefs south of the GBR in southern Queensland: the Gneering Shoals (26°S), Flinders Reefs (27°S) and northern New South Wales (28°S 30'S) on rocky reefs with populations of corals up to 50% coral cover.

WESTERN AUSTRALIA

Ashmore Reef, Scott Reef and Rowley Shoals

These isolated oceanic and shelf-edge reefs are far from any mainland influence, being scattered between 12°S and 18°S along a line 400km from the northwest Australian coast. Ashmore Reef (12°S) is closer to Indonesia than Australia and is regularly fished by Indonesians under a joint agreement. Scott Reef (14°S) is visited less frequently by Indonesians, and is a prospective site for extraction of liquid natural gas. The three reefs of the Rowley Shoals (17°S) are protected as State and Commonwealth marine parks. Trochus, trepang (sea cucumbers), shark and other fish are harvested, but the effects on the reefs are unknown. All reefs have hard pavements of coralline algae and low and stunted corals in exposed high wave-energy areas, and high coral cover and a large variety of growth forms in sheltered habitats. The NorthWest Shelf and coastline endures many large cyclones. Coral communities at all reefs were in good condition through most of the 1990s. In 1995, parts of the Rowley Shoals suffered significant cyclone damage. Scott Reef and the Rowley Shoals have been monitored by AIMS since 1995. Average cover on reef slope sites before the cyclone was 47%. In sites at Imperieuse Reef cover of branching *Acropora* spp. dropped from 50% to less than 1% after the storm. Then in 1998, Scott Reef was bathed in warm water for several weeks, resulting in extensive bleaching and subsequent coral mortality at monitoring sites at 9m. Extensive bleaching was observed to depths of more than 30m. Average coral cover in exposed sites prior to the bleaching was 54%, but after there was less than 10% cover of live coral. Recent surveys of the extensive lagoon floor showed that large areas of live coral survived at 25-60m (see Case study).

Cocos-Keeling Islands

These are atolls south of Sumatra at about 12°S. The human population is very low (600-700 people) with low impacts on the reefs, except for some over-harvesting of some molluscs (*Lambis lambis*). The Cocos-Keeling Islands are an Australian protected area: To quote from the Integrated Marine Monitoring Plan: "The waters surrounding Cocos are the responsibility of the Federal Government. ...The Federal National Parks and Wildlife Conservation Act 1975 was applied to Cocos in 1992 to provide for the protection of all wildlife (both terrestrial and marine). This legislation is unique in automatically protecting all species unless specifically unprotected by Ministerial Declaration." "Commercial fishing is of such a small scale (and low economic value), that the Australian Fisheries Management Authority (AFMA) is also unrepresented on the Islands. An AFMA officer visits once a year, and Parks Australia staff conduct any surveys and monitoring on their behalf." Thus these reefs are in virtually pristine condition, but there were reports of two large-scale coral die-offs in the past, followed by significant recovery of the corals.

The Dampier Archipelago and Monte Bello Islands

These reefs and islands (21°S) are on inner-shelf and middle-shelf parts of the continental shelf, respectively. There is high diversity and abundance of corals and fish, and despite increasing pearl oyster farming, petroleum exploitation, shipping, fishing and tourism activity, there is no evidence of significant human damage to the reefs. The Monte Bellos were used for British Nuclear tests from 1952-1956. A program of assessments of marine resources of each of these areas is in progress.

Ningaloo Reef
The Ningaloo Marine Park extends from 22°S, southwards for about 230km as a long fringing reef, separated from the shore by a sandy lagoon up to 3km wide. Land impacts are minimal as the adjacent land is arid. Coral and fish communities are diverse and abundant and fishing is only permitted in some parts of the Marine Park. The reef front receives the full force of swells from the Indian Ocean so corals tend to be compact, although cover may reach 40%. Coral cover behind the reef crest and in channels varies from 5-40% with an increase in coral cover to the south. Outbreaks of a coral-eating snail (*Drupella*) killed much of the coral in some areas in the 1970s and 1990s, and other areas have been damaged by low oxygen conditions when coral spawn decomposes. Now, new corals are recruiting and restoration of coral cover is progressing, but many areas are still dominated by dead coral and rubble. A broad-scale monitoring program has recently been established.

Abrolhos Islands
The islands and reefs are the southern limit (28°S) of reef development in Western Australia. These reefs have extensive areas of both kelp and corals, and support an important and well-managed rock lobster industry. The use of lobster traps is banned in areas with fragile corals, and there is no indication of any detrimental effects on the coral and fish communities. These reefs are in good to excellent condition and plans have been prepared for declaring the area as a marine protected area.

BIODIVERSITY

PNG
The reefs are part of the Western Pacific region of maximum marine biodiversity, where there is exceptional biodiversity on the coral reefs. Rapid ecological assessments of the coral reefs recently in Kimbe Bay, West New Britain Province; Lak Region, New Ireland Province; and Milne Bay, Mile Bay Province confirm this high biodiversity. Checklists for specific taxa have also been prepared for several other locations, including Bootless Bay on the south coast and Madang and Kamiali on the north coast. The diversity of reef fishes and corals at these locations ranks among the highest in the world. Furthermore, the number of species recorded on single dives is often among the highest recorded during such surveys anywhere in the Pacific. There is a low rate of endemism among fish, and probably other marine organisms. Milne Bay province has the majority of endemic species of fish, including several species known only from a single location. Biodiversity assessments on coral reefs in PNG have mostly concentrated on fish and corals. Consequently, the diversity of most other marine organisms is poorly documented. Comprehensive biodiversity assessments for vast areas of the reefs are entirely lacking.

Location	Fish	Corals
Milne Bay	1039	362
Kimbe Bay	837	347
Madang	804	--
Bootless Bay	730	284
Kamiali	577	--

Numbers of species of fish and corals known from sites around Papua New Guinea.

STATUS OF REEF FISHERIES

PNG

Reef fisheries in PNG are largely subsistence and artisinal. They provide a large proportion of the animal protein for coastal communities, however, data on the quantity and composition of the catch from these domestic fisheries are limited to one or two locations. The recorded export from reef fisheries is relatively low in comparison to some other Pacific countries, but has doubled to over 2,000 metric tonnes in the past few years, indicating increasing activity in this sector. Sea cucumbers are the largest component of the export catch, with over 680mt exported in 1998, followed by shark meat, molluscs and crustaceans. Reef fish are a relatively small component of fisheries exports but the volume has tripled in recent years to over 90mt in 1998. A number of live reef fish enterprises have started in PNG, but have failed because they did not take into account the complex social ramifications of their operations, or were found to be breaking fisheries regulations. There is currently a moratorium on the issuing of new licenses and fisheries authorities are developing management plans. The major concern about the live reef fish trade is how overfishing and destructive fishing techniques, such as the use of cyanide and targeting spawning aggregations, can be controlled.

EASTERN AUSTRALIA

Trawl Fishery

This is the largest fishery in the GBR targeting prawns (6,500mt worth US$60 million), scallops (200mt worth $18 million), Moreton Bay bugs (slipper lobsters, 500mt worth $5 million) and lesser quantities of blue swimmer crabs and squid. There are zoning restrictions on trawling in the GBR Marine Park and a variety of other management measures. Trawling is destructive in terms of by-catch and of the physical impacts of trawling gear on the seabed. Recent research indicates that a single pass of a trawl net can remove 5-25% of bottom living organisms. By-catch reduction devices (BRDs) and turtle exclusion devices (TEDs) are mandatory in some areas. Effort is fairly widespread and only a few areas are fished intensively. There has been a history of illegal trawling in inappropriate zones, but penalties and enforcement efforts have been increased recently.

Coral Reef Finfish Fishery

This covers coral reef fish species throughout their range in Queensland waters, although the fish are found predominantly within the GBR Marine Park. The fishery is second only to the trawl fishery for both the economic value and potential impact on Marine Park ecosystems. The main issues of concern to GBRMPA in this fishery are the sustainability of the target species and the potential secondary impact caused by the removal of high-order predators from coral reefs. The reef-line fishery includes commercial, recreational and indigenous fishers. The main targets are snappers (*Lutjanus* and *Lethrinus* spp.), groupers (Serranidae) with Coral Trout (*Plectropomus* spp.) making up about 40% of the commercial catch, and wrasses (Labridae). Between 3,000-4,000mt of fish worth about $25 million are taken by commercial reef-line fishers each year. There is much dispute as to how the recreational catch compares in size with the commercial line fishery and recreational fishers probably take similar amounts to the commercial fishers. The draft management plan and regulatory impact statement for the fishery was released by the Queensland Fisheries Management Authority (QFMA) in 1999, with the most significant feature an attempt to slow the continuing growth in fishing effort and remove the huge latent effort with over 1,000 little-used commercial line-fishing licenses. Also, the plan proposes reductions in recreational catches, improved protection of the breeding of key species, and a process to review the plan continuously with inputs of new information.

Trochus Fishery

Approximately 170mt of trochus (*Trochus niloticus*) are harvested annually in Queensland, with the regulation that the harvest must be by hand or hand-held implement. Wading or free-diving are used commonly in shallow waters (2-10m deep). Scuba and hookah can be used only by commercial fishers. Minimum and maximum size limits apply to all fishing (except by indigenous fishers collecting for traditional or customary purposes). Trochus is a limited entry quota-managed fishery with Total Allowable Catch (TAC) of 250mt for the East Coast. Trochus has been harvested commercially since 1912 from Torres Strait, and when the price for trochus shell peaked at $10.00 per kg during 1990, more than 600mt were landed. The present market for shell has collapsed as synthetic buttons are now indistinguishable from natural mother-of-pearl. The commercial market for trochus meat continue to grow. A harvest fishery management plan for trochus is due for release in October 2000, and there is a separate Torres Strait trochus fishery with a quota of 150mt managed by the Protected Zone Joint Authority.

Beche-de-mer Fishery

There are four main species of sea cucumber harvested; black teatfish, white teatfish, sandfish and prickly redfish. Approximately 200mt are harvested annually in Queensland as a quota managed fishery with a TAC for the East coast of 380mt allocated to 19 collectors. Management is hindered by illegal catches and unreliable catch returns data. Catch per unit effort (CPUE) of black teatfish peaked in early 1996 and steadily declined until 1999, so the fishery was closed to protect breeding stocks. Quotas for the other species are also being adjusted to avoid over-fishing. There is little information on the biology of holothurian recruitment rates and research to ensure a sustainable harvest. Another species (greenfish) may become more valuable with the recent discovery of pharmaceutical properties.

Tropical Rock Lobster

The GBR commercial tropical rock lobster fishery operates on the east coast from Cape York to 14^0S, harvesting 50-200mt annually. A separate fishery operates in the Torres Straits managed by a Joint Authority with PNG. One species, *Panulirus ornatus*, makes up over 90% of the catch, with 5 other species along the coast. Management sets both quotas and minimal size limits in Torres Straits, but there is no restrictions on Queensland waters. However plans are being considered to control the fishery to limit entry and impose bag limits on recreational fishers, but only hand spear, spear gun and hand collection is permitted now. Anecdotal information suggests that the maximum size of individuals on the east coast is decreasing, a classic symptom of growth overfishing. By contrast the Western Australian rock lobster fishery is well managed and sustainable, especially on the Abrolohos reefs and is a major export fishery of tails.

Marine Aquarium Fish

This is limited to the east coast of Queensland and is managed by input controls (on apparatus, number of participants, number of divers, area of operation). Commercial and recreational fishers are limited to collection of fish by hand or by using lines and cast, scoop and mesh nets, and scuba may only be used by commercial fishers. Limited catch and effort information is available and interim management arrangements include cost recovery through industry fees; transferability of authorities; zoning of the fishery; eligibility criteria for zones; and amendments to the application process, including entry criteria. A draft management plan is due for release in 2000.

Coral Collecting

The coral fishery in Queensland is both license and quota managed fishery, with 55 designated 'coral areas' along the east coast and a TAC of 200-250mt among 39 authority holders. 'Coral areas' of 200-500m of reef front to a depth of 6m may be either 'exclusive' (accessible by only one coral authority holder) or non-exclusive (accessible by more than one coral authority holder). The actual harvest is below 50mt and unlikely to reach the quota limit. These authorised 'coral areas' represent less than 0.0003 % of the total GBR World Heritage Area, with the main target being common, fast growing species, primarily *Pocillopora* and *Acropora*. Conflict and competition for use of some 'coral areas' has developed as a result of rapid growth of tourism, but monitoring is required to ensure that the harvest is sustainable.

Marine Shells (specimen shells/collecting)

There is a limit of specify 10 of any species allowed to restrict collection to specimen collectors rather than large volumes for export.

NORTHERN TERRITORY

There is a coastal line fishery, including some reef species and a total commercial catch per annum from 60–130mt over the last few years, with recent catches being lower. The recreational catch may 6 times the commercial landings and the catch by the traditional sector is not assessed.

ANTHROPOGENIC AND NATURAL THREATS TO REEFS

PNG

Some of the most serious threats to coral reefs in PNG come from terrestrial activities, such as large-scale forestry and agriculture as extensive tracts of coastal forest have been allocated for logging. There are insufficient mechanisms to prevent widespread damage to reefs from sedimentation as a consequence of this logging. Most of PNG's reefs are nearshore and significant inputs of sediments will impact directly on the reefs, but there are almost no data on elevated sedimentation from past or current logging operations, nor any monitoring data on reefs affected by logging activities. Increasing stresses on reefs is also coming from the growing coastal populations, though increased fishing pressure and pollution from sewage. Fish stocks are already overfished around the largest coastal towns and high levels of microbial contamination occur in the waters around Port Moresby.

Crown-of-thorns starfish (COTS, *Acanthaster planci*) are found on many PNG reefs, usually at low densities, with isolated instances of high population densities, but not as outbreaks. Consequently, COTS are not yet a major problem on PNG reefs. Blast fishing could be widespread in PNG and there are examples of substantial damage to some reefs. However, surveys in other areas where blast fishing is reported have not detected significant damage, therefore, the effects are relatively localised. Other fishing and collecting activities probably cause relatively minor physical damage to reefs unless the intensity of activity is high. Coral mining for lime, oil spills, industrial pollution, mine waste, land reclamation, ship groundings are probably localised threats.

Attempts to assess anthropogenic impacts to coral reefs in PNG are severely limited by a lack of data on the patterns and abundances of reef organisms and few data on the physical and chemical characteristics of the reefs. Thus, the apparent good condition of PNG reefs must be considered acknowledging this lack of information. Reliable monitoring programmes are needed, particularly in areas of increasing population pressure and where anthropogenic impacts are increasing.

Australia

Human pressures are low on Australian reefs as the population density is low, the reefs are mostly remote from the coast, fishing pressures are moderate to slight and in some areas virtually non-existent. In particular, the offshore atoll reefs on both coasts are so remote from the mainland influences that they are only subject to occasional fishing, although the extent of illegal and international fishing is poorly known. Also many reefs are adjacent to areas with low rainfall and minimal runoff. Increased levels of nutrients and sediment entering the GBR system in river discharges are potentially a significant threat to the GBR, particularly the inner-shelf reefs, however these impacts on coral reefs have never been measured directly. The major sources of nutrient pollution (nitrate and phosphate) on the GBR are increased runoff from large areas cleared for cattle grazing, enriched runoff from sugar-cane and banana farming and domestic sewage. Education and extension programmes run by the Department of Primary Industry and reef managers have raised awareness and lead to improved practices in the rural sector. Monsoonal flood events are the major vectors of sediment and nutrient pollution, particularly from grazing lands laid bare after long periods of droughts. Cane and banana farmers are regulating their use of fertiliser to minimise runoff loss. The practice of green tillage of crops and trash blanketing (leaving the trash on the ground as compost and not burning crops) is increasing, which reduces sediment and nutrient loss. All tourist resorts and cruise boats are required to either treat sewage or dispose of it so that there is no pollution (tertiary treatment).

There is potential for pollution from the extensive petroleum industry in Western Australia, however these industries are conscious of the public concern for the reefs and have strict procedures to manage any spills. There is some subsistence fishing on the remote reefs off Western Australia that is permitted under agreements between Indonesian and Australian governments, however there are continuing problems (mostly legal and political) of small-scale poaching of shark, fish, trochus, giant clams and sea cucumbers in other areas.

Although dugong are considered at risk globally, a large fraction of the world population is found in Australian waters, with about 15% of the Australian population in GBR waters. Aerial monitoring shows that the numbers of dugong have declined except in the northern section of the GBR with mortality attributed to nets, boat strikes and indigenous hunting. A major loss of numbers occurred after a major flood resulted in the death of inshore seagrasses in the early '90s. Following this decline in numbers, 16 dugong sanctuaries have been declared on the GBR, and the use of nets, particularly gill nets is restricted or banned in these areas. Permits for indigenous hunting have also been reduced.

Turtles are under similar global threats, with pressures in Australian waters from trawl nets, shark nets, traditional hunting, floating rubbish and boat-strikes as well as habitat loss and destruction of nests by feral pigs and foxes. Turtles migrate for thousands of kilometres and are probably exposed to intense hunting and nest predation in other countries. Many major nesting sites in Australia are protected, the use of turtle exclusion devices is spreading, and traditional hunting is regulated by permit.

STATUS OF MPAS AND REEF CONSERVATION

PNG provides a unique opportunity for the conservation of significant areas of coral reefs in the western Pacific region of maximum marine biodiversity, before they become severely impacted by local anthropogenic activities. Few other locations offer the combination of large areas of high diversity reefs mostly undamaged by human activity, relatively low populations in most coastal areas, a scientific and management community that is committed to sustainable use of marine resources, and a customary land tenure system that can be used to enhance conservation efforts. PNG has a number of legally designated protected areas that contain coastal and marine habitats, however, insufficient government resources for management means that the effectiveness of these areas for conservation is questioned. There are demonstrated successes in small-scale, community-based marine protected areas, which might provide a widely replicable model for other areas in PNG. These community based protected areas and conservation initiatives that link with social and economic development are probably most appropriate to the needs and land tenure realities of PNG. The success of protected areas and conservation projects in PNG is likely to rely heavily on continued NGO presence. Protected areas on both the south Papuan coast and the north coast and islands will be important for conserving the different faunal assemblages in these two regions. The Milne Bay province is a priority region for new protected areas, given that it contains major reef resources, the highest overall diversity of fishes and other marine organisms, and many of the endemic species.

The Australian Government's Oceans Policy includes a commitment by state and federal governments to implement a national representative system of marine protected areas. A plan of action has been published and bio-regionalisation schemes have been proposed that identify gaps. Many significant reefs are already included in protected areas. The vast geographic scale of Australia's reef resources means that capacity for enforcement is limited, but human pressures are generally light, economic status is high and the public supports conservation.

CORAL BLEACHING AND CLIMATE CHANGE

Coral bleaching has been reported from numerous locations around PNG since 1996. It appears that the 1998 bleaching event was less severe in PNG than in Australia and many other countries. In PNG, bleaching has also occurred outside El Niño periods, with the most severe and widespread bleaching in PNG during 1996-1997, centred apparently on Milne Bay with 54% of corals bleached. Most other areas had lower level of bleaching and good recovery. Bleaching was reported from some areas in 1999 and has again been observed in several locations during early 2000. The consequences of repeated bleaching events are unknown but may include changes in community structure or degradation of reef diversity. With the apparent increase in the frequency of bleaching events in PNG it is important that coral bleaching and associated physical parameters, such as sea temperature, be monitored in a coordinated manner. The likely effects of global climate change on reefs in PNG remain unclear due to uncertainty about appropriate models for predicting changes in temperature and sea level, the complex geological dynamics of PNG, and uncertainty about the effects that increasing temperatures will have on coral communities. Nevertheless,

---

PROTECTING BIODIVERSITY - REPRESENTATIVE AREAS IN THE GREAT BARRIER REEF MARINE PARK (GBRMP).

The GBRMP contains incredible biodiversity within the variety of habitats, which are protected through a number of management tools; zoning, education, permits, and management plans. To ensure all habitat types are adequately protected, the Great Barrier Reef Marine Park Authority (GBRMPA) is using a 'representative areas' approach to identify the different habitat types, assess threats and identify an appropriate level of protection. The 'Representative Areas Program' contributes to an Australian review of marine and terrestrial areas requiring protection for biodiversity. A **representative area** is typical of the surrounding habitats at an ecosystem scale, with physical features, oceanographic processes and ecological patterns representative of the full range of habitat and ecosystem types. Recently, there has been a growing realisation of the need to identify and protect representative examples of the diversity of habitats for species dependent on those habitats, rather than focus on specific habitats or individual species. Coral reefs are given higher levels of protection within the Marine Park, but other less-known and less-spectacular habitats are also important and should be represented within highly protected zones. Consequently, GBRMPA has started a program to classify representative habitats, review existing protected areas and identify candidate areas to ensure protection of all representative habitats, with the process scheduled for completion in late 2002. A major feature of the process is the involvement of a broad base of stakeholders, who will provide input through meetings, workshops, written documents and invitations for submissions.

increased frequency of coral bleaching and inundation of coastal areas can be expected if increases in sea surface temperature and associated changes in sea level occur as predicted. Bleaching in Australia was focussed on inshore areas of the GBR in early 1998 and on offshore reefs off Western Australia (see Chapter 1).

GAPS IN REEF MONITORING AND MANAGEMENT

PNG

Scientists and technicians capable of coral reef monitoring and the support infrastructure required for monitoring programmes are scattered among a few government departments, academic institutions, NGOs, mining companies, and the dive tourism industry. However, financial constraints and personnel shortages have consistently limited the ability to continue monitoring programmes that have started from time to time. There is an urgent need for baseline monitoring in areas currently or likely to be subjected to increasing anthropogenic stresses. The University of Papua New Guinea (UPNG) instigated a Reef Check program in 1998 and students were trained as divers and schooled in survey techniques. Surveys were conducted in many locations around PNG with the assistance of local dive operators. Training in AIMS/GCRMN standard monitoring techniques has also been conducted through the Motupore Island Research Department of UPNG. Further development and ongoing support of these programmes is essential to ensure they can continue and be expanded. In addition to training more local divers and building a regular monitoring program, there is a need for facilities and capacity for data analysis and the dissemination of results.

Dive tourism is one of the biggest non-destructive commercial activities in PNG. The PNG Divers Association plays a significant role in marine conservation and could also assist in ongoing monitoring programmes and skills training for marine scientists. Dive operators clearly have the best infrastructure for visiting reef areas and monitoring programmes that effectively harness this capacity are likely to be most successful. Government departments with responsibilities for the management of reef systems are hampered by staff shortages and severe funding cutbacks. Much of the capacity for biodiversity assessment, the establishment and management of marine protected areas, and the development of community education programmes now rests with NGOs. There appears to be sufficient capacity for the development of policy and regulations for fisheries and marine protected areas. Unfortunately there is little capacity for enforcement of laws, quotas and regulations. Local communities could play a greater role in enforcement of fisheries regulations and marine protected area, through the expansion of community based management programmes. As the anthropogenic stresses on the reefs increase, the need for substantially greater capacity to assess and deal with threats to reefs will become critical. A high priority must be protecting and enhancing the capacity for impact assessment, monitoring and management of reefs within PNG.

Australia

Australia's largest reef area, the Great Barrier Reef is the subject of several monitoring programmes, notably the AIMS Long-term Monitoring Program (crown-of-thorns starfish, benthic organisms and reef fishes) and the Long-term water quality monitoring programme of the GBR Marine Park Authority (Chlorophyll). Other agencies have programmes to monitor particular groups of organisms e.g. seabirds, turtles, dugongs in limited areas. Other programmes monitor physical variables, GBRMPA monitors water temperature over a

MONITORING OF THE GBR

As a federal government research institute adjacent to the Great Barrier Reef, the Australian Institute of Marine Science (AIMS) is concerned with studying many aspects of coral reefs, but one specific task is the continued assessment of the health of the reefs: the AIMS Long-term Monitoring Program on the GBR. The objective is to monitor regional status and trends of selected organisms on the reefs with the results contributing significantly to fulfilling the reporting requirements associated with World Heritage listing of the GBR. Each year, 48 'core' reefs are visited, selected on their position across the continental shelf (there are very clear differences in reef communities on the coast – ocean gradient) and by latitude. Marked permanent transects are placed in one habitat, the NE slopes of the reefs. These are surveyed for benthic organisms using underwater video and 191 species of reef fishes are counted on the transects. The perimeters of the core reefs and of about 50 other reefs are surveyed by manta tow to record densities of COTS and to estimate reef-wide coral cover. Extensive quality control procedures are in place, and outlined in Standard Operating Procedure documents (SOPs) that are reviewed regularly. Observers are cross-calibrated at frequent intervals, and specialised data-entry programmes are used to help extract information from the video records and to check the data as they are entered. These include statistical data checking for unlikely values. Methodological studies are also an important part of the program. Much emphasis is given to reporting via the Internet [www.aims.gov.au/reef-monitoring]. Preliminary reports emphasising COTS are posted within a few days of each survey, and interactive data summaries are available, which include the most up to date information. There are pages dealing with particular issues such as COTS and coral bleaching. The program produces annual Status Reports, which are also available via the Internet along with SOPs and other documentation. Many aspects of reporting, particularly statistical analyses and graphs, are generated automatically, and the program produces scientific papers describing the large scale patterns revealed in the monitoring data. The monitoring team have also been involved in considerable training of people in many countries for the GCRMN. Staff of the program includes 6 graduate biologists in the field team, a database administrator, a biostatistician and a coral reef biologist; all this costs more than US$1.5 million per year.

large area and there is a project linking real-time records from automatic weather stations run by AIMS with NOAA-NESDIS satellite SST data as part of the 'Bleaching Hotspots' program to identify locations at risk of coral bleaching.

There are also monitoring programmes at Ningaloo Reef, Scott Reef and the Rowley Shoals, Lord Howe and the Solitary Is. There is a wide gap across much of far northern Australia from the northern quarter of the GBR to the Pilbarra in Western Australia where there is no monitoring and little baseline information. Australian government funding policies favour community-based monitoring programmes in both terrestrial and marine environments. While this approach has a number of potential benefits, it remains untested as a method of producing reliable long-term data on reef ecosystems.

LEGISLATION AND REGULATION

PNG
There appear to be adequate laws and legislation for the conservation and management of coral reefs in PNG. However, most legislation does not specifically refer to marine systems and this has caused uncertainty about how it should be applied to coral reefs. Also, the laws relevant to different sectors (e.g. fisheries, mining, environmental protection) are not fully integrated which has lead to confusion over which laws have priority, who is responsible for management, and the rights of the various interest groups. There is little government capacity for enforcement of laws, quota and regulations. A national surveillance strategy has been suggested which would involve participation from all levels of government, NGOs and local communities. Local communities could play a greater role in enforcement of fisheries regulations and marine protected area, through the expansion of community based management programmes.

Australia
Management of the coral reefs are now covered by substantial legislation, with the authority vested in a single authority for the GBR (GBRMPA) in cooperation with the Queensland State Government. This was reviewed in the 1998 report.

Fisheries resource management is a complex mix of Commonwealth and State/Territory responsibility, with the States or Territories responsible out to 3 nm and the Commonwealth managing fisheries beyond that to 200nm (EEZ). Most commercial fisheries are managed mainly through input controls limiting the number of vessels, the time and place of fishing, and the type gear. Progressively, the management focus for the coral reef fisheries has moved from considering the fishing impacts on target species to consideration of non-target species and the environment, especially the effects of trawling on benthic habitats. The release of the Australian Ocean Policy in 1998 listed the principles, and the planning and management approaches necessary to achieve the ecologically sustainable development of the ocean resources. In 1999, major changes were made to national environment legislation via a new Act (the *Environment Protection and Biodiversity Conservation Act 1999*, EPBC Act) by replacing 5 existing Commonwealth Acts and introducing an assessment and approval process for activities likely to have significant impacts on: Commonwealth marine areas; nationally threatened species and ecological communities; the marine environment; and on internationally protected migratory species. This act is particularly relevant to fisheries within the GBR World Heritage Area.

The Endangered Species Protection Act 1992 seeks to limit the effects of unintended marine catch (by-catch) on endangered species, by controlling threatening processes. A recommendation in 1999 to list otter trawling as a 'key threatening process' for sea turtles is being reconsidered, and another is to list gillnetting because of its effects on cetaceans (particularly the Indo-Pacific humpback dolphin and the Irawaddy dolphin) and dugong. The effect of fishing on non-target species is a major issue within some reef fisheries as the amount of discarded by-catch exceeds targeted catch in some trawl fisheries by up to a factor of 10 or more. The scarcity of information makes it very difficult to assess what fisheries management procedures are required to address the legislated objectives for non-target species. This requires further monitoring and assessment.

RECOMMENDATIONS

PNG
There is an immediate need to build scientific and management capacity related to coral reefs in PNG. Programmes are urgently needed to retain the current capacity for assessment, monitoring and management with long-term employment, and support the training of future marine scientists and managers.

An integrated approach is needed between environmental planning and any economic developments that will impact on the reefs. Effective management of urban development, watershed degradation and large-scale commercial activities requires much greater capacity in provincial and national government agencies.

Monitoring initiatives need to be supported and developed to provide an effective and ongoing assessment of reef health. An integrated monitoring program involving UPNG, PNGDA, and the Office of Environment and Conservation (OEC) could provide an effective monitoring network with links to management objectives. Monitoring is urgently needed in areas likely to come under stress from coastal development and other terrestrial activities. Support is needed for diver and technical skills training, routine monitoring trips, quality control, data assessment and dissemination.

High quality mapping of PNG reefs is important for conservation initiatives and appropriate management of fisheries resources. Continued support for biodiversity assessments conducted by appropriate experts and basic biological and ecological research is important for the long-term sustainability of PNG's coral reef resources.

Community-based programmes could be established within a network of marine protected areas. Addressing the needs of local communities and integrating them in the development, management and enforcement of protected areas is likely to yield success. Continued community support is important for the success of marine protected areas; and Effective methods of enforcing fisheries regulations are urgently needed. This is particularly important for any re-introduction of the live reef-fish trade. Community education and alternative income programmes can help reduce destructive fishing practices. Increased commitment is also needed at provincial and national levels.

Australia
- Most Australian reefs are well protected against the pressures that they are subject to, both legally and in terms of capacity to enforce those regulations. This situation needs to be maintained;
- There are a growing number of monitoring programmes covering many reef areas; maintenance of effective monitoring is important, given potential effects of changing human pressures and climate change;
- The long-term effects of chronic pollution from river runoff and coastal activities are poorly understood and closer attention is required to managing these to ensure the sustainability of the GBR;
- Chronic fishing pressures, both commercial and recreational, have the potential

to alter the fish populations on coral reefs by selectively targeting key predator species, such as groupers and snappers. Such changes conflict with the aesthetic values requested by the international tourism market and this will require management attention; and
- The GBR is relatively well studied. Major pressures are identified as fishing (trawling and line fishing) and changing water quality due to the development of agriculture in the catchments of rivers that flow into GBR waters. Major research programmes investigating these impacts are in progress, coordinated by the CRC for the GBR World Heritage Area. Their findings need to be included in future management plans.

SUPPORTING DOCUMENTS AND CONTACTS

Wachenfeld DR, Oliver JK, Morrisey JI (ed) 1998 State of the Great Barrier Reef World Heritage Area 1998. GBRMPA Townsville 139 pp. [available on the web or contact: davidwa@gbrmpa.gov.au)

Sweatman H, Cheal A., Coleman G, Fitzpatrick B, Miller I, Ninio R, Osborne K, Page C, Ryan D, Thompson A, Tomkins P (2000) Long-Term Monitoring Of The Great Barrier Reef. Status Report Number 4. Australian Institute Of Marine Science, Townsville 127 pp. [available on the web or contact: h.sweatman@aims.gov.au]

Caton A, McLoughlin K (eds) (2000) Fishery Status Reports 1999: Resource Assessments of Australian Commonwealth Fisheries. Bureau of Rural Sciences, Canberra.

USEFUL WEB ADDRESSES FOR FURTHER INFORMATION OR COPIES OF THESE REPORTS:

www.environment.gov.au – Environment Australia
www.gbrmpa.gov.au – Great Barrier Reef Marine Park Authority
www.aims.gov.au – AIMS
www.calm.wa.gov.au - WA Dept of Conservation and Land Management
www.nt.gov.au/paw/ - Parks & wildlife commission of NT
www.brs.gov.au/fish/ - Bureau of Rural Sciences, Fisheries and Forestry Sciences Division

Thomas Maniwavie is at the Motupure marine station of the University of Papua New Guinea (motupore@upng.ac.pg), Hugh Sweatman leads the Long Term Monitoring Program at AIMS (h.sweatman@aims.gov.au), Paul Marshall manages research and monitoring projects at the Great Barrier Reef Marine Park Authority (p.marshall@gbrmpa.gov.au), Phil Munday conducts research on fishes of PNG from James Cook University (Philip.Munday@jcu.edu.au), and Vagi Rei works with The Nature Conservancy in PNG.

10. Status of Coral Reefs in the Southwest Pacific: Fiji, Nauru, New Caledonia, Samoa, Solomon Islands, Tuvalu and Vanuatu

Robin South and Posa Skelton

Abstract

The IOI-Pacific Islands GCRMN Node covers 7 Pacific Island countries including: Fiji, Nauru, New Caledonia, Samoa, Solomon Islands, Tuvalu and Vanuatu, which together have over 2100 islands and islets. Collectively these countries illustrate a wide range of island and reef biogeographies, including: high and low lying islands; atolls with fringing, barrier, submerged, platform, oceanic ribbon, and near atoll formations. The diversity of marine species declines eastwards from the Indo-Pacific centre of concentration. Finfishes decline from 1168 species in Fiji to 991 in Samoa and just over 400 in Tuvalu. Marine benthic algal species decrease in number from 422 in Fiji to 287 in Samoa and 40 in Nauru. Extinct species include the giant clams: *Tridacna maxima* and *Hippopus hippopus* from Nauru and Samoa respectively. Endangered species include: marine turtles, giant tritons, mangrove crabs, bêche-de-mer, trochus and turban shells, and highly targeted reef fishes. Coral reefs are in good condition in most countries, although significantly degraded in urban areas. The reefs are becoming increasingly vulnerable from over-fishing, pollution, sedimentation, environmentally un-sound development, crown-of-thorns starfish outbreaks and habitat loss. Climatic factors such as cyclones, sea level rise, coral bleaching (such as the early 2000 bleaching event in Fiji) are real or potential threats. Effective long-term monitoring of biotic reef systems is not in place, although the development of marine protected areas has been implemented in Samoa, Solomon Islands, Tuvalu and Vanuatu, and identified in Fiji and Nauru. Only New Caledonia has active monitoring, having established 18 stations since 1997. There is a need for integrated coral reef management within a broader context of coastal and island resources management. Basic technical knowledge exists in most government departments, but there remains a need for more trained biologists, taxonomists, ecologists and managers in all countries at local and national levels.

Introduction

All of these countries of this South West Pacific region were assisted by the International Ocean Institute of the University of the South Pacific in Fiji to produce this report. The first stage was a series of reports presented by national experts of these countries at the International Coral Reef Initiative (ICRI) Regional Symposium on Coral Reefs recently held in Noumea, New Caledonia during May 2000. These countries have now formed a GCRMN Node to address the gaps that have been outlined in this report, in particular the lack of monitoring capacity and support to assess the status of coral reefs in these countries. This

Status of Coral Reefs of the World: 2000

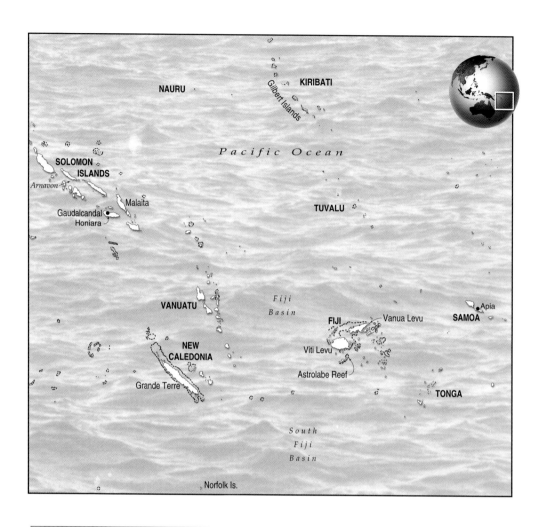

No. islands, islets & coral cays	Land Area (km²)	EEZ (km²)	Mangrove Area (ha)	Reef Type
Fiji (1,000)	18,500	1,290,000	35,000	barrier, fringing, atoll, platform
Nauru (1)	21	430,000		fringing
New Caledonia 4 (Main Islands)	18,585	1,740,000	20,000	barrier, fringing, atoll, platform
Samoa (16)	2,935	120,000	~100	fringing
Solomon Islands (1,000)	28,370	1,340,000	52,000	barrier, fringing, atoll
Tuvalu (9)	26	900,000	<20	fringing, atoll, platform
Vanuatu (80)	12,190	680,000	25	barrier, fringing, platform

Basic geographic details of the 7 countries covered by this report.

	Population (increase %)	Population per km²	Per capita GDP in US$	Population Distribution
Fiji	775,077 (2.0)	42	2,118	15% (Suva)
Nauru	12,460 (2.9)	593	17,486	N/A
New Caledonia	193,386 (2.0)	10	12,753	60% (Noumea)
Samoa	170,000 (0.5)	58	1,018	20% (Apia)
Solomon Ids	408,358 (3.4)	14	739	High (Honiara)
Tuvalu	10,144 (1.7)	390	7,053	40% (Funafuti)
Vanuatu	177,400 (2.8)	15	1,309	25% (Port Vila & Louganville)

Basic demographic and economic data for countries in the Southwest Pacific.

need became particularly evident in early 2000 when large-scale coral bleaching affected many of the coral reefs in the region, but there was little information on the condition of the reefs beforehand to determine the scope and scale of this bleaching event. Hopefully better monitoring will be in place to document and compare any future bleaching events, as well as other stresses.

REEFS AND PEOPLES IN THE SOUTHWEST PACIFIC ISLANDS

The reefs and islands in the 7 southwest Pacific countries and states vary from tiny atolls to large, high islands with all coral reef types. The smallest nation is Nauru, a single island of 21km² with a fringing reef of 7.4km². The largest country is the Solomon Islands which has 28,370km² of land area an extensive EEZ. New Caledonia consists of a large continental island (Grande Terre) surrounded by one of the longest barrier reefs in the world enclosing one of the region's largest lagoons. Most of Fiji, Solomon Islands, Samoa and Vanuatu are high volcanic islands, whereas Nauru is a raised calcareous platform and Tuvalu is comprised of a series of atolls. There are more than 2,100 islands, islets and cays, almost all with coral reefs. Together, these countries control large Exclusive Economic Zones (EEZs) totalling over 6.0 million km². Therefore oceanic fisheries (especially tuna) are a primary foreign exchange earning source for most of these countries.

Most of these southwest Pacific countries have large concentrations of people clustered into a few urban centres, with the remainder living in rural, remote, coastal communities. Population densities in Nauru and Tuvalu are among the highest in the world. New Caledonia has a large expatriate population in the urban area of Noumea. The GDP of most countries is mostly very low. Population growth rates are very high and there is a heavy reliance on reef resources for food security. These countries include a diversity of cultural groups: Polynesian (Samoa, Tuvalu), Micronesian (Nauru) and Melanesian peoples (Fiji, Solomon Islands, Vanuatu, New Caledonia). Fiji is a mixture of Melanesian and Polynesian people, with a large Indo-Fijian population (40% of total). Collectively more than 100 languages are spoken throughout the region.

STATUS OF CORAL REEFS

Narrow fringing reefs border most high island shorelines and are frequently separated from the shore by a shallow channel. Much wider barrier reefs are separated from the shore by deep lagoons up to 70m deep in Fiji, Vanuatu, Solomon Islands and New Caledonia. Many Samoan reefs grow on recent lava flows with virtually no lagoons. The Great Sea Reef of Fiji is one of the largest reefs in the South Pacific (370km long), and the Great Astrolabe Reef, also in Fiji, has been proposed as a World Heritage Site. Double barrier reefs (two reefs in parallel formation) are rare and occur only in the Solomon Islands and New Caledonia, the latter has a reef area of 40,000km^2. Atolls, formed on drowned sea mounts and volcanoes, occur in Tuvalu, at a few sites in Solomon Islands, Fiji (northern Lau Group), and New Caledonia. Low-lying coral islands (cays) form on the rim of the reef, and rarely rise more than 3m above sea level. Most atoll lagoons are moderate in size, unlike some elsewhere in the Pacific (e.g. Kwajalein Atoll in the Marshall Islands is over 100km in diameter). Patch reefs occur in most countries in the lagoons behind barrier reefs, but are rare in Samoa. Submerged or drowned reefs occur in the open ocean, some as shallow as 20–30m depth, which can be a challenge to navigation.

The majority of reefs in the southwest Pacific region are in good condition; however, there are increasing pressures from anthropogenic impacts with some reefs being severely degraded. Most people in this region are strongly dependent on coral reefs for food resources, cash income from reef fisheries, coastal protection, sand and rock for building of roads, housing, as well as healthy reefs to support growing tourism industry. The sustainable development of marine and coastal resources is one of the most critical issues for the Pacific Islands for the following reasons and formed the core findings of the South Pacific Regional Environment Programme (SPREP) South Pacific Environment Outlook on developments in coastal areas of the Pacific Islands over the last 100 years. The report also noted that few data exist on historical pressures on coastal environments in the Pacific Islands, but the pressure on these resources is now considered as a serious issue:

- marine resources are the major source of protein for most Pacific islanders;
- most Pacific islanders live in the coastal zone;
- parts of the Pacific have the highest marine biodiversity in the world;
- the development of marine resources represents critical opportunities for substantial economic advancement for many Pacific Islands, especially in atoll states like Tuvalu; and
- commercial fishing is one of the fastest growing industries in many Pacific Island countries.

The following summarises the current state of coral reefs for each country in this region, but even today there is still a reliance on anecdotal information as there is still an urgent need for more quantitative data across temporal and geographic scales.

Fiji
- reefs around urban centres are significantly degraded through eutrophication, pollution, crown-of-thorns starfish (COTS), coastal development and siltation;

- most commercial species are depleted and some species are extinct;
- mass coral bleaching severely affected many reefs in February to April 2000;
- destructive fishing, mining, forestry, agriculture, and poor tourism developments are impacting reefs in some areas;
- periodic severe cyclones cause extensive reef damage in localised areas;
- a lack of coordinated monitoring prevents adequate assessment being made on the status of the reefs;
- a lack of Marine Protected Areas is hindering conservation efforts;
- biodiversity studies have only occurred on Viti Levu, particularly near Suva, and many reefs remain unstudied; and
- reef degradation is exacerbated by coastal development, mangrove destruction and other anthropogenic impacts.

Nauru
- reefs are over-fished with some target species being depleted or extinct (e.g. giant clams - *Tridacna maxima*);
- no Marine Protected Areas have been established;
- no reef monitoring exists and there is no expertise at present;
- the status of reef biodiversity is poorly documented (except for fishes);
- blasting of reef channels has caused some damage; and
- ciguatera fish poisoning is a moderate problem.

New Caledonia
- the state of coral reefs is not generally quantified and not well known;
- the majority of reefs are in good condition, except some fringing reefs around Noumea and on the east coast of Grande Terre, where nickel mining runoff is prevalent;
- 23-28% of mangroves have been lost around Noumea since the 1960s;
- 200ha of other coastal habitats (including seagrass beds) have been lost through coastal development;
- biodiversity is not well documented except for some animal groups e.g. corals, sponges, echinoderms; and
- MPAs and conservation areas have been established since 1971, with effective enforcement since 1991, and more are being developed.

Samoa
- most reefs are degraded by over-fishing and land-based activities;
- there has been severe damage from cyclones (most recent in the early 1990s);
- some species are extinct (e.g. the giant clam *Hippopus hippopus*); and
- marine biodiversity is poorly documented.

Solomon Islands
- rapid population growth is putting pressure on reef resources;
- biodiversity is poorly documented, and most reefs have not been investigated;
- industrial pollution is significant in Honiara and causing extensive reef damage;

	FIJI	Nauru	New Caledonia	Samoa	Solomon Islands	Tuvalu	Vanuatu
Corals (stony)	198	?	300	>50	?	?	?
Molluscs	478	?	5,500	?	?	?	?
Seagrasses	4	0	?	2	7	?	?
Mangroves	9	1	?	3	26	3	?
Benthic Algae	422	>40	350	>287	233	?	?
Fishes	1198	?	1,950	991	725	400	469
Turtles	3	?	4	3	5	?	?
Sponges	?	?	600	?	31	?	?

Current state of biodiversity knowledge in the region with a ? indicating that there is no reported information

- widespread logging is having a major impact on catchment areas and reefs, resulting in massive sediment increases; and
- monitoring programmes, and appropriate legislation to protect reefs is lacking.

Tuvalu
- most reefs and biodiversity are poorly documented, except Funafuti Lagoon;
- sand mining is causing significant degradation of reefs;
- ciguatera poisoning is a serious problem; and
- cyclones have caused serious damage.

Vanuatu
- the reefs have deteriorated since 1985, with 50% of reefs now considered degraded;
- there is a high incidence of ciguatera poisoning;
- cyclones have impacted on reefs, but the impacts were not assessed; and
- some reefs are still pristine with exceptional water clarity.

BIODIVERSITY PERSPECTIVES

Coral reefs of this region are part of the broader Indo-Pacific major biodiversity ecoregion, illustrating a general reduction in marine biodiversity from west to east. Small countries with few reef habitats, like Nauru, have fewer species than the Solomons and Fiji, with many more reef types. There is a serious lack of information on biodiversity of most trophic groups for most countries, with only fishes and benthic algae being studied to some depth. Many areas remain unexplored, and there is currently no regional database on coral reef biodiversity. This lack of information is hampering attempts to identify vulnerable or important species-rich areas for conservation, except for highly conspicuous animals like marine turtles, which are the focus of region-wide conservation efforts by SPREP and others. The rate of extinctions for most marine species is not known, but it is likely that many species have already been lost through reef degradation, even before they were named. There is probably a very low rate of endemism among the marine biota, although some sub-species may have developed because of the isolation of many reefs.

Stress & (score)		Fiji	Nauru	New Caledonia	Samoa	Solomon Islands	Tuvalu	Vanuatu
Coral Harvesting/ Fish Trade	(7)	2	0	2	1	2	0	0
Oil Spills	(8)	3	0	2	1	2	0	0
Destructive fishing	(9)	4	0	0	4	0	0	1
Acanthaster	(11)	4	0	2	1	2	0	2
Coral Bleaching	(11)	5	0	2	2	1	0	1
Tourism	(12)	4	0	3	2	2	0	1
Urbanisation	(12)	3	1	2	2	2	0	2
Extraction	(14)	3	0	0	3	2	4	2
Pollution	(16)	4	1	4	2	2	0	3
Climate change	(18)	4	1	1	3	2	5	2
Coastal Alteration	(17)	4	1	2	2	1	3	4
Cyclones	(18)	5	0	1	5	0	3	4
Overfishing	(19)	4	2	2	4	3	3	1
Sedimentation	(19)	5	0	4	3	3	0	4
TOTAL IMPACT RATINGS		54	6	27	35	24	18	27

National experts rated causes of coral reef degradation as follows: 0 = no threat; 1 = low threat; 2 = localised low threat; 3 = average threat; 4 = localised major threat; 5 = general major threat.

There is a serious lack of marine biologists and taxonomists among Pacific Island nationals. Therefore, most biodiversity studies have been carried out by outside experts, and the specimens deposited in collections all over the world. USP has established a Marine Biodiversity Collection in Suva, Fiji with important collections of corals, invertebrates, fishes and benthic algae. This could become the nucleus of an important regional biodiversity collection; however, much work remains to be done to document coral reef biodiversity of the region. This is essential as baseline information upon which degradation can be monitored and reversed. Such potential taxonomic information limits threaten the growth of bioprospecting as a potentially important new activity in the region, especially in Fiji. Specimens are collected, identified, and extracted to seek potentially useful compounds, such as beqamide, an anti-cancer substance found in Fiji sponges. While the bioprospecting success rate is extremely low, there are parallel challenges on intellectual property rights that must be addressed.

CAUSES OF CORAL REEF DEGRADATION

Most of the reefs in the area have been impacted by the same suite of natural and anthropogenic impacts but there is insufficient monitoring information to prepare detailed 'state of reefs' reports. Reef degradation is highest in Samoa, Fiji and Vanuatu with overfishing of nearshore resources being a common problem throughout. Some of the most important target species, like giant clams, are rare or extinct throughout the region. Coastal development and habitat destruction, like clearing of mangroves, is prevalent, and pollution is serious near urban areas where there is a lack of sewage treatment and proper waste disposal. Sand and rock mining causes reef degradation in atolls, and there is significant sedimentation and eutrophication in the larger countries (Fiji, New Caledonia and Solomon Islands) because of extensive logging, poorly planned agriculture, uncontrolled coastal development, and mining. The major threats to the marine environment throughout the Pacific Islands are summarised below:

- nutrients from sewage, soil erosion and agricultural fertilisers;
- increased sedimentation from poor land use;
- physical alterations through destruction of fringing reefs, beaches, wetlands and mangroves for coastal development and sand extraction; and
- over-exploitation of coastal food fisheries, particularly through destructive fishing methods.

Most of the population lives in the coastal zone, and in atolls like Tuvalu, all livelihoods are coastal based. Rapid growth of these coastal populations is responsible for much of the plundering of coastal resources, destruction of mangroves, pollution of lagoons and harbours, and loss of some marine species and ecosystems. Some of the damage is now irreversible, e.g. coastal modification and destruction of habitats, and this will increase in the future, if population growth continues unabated. Land based sources of pollution are among the greatest threats to marine biodiversity. Marine invasive species are poorly documented, but may pose a threat near ports, harbours and coastal habitats. Ship-sourced marine pollution is also a potential threat.

Tourism is the most rapidly growing industry in parts of the region (e.g. Fiji, Samoa and Vanuatu), with a large proportion of these tourists being divers. Construction of tourism

CROWN-OF-THORNS STARFISH

Population explosions of the coral eating crown-of-thorns starfish (COTS, *Acanthaster planci*) have been a major problem on many coral reefs, such as the Suva Barrier Reef, Fiji since 1979. While human activities have been blamed for these outbreaks, they may also reflect long-term natural variations in the populations or consequences of climate events like El Niño events. However, there is evidence that eutrophication and siltation encourage larval survival and enhance the proliferation of the animals. Heavy infestations cause major changes in coral communities and have the potential to shift species diversity because they target the branching corals of the species Acropora and Pocillopora. Major outbreaks were observed in 1986, 1988 and 1997 on Suva reefs. In 1996, the Mamanuca group was affected and was followed by a moderate outbreak in the Kaba area in 1997. In 1998 the Wakaya and the Lau Group reefs saw a large aggregations of COTS, as were Taveuni and the Somosomo Strait the following year. Early this year Cuvu Reefs on the Coral Coast were affected by COTS and villagers collected over 2000 COTS for a one-km stretch of reef. Current reports from August to September 2000, indicate large numbers of COTS aggregating near the Solo Lighthouse Reef in the Great Astrolabe Reef. There has been a recent outbreak around Noumea, New Caledonia, and there was a clean up at major dive sites; 1,500 were collected from Tabu reef, and 1000 from Ilot Maitre. These are the first outbreaks for at least the last 10 years.

	Subsistence		Commercial	
	Mt per yr	$US (x 1000)	Mt per yr	$US (x 1000)
Fiji (40kg yr-1)	16,600	45,800	6653	18,300
Nauru	98	220	279	630
New Caledonia	2,500	9,000	981	7,970
Samoa (26kg yr-1)	3,281	5,000	208	320
Solomon Islands (34kg yr-1)	10,000	8,410	1150	4,340
Tuvalu	807	660	120	980
Vanuatu (14kg yr-1)	2,045	1,950	467	1,510

A summary of the mean annual subsistence and commercial fisheries production between 1989 – 1994 with the annual fish consumption listed in parentheses.

facilities has accelerated erosion in some areas and caused sedimentation on reefs and in lagoons. Some areas with intensive development, such as the Coral Coast on Viti Levu (Fiji), have suffered from increased eutrophication and sedimentation and there has been an increase in the growth of benthic macroalgae, such as *Sargassum* on reefs and reef flats.

CORAL REEF FISHERIES

Coastal and offshore fisheries are the sources of most animal protein and the backbone of Pacific Island economies, particularly those with small land areas and large EEZs. The Pacific offshore fishery is of global significance and was valued at more than US$2.0 billion in 1998. The Pacific Island region contributes over 50% of the world's tuna catch of 3.4 million mt. The offshore fishery is important to the economy of Pacific Island countries such as Samoa, where annual fisheries exports are estimated at US$14.95 million, or 70% of total merchandise exports for 1999.

Coastal and reef fisheries play a significant social and economic role for Pacific peoples, and constitute 80% of the total fishery. Reef fisheries are predominantly subsistence, involving traditional, simple non-mechanised and non-technical gear. Fishing is carried out mostly by rural dwellers, especially women, and provides the bulk of animal protein, as well as supporting traditional rituals and culture. Most (about 80%) of harvested marine resources are consumed by the village, with the remainder sold at local markets or exported. The total value of the subsistence fishery in this region is US$71.04 million, while the commercial coastal fishery earnings are US$30.05 million. But the true quantity and value of the subsistence fishery in most Pacific countries is poorly documented, and estimates of marine resources harvested range from 17,000mt/yr (Fiji), to 4,600mt (Samoa) and 2,000mt in Vanuatu. The annual consumption per capita ranges from 14kg/capita/yr (Vanuatu) to 40kg/capita/yr (Fiji). It is very much higher in atoll countries such as Tuvalu, where up to 1kg of fish may be consumed daily.

In most of the southwest Pacific countries, management of the coastal fisheries had been essentially traditional since the islands were inhabited, but management of marine resources became the responsibility of the 'State' during colonial rule. Since these periods,

> ## CIGUATERA FISH POISONING
>
> Ciguatera fish poisoning is a problem in Vanuatu and Tuvalu. Ciguatera is caused by a toxin from a microscopic marine dinoflagellate alga *(Gambierdiscus toxicus)* which lives on the surface of larger algae and coral fragments. It is bioaccumulated up the food chain, starting in herbivorous fishes and ending up in carnivores, which are the targets for human consumption. The toxin also accumulates in people and is difficult to dilute out of the body if fish is a major dietary component. Symptoms range from mild to serious, and occasionally death. The creation of favourable habitats for the toxic algae through dredging of lagoons and channelisation can increase the incidence of ciguatera poisoning.

'western style, open access' legislation was introduced and fishing activities were regulated by gear and size rather than by tenurial areas or histories. Most fisheries departments have policies that encourage maximum rather than sustainable fish catches, which usually leads to overfishing. Conservation is typically the role of 'environment' departments. In addition, there is lack of information on stock estimates and the basic biology or natural history of many of the species. Therefore regulation of the subsistence fishery is trivial and catch and size limits are rarely set or enforced at the village or market level through governmental mechanisms.

Regional and national attempts have been made to re-stock important reef species like giant clams, trochus and green snails, but these have had limited success. There are some projects in Fiji and Solomon Islands to replenish coral stocks (e.g. *Acropora*) by seeding fragments in lagoons under controlled conditions. Mangrove trees are being re-planted in other pilot projects. A general strategy to supplement declining reef fish stocks through aquaculture or stock enhancement has been adopted by all countries. *Tilapia* cultivation in village projects has been successful in Fiji, but other aquaculture projects have been less beneficial, and had adverse consequences for coral reefs. One example is the widespread cultivation of the introduced seaweed *Kappaphycus*, which has spread to areas of Fiji well away from the original farm sites.

THREATS TO CORAL REEF FISHERIES

Coastal reef fisheries face a number of threats and challenges, with the most important being the need for sustainability to maintain food security of the people. But the lack of data, and few qualified personnel and resources to regulate the subsistence fishery, plus the lack of properly managed conservation areas hinder efforts towards sustainable management of reef fisheries for the future.

Traditional subsistence fishing methods have rapidly been displaced by modern methods, which are often biologically and environmentally unfriendly. This has been accelerated with the advent of local cash economies. Destructive fishing methods such as the use of dynamite, plant (e.g. derris) and chemical poisons, have degraded large areas of coral reefs, and interrupted the biological cycles that support the coastal fisheries. Selective fishing of the top predators has also changed the structure of reef fish communities, with

	Harmful fishing methods	Endangered or threatened species
FIJI	Scuba; hookah; poisons (e.g. derris)	Tridacna, Pinctada marginifera, Holothuria fuscogilva, H. scabra, H. nobilis, Charonia tritonis, Bolbometopon muricatus
NAURU	Scuba; Small mesh sized seine and cast nets	Mullets; Kyphosus cinerascens, Cephalopholis miniata, Cheilinus undulatus, spiny lobsters Tridacna
NEW CALEDONIA	Reef walking, bag nets, crowbars, small-mesh nets, Poisons,	?
SAMOA	Poisons	Tridacna squamosa, Hippopus hippopus, Mugil cephalus, Chanos chanos, Charonia tritonis, Scylla serrata.
SOLOMON ISLANDS	Explosives, poisons (e.g. derris, Barringtonia, cyanide)	Green snail (Turbo mamoratus), lobsters, dugongs
TUVALU	scuba, hookah	Tridacna spp.
VANUATU	Poisons	?

Summary of harmful fishing methods and endangered or threatened species.

unpredictable long-term implications. Pacific Island populations doubled between the 1960s and 1990s, which exerted increasing pressure on inshore and reef subsistence fisheries. Over-fishing was inevitable, and fishers now spend more hours at sea and travel further to sustain the same level of catch.

The life histories of marine organisms are often disrupted by encroachment (reclamation for land development) and habitat destruction (mining, rubbish dumping). Target species (e.g. grey mullets) have become endangered in Samoa and Fiji because they are fished by more fishers using overly efficient methods (e.g. gill and seine nets). As more target marine animals are threatened, fishing methods and activities will continue to increase in efficiency and destructiveness until the biological diversity will be compromised and the food security of many Pacific Island countries will be threatened.

Many of the same natural and anthropogenic threats to coral reefs also affect the fisheries. Additional threats include rapid population growth with consequent increases on reef resource exploitation, the incentive from the cash economy to catch more fish than needed for home consumption, and the loss of traditional systems controlling the use of resources. When these threats are combined with the other threats to reef health, the prognosis for sustainable reef fisheries is poor.

CLIMATE CHANGE IMPACTS

Cyclones are the major factors that determine the geomorphology of many coral reefs; they build up rubble and sediments, to create new habitats and islands; and also can remove coral cover from large areas of reefs. Cyclones can also cause increased sediment and pollutant loadings in rivers, which transport sediment to coral reefs. The reefs around Fiji, Vanuatu and Samoa have all suffered badly from cyclones in the early 1990s, with recovery anticipated to require 10 to 20 years. Therefore an increase in the frequency of cyclones, often linked to climate change, will have serious impacts and prevent natural recovery of damaged reefs. Sea level rise and increased storm surges are also other major impacts anticipated particularly on low atoll countries like Tuvalu.

A serious coral bleaching event occurred in early 2000, and affected Fiji and Solomon Islands. In some areas of Fiji, up to 90% of corals were killed (e.g. parts of Beqa Lagoon, Somosomo Strait and the Yasawa Islands), while in other areas damage was less severe and recovery was taking place several months after the event. The longer-term impacts of this event are being studied with coordination by the IOI-Pacific Islands GCRMN Node. Unfortunately there was

Mass Coral Bleaching in the Fiji Islands 2000

A HotSpot of sea surface temperatures 1-1.5°C above average was centred on the Fiji Islands in 2000, accompanied by unusually calm conditions. Massive coral bleaching was seen in early surveys of southern Viti Levu with 50-100% of corals bleached to 30m depths. Aerial surveys and spot checks showed more intense bleaching on the southern coasts of Vanua Levu and Viti Levu with less intense bleaching on the northern coasts. The important reefs in Beqa Lagoon, Great Astrolabe Reef, Taveuni, Ovalau, the Somosomo Strait and the Yasawa Islands were seriously affected. The HotSpot dissipated abruptly in mid-April with cooler weather and the onset of southeast trade winds.

The first reports from the National Tidal Facility (Flinders University, Australia) indicated elevated sea surface temperatures from December 1999-January 2000. NOAA distributed satellite images of the South Pacific HotSpot in early February. Soon there were coral bleaching reports from Solomon Islands and Easter Island. Staff from USP, the aquarium trade, dive operators, and the Waikiki Aquarium (Hawaii) surveyed reefs throughout Fiji coordinated by the IOI-Pacific Islands GCRMN Node. By July 2000, there were observations that some reefs were largely dead (Taveuni and Yasawa Islands), whereas others (Beqa Lagoon) were recovering.

The bleaching could have serious impacts on the dive tourism industry in Fiji, which brings in more than US$100 million annually. The USP Marine Studies Programme started medium to long-term experiments to assess the effects of bleaching and recovery rates, potential impacts on reef fish populations and effects on food fisheries. They are now establishing long-term monitoring sites on the Suva Barrier Reef, the Great Astrolabe Reef and in western Viti Levu, to avoid the problem of having no baseline data to assess the impact of the bleaching. The 2000 bleaching event in Fiji, recorded by satellite imagery, was the most serious ever experienced and reinforces the links between elevated sea surface temperatures and coral bleaching and may be a harbinger of future trends.

not a good set of 'before bleaching' data, but current monitoring will make Pacific islands better prepared if bleaching happens again. Bleaching has previously occurred in Fiji, although not to the same degree as it did in early 2000. Bleaching is a regular but less dramatic event in Samoa. There is little documented evidence of repeated bleaching in Tuvalu, Solomon Islands, or Vanuatu. New Caledonia experienced some bleaching in 1995-96, but recovery was strong with minimal mortality. Current evidence suggests that coral bleaching will become a more frequent and severe phenomenon in the future.

Changes in sea level are frequent from a geological perspective in the Pacific Islands, and the coral reefs have been able to adapt. Therefore sea level rise is not a major threat in itself to coral reefs, but a combination of sea level rise and increases in atmospheric CO_2 concentrations, which will reduce the rate of calcification of coral skeletons, could seriously threaten coral reef ecosystems. Thus, if predicted increases in sea level (15 - 95cm by 2100) occur in combination with increased CO_2 levels, coral reef growth may not be able to keep pace with sea level rise. Other factors such as increased reef degradation from human sources and increases in mean sea surface temperatures, could affect the geographic distribution of coral reefs.

STATUS OF REEF CONSERVATION

All Pacific Island countries recognised the need to conserve marine habitats in the National Environment Management Strategies (NEMS), coordinated by SPREP in preparation for the 1992 UNCED Conference in Rio. SPREP now assists these countries in their compliance with Conventions and Agreements on marine conservation and sustainable development, such as the Convention on Biological Diversity, the Ramsar Convention on wetlands, and the United Nations Framework Convention on Climate Change (through the Pacific Islands Climate Change Assistance Programme). Reef conservation is specifically reinforced in the Activity Plan for the Conservation of Coral Reefs in the Pacific Islands Region (1998-2002). The Plan is a partnership between countries and regional organisations, and focuses on five areas: 1) education and awareness, 2) monitoring, assessment and research, 3) capacity building, 4) legislation; and 5) networking and linkages across people and programmes.

SPREP has also assisted in coral reef monitoring training in several countries to assist the formation of GCRMN Nodes. Likewise Reef Check has also been active in Fiji. Both of these monitoring programmes are now the responsibility of GCRMN Nodes to implement, building on Fiji's example. While there are basic databases, these need reinforcement, appropriate coordination, and greater commitment.

Most countries have a strong tradition of customary marine tenure frequently aimed to conserve marine resources for future use. In that most tenure mechanism are ancestral and exist along village and family rights and owned areas, it is a challenge to integrate these with more 'western' governance and zoning structures that are more 'public' or common property based in design. Thus the establishment of both MPAs and coastal zone management (CZM) plans (which are lacking for most countries) can made more effective by enhanced incorporation of customary marine rights. However, there is still a need for underpinning marine conservation policy dialogue, economic incentives and conservation legislation for both traditional and governmental application of MPAs and CZM plans. The following provides an overview of current conservation efforts for each country, followed by a case study on Samoa for village-based monitoring and legislation.

Fiji
- Extensive traditional systems exist, but are being challenged by urban values and pressures to maximise commercial opportunities in the fishery, tourism and other coastal resources sectors;
- No established system of MPAs exists to date, although the Fiji National Historic Trust is charged to develop Marine Reserves in Fiji. The Sustainable Development Bill is now on hold with the current political disturbances;
- Some privately owned marine sanctuaries are associated with tourist resorts (e.g. Turtle Island);
- Some villages (e.g. Waisomo on Kadavu) are establishing village-based and managed MPAs;
- Various local NGOs and private sector dive operators are actively raising marine awareness and promoting the establishment of MPAs.

Nauru
- No system of MPAs or relevant legislation exists;
- Traditional marine tenure is lacking.

New Caledonia
- There are 37,500ha of MPAs, possibly second only to Hawaii;
- A 1km wide conservation zone has been declared around all land areas where commercial fishing is prohibited;
- The Environment Plan for the North Province proposes to designate 10% of coral reefs for strict protection and 10% for subsistence use only;
- Customary reserves and traditional fishing zones have been recognised;
- Fishing regulations prohibit the following: fishing with scuba; spearfishing at night; nets set in estuaries and mangroves; explosives and poisons; and there are specific regulations on turtles, dugong, corals, aquarium fish, rock oysters, lobsters, mangrove crabs, trochus and rabbitfish.

Samoa
- Strong traditional system exists and the Village Fisheries Management Plan in the Fisheries Division under an Australian (AusAID) project has resulted in 54 fish reserves being established;
- One national marine reserve, the Palolo Deep Marine Reserve, was established in 1979;
- IUCN (The World Conservation Union), with Global Environment Facility support, and the Samoa Department of Environment and Conservation are establishing two new reserves in Aleipata and Safata, on Upolu Island;
- The Siosiomaga Society works at the village level on conservation issues;
- The SPREP-funded Pilot Village Level Coral Reef Monitoring project trained over 40 villagers in coral reef monitoring;

- IOI-Pacific Islands organised a Marine Awareness Workshop in 1997 to promote marine conservation in Samoa.

Solomon Islands
- Extensive traditional conservation practises still dominate nearshore marine and island resources management;
- The Arnavon Marine Conservation Area (Isabel and Choiseul Islands area of 1000ha) has been established with an important turtle nesting area, but needs increased management and training capacity at the village level;
- There are plans to establish other community-based marine conservation areas.

Tuvalu
- The traditional marine tenure system has broken down, but a conservation ethic remains;
- The first marine park was established at Funafuti Lagoon.

Vanuatu
- Extensive traditional conservation practices remain, utilising a network of customary marine tenure protected areas that are established jointly with the government;

VILLAGE LEVEL CORAL REEF MONITORING IN SAMOA

Most surveys and monitoring of marine resources are done by scientists and government officials in the Pacific to develop management policies, with only limited information reaching village users of coral reefs. There is also no coordinated database on coral reef monitoring in the Pacific. The GCRMN IOI-Pacific Islands node collaborated with the Fisheries Division, the Division of Environment and Conservation of the Government of Samoa and SPREP in a 1998 pilot study to train villagers to monitor their reefs. This complemented the successful Village Level Fisheries Management Project (VLFMP) by the Fisheries Division's Extension Unit with assistance from Australia (AusAID). Ten Fisheries and Environment staff were trained in the local language to monitor coral reefs, and they trained over 40 villagers who had set up fish reserves in 6 villages. They were trained to use the 'Manta Tow Technique, Under-Water Visual Census (UVC) and Line Intercept Transect (LIT)' with the focus on marine organisms that the villagers could recognise and can be used to show changes to the ecosystem over time. The aim today is to enable village chiefs and people to make informed management decisions to protect their fisheries. This 18 month project is continuing with Fisheries staff training more villagers and expanding efforts to raise awareness by distributing coral reef monitoring forms to the villagers. These data are being collated as an IOI-Pacific Islands contribution to GCRMN global reports. Projects of this type are an important step towards empowering coral reef users to take control of managing their environment through understanding how their actions can be transferred to other islands in the near term and passed on through generations.

> **VILLAGE LEVEL BY-LAWS IN SAMOA**
>
> Samoa is leading the Pacific Islands in the involvement of local stakeholders in coastal fisheries management by taking something 'old', and combining it with something 'new' to develop Village Level By-Laws. The successful Village Level Fisheries Management Project (VLFMP) in 1995 was implemented in coastal villages to use the traditional custom (fa`asamoa) to establish village 'fish reserves'. However this contradicted the 1960 Samoan Constitution, which stated that 'all lands lying below the high water mark belong to the State', which in effect gave outsiders the right to fish in other village 'fish reserves' and became the main impediment for villages to set aside their own traditional fishing ground for conservation. Therefore the Fisheries Division lead the establishment of Village Level Bye-laws which legally recognised village level conservation efforts, allowing chiefs to declare conservation sites, restrict fishing activities including fishing gear and impose penalties. All these By-Laws were drawn directly from the Fisheries Act (1988) and the Fisheries Regulations (1996), but the penalties are at the discretion of the Village Council. These may include traditional punishment such as provision of fine mats, pigs, or cash payments.

- The draft Environmental Resource Management Bill (1999) provides for the establishment of Community Conservation Areas;
- Many reefs were surveyed by scientists from the Australian Institute of Marine Science.

MONITORING AND GAPS IN REEF MONITORING AND MANAGEMENT

There is a range of coral reef monitoring capacity in the region: there is none in Nauru, Solomon Islands and Vanuatu; some in New Caledonia, Samoa and Tuvalu; while there is reasonable capacity in Fiji. Training in coral reef monitoring (GCRMN and Reef Check recommended methodologies) has been conducted in Samoa and Fiji, with the cooperation of the Fiji dive industry and the USP (University of the South Pacific) Marine Studies Programme. Countries have considerable interest in setting up monitoring programmes and improving coral reef management, but there is a lack of expertise and funding. Most importantly, there is a strong need to involve the principal resource owners in monitoring and management of their local reef areas. Awareness raising is a paramount necessity to raise their understanding of conservation and management ethics to strengthen current government activities.

Fiji

Reef Check monitoring in Fiji is coordinated by Ed Lovell, and USP has actively surveyed the Suva Barrier Reef since 1987, as well as Great Astrolabe Reef. Surveys have also been conducted to assist the establishment of MPAs (such as at Waisomo, Kadavu), and for environmental impact assessments. Extensive surveys are now monitoring the impact of the February-April 2000 mass coral bleaching event, with assistance from the tourism industry. Some aquarium trade companies have extensive data on reefs where they collect, but these data are not widely disseminated. The main problems in Fiji are a lack of coordination, no properly identified long-term monitoring sites, and an ineffective database.

Nauru
Surveys are limited to fisheries-related needs, although there are plans for a reef monitoring programme in the future.

New Caledonia
Coral Reef Observatory (1997) trains experienced divers to take samples at regular intervals at pre-defined sites (undisturbed, disturbed, mining, agricultural and managed). Local consultants are responsible for collection and analysis of data for Reef Check and GCRMN. IRD conducts long-term lagoon research programmes focussing on marine resources and impacts of land based sediments and anthropogenic activities. The Southern Province established a coral reef monitoring network with volunteer divers in 1997, and maintains 18 stations twice a year as part of Reef Check. The divers created an NGO in 1999 to receive public funds and expand activities to two other provinces. All MPAs in the Southern Province were surveyed in 1994 and 2000 (aerial photos and ground truthing) to assess reef status, with all reefs being in better shape today, except one damaged by COTS. Fish populations have shown a dramatic increase since the purchase and operation of patrol boats.

Samoa
An extensive monitoring programme by the Fisheries Division focuses on 54 fish reserves established under the Village Fisheries Management Plan project. A village level coral reef monitoring project has been coordinated by the IOI-Pacific Islands since 1998-1999 with the training of 42 villagers in 6 villages. A monitoring programme is planned for 2 new reserves in Aleipata and Safata by the IUCN in collaboration with the Division of Environment and Conservation. Despite these activities, improvements in monitoring capacity and expertise are needed and a national or regional monitoring database is required to house the data.

Solomon Islands
There is no coordinated monitoring programme, and a serious lack of monitoring capacity and expertise. Surveys are limited to fisheries-related issues.

Tuvalu
Monitoring by the Fisheries Department is only for fisheries related issues, and some monitoring of coastal erosion is carried out by the government, in collaboration with the South Pacific Applied Geoscience Commission (SOPAC). Responsibility for monitoring the new Funafuti Lagoon reserve is vested in the Funafuti Town Council, the Environment Unit, and the Fisheries Department. But Tuvalu faces great difficulties in periodic monitoring and conservation of marine resources, which has resulted in difficulties in evaluating the damage to marine communities by shipwrecks, spills, cyclones or COTS infestations.

Vanuatu

There is a strong belief that reef monitoring and management should be devolved to the communities as principal resource owners, as well as to NGOs and the private sector (such as tourist resorts). The Marine Biodiversity Strategic Action Programme was set up to monitor MPAs, but there is a need to reinforce monitoring capacity, and extend the results of pilot reef conservation projects to other areas.

LEGISLATION AND REGULATION

Only 4 fisheries departments have been established within Pacific Island governments since the 1960s. Colonial governments did not administer fisheries separately, irrespective of the importance in these islands. There were no licensing schemes, with only passive fisheries regulations e.g. minimum size limits and reservations on certain areas. These were enforceable with the minimum of effort and designed to prevent major damage to fisheries, not towards the maintenance of a sustainable yield.

There is no legislation in the Pacific that is dedicated to conserving coral reefs, although some governing fisheries resources may include coral reefs. These are usually covered by legislation under several government departments in some countries. Traditional marine tenure systems have been eroded in many countries, however the indigenous system is strong and still effective in others (e.g. Vanuatu and Fiji). Customary systems are being revived to complement state legislation for better management of marine resources in Samoa, Solomon Islands and Tuvalu. The following is pertinent legislation in the region:

Fiji
- State Lands Act 1946 - governs littoral zone, foreshore and submerged seafloor;
- Fisheries Act 1942 - prohibits destructive fishing and imposes minimum sizes on a number of reef species;
- Sustainable Development Bill 1999 (currently under review) provides codes for sustainable practice, national management plan, offences and penalties.

Nauru
- Fisheries and Marine Resources Authority Act 1997 - enables the management and sustainable utilisation of the fisheries and marine resources;
- Fisheries Act 1997 - provides for the management, development, protection and conservation of the fisheries and marine resources.

These two are inadequate to address conservation programmes, and a Marine Conservation Bill has been tabled in Parliament. A draft was written, but widespread consultation with stakeholders should be undertaken first.

New Caledonia

Territorial and Provincial regulations cover: Protected Areas; Biological Resources; and Noumea Agreements (improved recognition of customary laws and administrative structures). The enforcement of laws and regulations is complex because of overlapping powers of State, Territory and Provincial bodies, and this is especially acute in marine affairs where there is a lack of enforcement. The Southern province has two patrol boats, which are also used for management, and several Gendarmerie Precincts now have small boats, although this is insufficient for adequate enforcement.

Samoa
There are 10 pieces of legislation, involving 8 different government departments, with the most relevant being:

- The National Parks and Reserves Act 1974 - provides for the establishment of Marine Parks and Reserves;
- The Fisheries Act 1988 - a new bill is currently proposed to repeal this Act to govern issues relevant to marine conservation and monitoring, prohibition of fishing activities, research and establishment of Regulations and Bye-laws;
- Fisheries Regulations 1996 - regulate exploitation of certain marine species, fishing practices and fish aggregating devices;
- The Village Bye-laws 1999 - this new legislation allows villages to regulate, manage and enforce rules within their customary fishing areas using traditional customs e.g. a penalty of fine mats, pigs or boxes of tinned fish can be imposed on the offender.

Solomon Islands
- Fisheries Act 1998 - gives power to the Minister of Fisheries to manage, develop and conserve fisheries resources to ensure that the resources are not endangered by over-exploitation;
- Environment Act 1998 - makes provision for and establishes integrated systems of development control, environmental impact assessment and pollution control.

Tuvalu
- Fisheries Act 1978 - (revised in 1990) promotes the development of fisheries and ensures that fishery resources are exploited to the full for the benefit of Tuvalu;
- Marine Zones (Declaration) Act 1993 - refers to sovereign rights to explore, exploit, conserve and manage living and non-living resources within the area of its jurisdiction;
- Local Government Act - gives local governments the power to improve and control fishing and related industries and prohibit, restrict or regulate hunting, capture, killing or sale of fish;
- Local Government Bye-laws - several island councils have exercised their right under the Local Government Act by establishing by-laws that ban fishing practices, prohibit certain areas from fishing activities, license commercial fishers etc.

Vanuatu
Traditional custom ownership still prevails in many parts of Vanuatu.

- Fisheries Act 1982 - development and management of fisheries including provisions to prohibit the use of explosives, poisons and noxious substances for fishing;
- The Environmental and Resource Management Bill 1999 - provides for the establishment of a Coastal Management Committee to initiate and develop a coastal resource inventory at the local, island, provincial and national level.

Despite this array of legislation, there are numerous problems with compliance and enforcement. Foremost, is that fisheries legislation affecting people in remote areas, is often developed in urban centres, and a lack of community support can render such legislation ineffective. Legislation must be enforced to be effective, and the most important aspect of enforcement is education and awareness. Resource users need to be made familiar with legislation and why it is imposed before they will support the legislation, and apply peer and community pressure to deter those breaking the law.

Penalties and prosecution should be the last resort, however, legislation which is not enforced, either due to insufficient enforcement staff, or overly complex and impractical rules, will fail leading to more destruction of reef resources. Therefore it may be better to have less direct legislation that is cheaper and easier to police, than ideal laws. The imposition of traditional punishments must be considered as a deterrent to offenders, provided that these do not contravene constitutional laws or human rights provisions.

RECOMMENDATIONS

- Capacity building is a high priority for the region to set up coral reef monitoring in a series of long-term monitoring sites and this should be addressed at all levels, including stakeholders, NGOs, governments and the private sector;
- More attention needs to be focussed on the most highly stressed coral reefs areas in the region, particularly around urban and coastal areas where anthropogenic stresses are concentrated;
- There is an urgent need to establish community based marine protected areas throughout the region;
- The development of appropriate national coastal zone management plans and policies is required for all countries in the region;
- All countries should incorporate coral reef issues into national climate change strategies under the UN Framework Convention on Climate Change (UNFCCC);
- There is an urgent need to document the biodiversity of coral reefs in all countries, which will require training of marine taxonomists and strengthening of the marine Biodiversity Centre at the University of the South Pacific e.g. the biology of the important reef food fishes is poorly known, which impedes the introduction of sound sustainable fisheries management;
- Regional and national strategies for the preservation of intellectual property rights on marine biodiversity must be developed;
- Legislation and regulations for the management of coral reefs need urgent upgrading, especially the incorporation of integrated coastal management, and sustainable fisheries;

- Coral reefs should be mapped using advanced technology such as the use of Remote Sensing and Geographical Information Systems.

ACKNOWLEDGEMENTS

G. Robin South and Posa A. Skelton were lead authors for this regional report, and together they coordinate the GCRMN IOI-Pacific Islands Node from the Marine Studies Programme of the University of the South Pacific in Fiji. The following persons contributed to this regional status report: FIJI: Veikila Vuki, Milika Naqasima-Sobey, Ron Vave. NAURU: Peter Jacob. SAMOA: Lui Bell, Atonio Mulipola, Anne Trevor, Posa Skelton. SOLOMON ISLANDS: Cameron Hay, Reuben Sulu, Michelle Lam, Peter Ramohia. TUVALU: Samasoni Sauni. VANUATU: James Aston. Information for New Caledonia was drawn from Garbrie, C. (1999). The authors wish to thank the Conservation Action Fund of the New England Aquarium, Boston, USA, the International Ocean Institute Headquarters, Malta, the IOI-Pacific Islands Operational Centre, Fiji, the South Pacific Regional Environment Programme, Samoa and the Marine Studies Programme, the University of the South Pacific for financial and in-kind support.

SUPPORTING DOCUMENTATION

Fiji
Vuki, V., Naqasima, M., & Vave, R.D. 2000. Status of Fiji's Coral Reefs. Pp. 1-16.

Nauru
Jacob, P. The Status of Marine Resources and Coral Reefs of Nauru. Pp.1-10.

New Caledonia
Gabrie, C. (1999). State of Coral Reefs in French Overseas Departments and Territories. French Coral Reef Initiatives. Ministry of Spatial Planning and Environment: States Secretariat for Overseas Affairs. Government of France for New Caledonia.

Samoa
Skelton, P.A., Bell, L.J., Mulipola, A., & Trevor, A. 2000. The Status of the Coral Reefs and Marine Resources of Samoa. Pp. 1-35.

Solomon Islands
Sulu, R., Hay, C., Ramohia, P., & Lam, M. 2000. The Status of Solomon Islands' Coral Reefs. Pp. 1-58, 1 table.

Tuvalu
Sauni, S. 2000. The Status of the Coral Reefs of Tuvalu. Pp. 1-24.

Vanuatu
Aston, J. 2000. Status of the Coral Reefs of Vanuatu.

THE GCRMN SOUTHWEST PACIFIC NODE

Countries of the southwest Pacific are now being assisted by the International Ocean Institute Operational Centre for the Pacific Islands (IOI-Pacific Islands) as one of 4 GCRMN Nodes for the Pacific Islands. This was initially decided at the VIIth Pacific Science Association's Inter-Congress, in Fiji in July 1997, but membership has changed since to match campus locations of the University of the South Pacific and the realities of finding sustainable funding. The countries now included are: Fiji, Nauru, New Caledonia, Samoa, Solomon Islands, Tuvalu and Vanuatu, with all needing funding assistance, except New Caledonia. This node is coordinating both GCRMN and Reef Check level training and monitoring in all countries. These countries prepared their first National Status of Coral Reefs for a meeting in Noumea, May, 2000 and in future will input data into the Pacific Global Ocean Observing System (Pacific GOOS), coordinated by the South Pacific Applied Geoscience Commission (SOPAC). But there are large information gaps for many important topics in this report, which is setting the agenda for the node – to provide the basic information to develop projects to achieve conservation and sustainable use of our reefs.

Node regional contacts are:
Robin South (south_r@usp.ac.fj), Posa Skelton (skelton_p@student.usp.ac.fj), Ed Lovell (lovell@suva.is.com.fj)

IOI-Pacific Island Node National Contact Points include:
Fiji: Dr Robyn Cumming (robyn.cumming@usp.ac.fj), and Mr Ed Lovell (lovell@suva.is.fj)
Samoa: Mr Atonio Mulipola (fisheries@samoa.ws)
Nauru: Mr Peter Jacob (peterjacob_nfmra@hotmail.com)
Tuvalu: Mr Samasoni Sauni (sauni_s@usp.ac.fj)
Solomon Is: Mr Reuben John (Sulu rjsulu@hotmail.com)
Vanuatu: Mr Moses Amos (fishery@vanuatu.com.vu)
New Caledonia: Dr Richard Farman (drn@province-sud.nc)

11. Status of Southeast and Central Pacific Coral Reefs 'Polynesia Mana Node': Cook Islands, French Polynesia, Kiribati, Niue, Tokelau, Tonga, Wallis and Futuna

Bernard Salvat

ABSTRACT

The Cook Islands, French Polynesia, Kiribati, Niue, Tokelau, Tonga and Wallis and Futuna are 7 independent or autonomous countries or territories including 347 islands with coral reefs and an EEZ of more than 12 million km^2. The total population of this region is approximately 450,000 inhabitants throughout half of the islands, with 80% concentrated in the urban centres of 7 islands.

Coral reef habitats and biodiversity decline in scale and scope along a west to east gradient in the Pacific. Cyclones are rare and only impact on the reefs at decadal scales. Crown-of-thorns starfish have appeared with varying levels of intensity throughout this area, with most major outbreaks occurring in the 1970s in French Polynesia. Bleaching events with high mortality were also reported there in 1991, as well as some localities throughout the region during 1999 and 2000. Pollution, primarily from eutrophication and sedimentation is concentrated near urban centres. Although coral reef resources are important for subsistence and local commercial exploitation, documented information on reef stocks and exploitation of resources is limited, except in French Polynesia. Tourism (about 300,000 tourists per year) is still in the early stages of development in the region with the

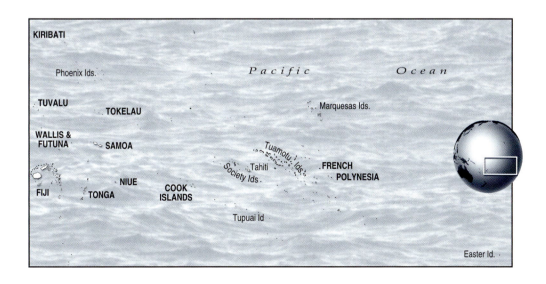

exception of a more intense industry in French Polynesia. Black pearl oyster culture plays a significant role in local economies in the Cooks and French Polynesia. Threatened reef species include giant clams, however there are plans for reintroduction and commercial exploitation. Regular coral reef monitoring programmes exist in French Polynesia, although programmes are now getting underway elsewhere. The level of supporting conservation laws and legislation varies throughout the region, yet enforcement is weak in most cases. The major needs for reef conservation and science include increased funding levels across all scales, broader and deeper knowledge bases, education materials, economic resource evaluations, on-site and networked monitoring programmes, as well as increased interest at various political levels about the real value of coral reefs to subsistence-dependent users.

INTRODUCTION

Polynesia Mana Node

This report covers the Polynesian countries of the Cook Islands, French Polynesia, Kiribati, Niue, Tokelau, Tonga and Wallis and Futuna. These countries form the 'Polynesia Mana Node'of the GCRMN (Global Coral Reef Monitoring Network) with co-ordination from the CRIOBE-EPHE Research Station on Moorea. 'Mana' is a Polynesian term meaning supernatural power and is used here to reflect a sense of authority and symbol for the Pacific peoples and their desire to conserve valuable reef resources. The Polynesia Mana Node was first initiated at the VIIth Pacific Science Association Inter-Congress in Fiji in July 1997. Membership has evolved to embrace cultural alliances and funding opportunities, rather than on biogeography arrangements. This node interlinks with the 'IOI-Pacific Islands' one covering the southwest Pacific countries of Fiji, Nauru, New Caledonia, Samoa, Solomon Islands, Tuvalu and Vanuatu.

National Coral Reef Status Reports

Most Polynesia Mana countries presented national coral reef status reports at the ICRI Regional Symposium on Coral Reefs in Noumea, New Caledonia during May 2000. This regional report is based on these national reports, and specific acknowledgements to the following authors: *Cook Islands* by Ben Ponia; *French Polynesia* by Bernard Salvat, Pat Hutchings, Annie Aubanel, Miri Tatarata and Claude Dauphin; *Kiribati* by Ed Lovell, Taratuu Kirata and Tooti Tekinaiti; *Tonga* by Ed Lovell and Asipeli Palaki; *Wallis and Futuna* by Paino Vanai. Reports from Kiribati and Tonga were produced with the help of IOI-Pacific Islands. Although no reports were prepared for *Niue* and *Tokelau*, information has been drawn from existing literature.

Although some data have not been updated since the early 1990s and major information gaps continue to drive an urgent need to build better capacity in reef resource assessment in some countries, a substantial process has started with the Polynesia Mana Node to rectify these gaps and provide better information in the near to long-term. For example, a seminar is planned in Moorea 2001 with authors of national reports and input by the South Pacific Regional Environment Programme (SPREP) to improve capacity to monitor and report on coral reef status and foster increased coordination and collaboration between reef scientists and users across these countries.

Reefs in the South-East and Central Pacific

Vast Oceanic Zone
The 7 countries of this node lie between 135° West (French Polynesia) to 170° East (Kiribati) from the Equator to 25° South. Collectively the area has a combined EEZ of 12 million km^2, 60,000km^2 of lagoons, and 344 islands with a total land area of less than 6,000km^2. Most of the islands in this region originated as volcanic hot spots on the ocean floor. They now move north-west at a mean speed of about 12cm per year along the Pacific tectonic plate. In this alignment, the oldest islands are in the north-west, and the most recent are in the south-east. As these volcanic islands subside and erode, low coral islands (atolls) often form on them. However, the Tongan island group did not originate from hot spots, but from uplift as the Australian and Pacific tectonic plates converged.

Island Geomorphology
There are a wide variety of island types in the region, including: 126 atolls (out of a total of 420 in the world); 8 uplifted atolls; and many other low coral islands and reefs without lagoons growing on platforms (e.g. in Tonga). Kiribati, Tokelau and Tuamotu are atoll archipelagos with no high islands, and Niue consists of one raised island atoll. Most of the atolls are closed and do not have passes, unlike most other atolls in Micronesia and Melanesia. All other islands in this region are high volcanic islands (2 in Cook Islands; 34 in French Polynesia; 3 in Wallis and Futuna; and many in Tonga). French Polynesia includes the large atolls of Rangiroa and those of the Tuamotu Archipelago, as well as Tahiti the second largest high volcanic island in the world.

Biodiversity
The countries of the Polynesia Mana node are at the eastern end of a Pacific-wide biodiversity gradient with genera and species decreasing from west to east. There are no natural mangroves in the Cook Islands or French Polynesia. Bays are rare on most islands so there are only a few estuaries with mixed fresh and seawater, further limiting species diversity. For example, many soft and gorgonian corals (Alcyonacia), and sponges play only a minor role on French Polynesian reefs, and crinoids (feather stars) are absent. The number of coral species varies between 115 on Kiribati and 192 on Tonga, as compared to over

	Land area sq.km	EEZ mil sq. km	Islands	Atolls	Raised atolls
KIRIBATI	823	3.5	33	33	0
WALLIS FUTUNA	215	0.3	3	0	0
TONGA	699	0.7	174	0	0
TOKELAU	11	0.23	3	3	0
NIUE	258	0.32	1	0	1
COOK ISLANDS	237	2.0	15	7	6
FRENCH POLYNESIA	3430	5.5	118	82	1
TOTAL	**5662**	**12.3**	**347**	**125**	**8**

Biogeographic overview of 'Polynesia Mana' node countries (EEZ: Exclusive Economic Zone)

	Population	Density	Inhabited Total islands	Capital	Urbanised Island	GNP in $ US
KIRIBATI	80,000	112	21/ 33	Tarawa	Gilbert - 72.000	858
WALLIS AND FUTUNA	15,000	55	3 / 3	Alofi	Wallis - 10.000	3,333
TONGA	105,000	130	44 / 174	Nuku Alofa	Tongatapu - 67.000	1,848
TOKELAU	1,700	168	3 / 3	Fakaofo	Fakaofo - 1.000	5,282
NIUE	2,500	10	1 / 1	Alofi	Niue - 2500	4,000
COOK	20,000	78	12/15	Avarua	Rarotonga - 10.000	4,737
FRENCH POLYNESIA	220,000	53	76 / 118	Papeete	Tahiti - 115.000	10,254
TOTAL	444,200		160/347		7 islands - 277.500	

Basic demographic and socioeconomic overview of Polynesia 'Mana' Node Countries (GNP: Gross National Product in $US per capita)

350 on the Great Barrier Reef. French Polynesian reefs also support an impressive 800 different fish species, about half the numbers found in Australia.

DEMOGRAPHIC AND CULTURAL PERSPECTIVES

Population
While the Polynesian Mana node region has 440,000 inhabitants, the density per country is highly varied - French Polynesia has the largest population of 220,000 people, while Tokelau has less than 2,000 inhabitants. However, in many countries there are large non-resident populations living in countries with historical ties, e.g. New Zealand and France. Population increases are between 0.5% (Tonga) and 2.3% (Kiribati), with life expectancy over 60 (and as high as 72) years. All countries have only one urban concentration as the capital, with more than half the total population living in these 7 towns. More than half the 347 islands are uninhabited, for example, only 37 of the 174 Tongan islands are inhabited. Although all peoples are originally Polynesian, there are many distinct Polynesian languages, but most can communicate using a common language base.

Political Status and Links
The present political alliances, international funding and predominant languages spoken reflect previous histories. In this region, only Kiribati, Niue, and Tonga are fully independent countries. Tokelau and Cook Islands are associated with New Zealand, while French Polynesia, and Wallis and Futuna have autonomous links as part of the French Overseas Territories. All countries are members of SPREP and the South Pacific Community (SPC). Some attend the Pacific Forum (Cook Islands, Kiribati, Tonga) and others have links to the European Union (French Polynesia, Kiribati, Tonga, and Wallis and Futuna).

Economies
The Gross National Product (GNP) varies considerably between countries with Kiribati and Tonga at the low end of the spectrum and the Cook Islands, Niue, Tokelau, Wallis and Futuna, and French Polynesia at the higher end. The differences in the economies reflect

limited industrial options beyond tourism and black pearl oyster cultivation, each of which are most prevalent in French Polynesia and the Cook Islands. The economic inputs include fishing and agriculture, and most phosphate mining has ceased over the past several decades. Therefore, all countries presently rely on foreign assistance, which can constitute up to 50% of GNP in some cases. In this region, expatriate workers also provide substantial contributions to the GNP (e.g. Niue, Tonga, Wallis and Futuna).

Traditional Reef Reliance

Many Pacific islanders today still rely on the reef for basic food security and limited cash resources, especially on low populated atolls. Therefore, most Polynesian peoples maintain essential relationships to coral reefs as they did in the past. Such linkages have fostered a strong and respected traditional knowledge base of reef-based resources.

STATUS OF THE REEFS

Over the past 10 years, coral reef surveys have provided percent coral cover data on different reef systems throughout the region, which is making it easier to discriminate between natural and anthropogenic impacts. Coral cover in lagoons is very variable due mainly to the high variability of the environment and conditions, therefore comparisons between countries are difficult to interpret. Comparisons using outer reef slopes are more reliable because they occur in generally similar environmental conditions with dominant ocean influences and fewer sea-based human impacts, with the exception of varying degrees of cyclone damage on reefs in this area. Outer slope coral cover on Kiribati and French Polynesia are typically about 50%, whereas cover on Wallis and Futuna is less than 20%. Monitoring programmes on the outer slopes of 14 islands in French Polynesia

ISLAND	1993	1994	1995	1996	1997	1998	1999	2000	Causes of variations
Marutea Sud	-	37	-	-	49	-	54	-	Increase after 1991 bleaching
Moorea Entre 2 Baies	16	25	-	-	29	33	-	43	Increase after 1991 bleaching
Tahiti Faaa	14	19	-	31	44	-	37	-	Increase after 1991 bleaching + new bleaching mortality 1998
Tetiaroa	47	45	-	-	38	-	31	-	Anthropogenic damage (?) bleaching mortality in 1998
Tikehau	-	40	-	-	-	4	-	5	Destructive cyclone VELI January 1998

Coral cover (percent) on the outer slopes of some islands of French Polynesia.

conducted since the early 1990s show declines in coral cover which are probably due to cyclones and/or bleaching events.

Coral Reef Knowledge
For large areas of this Polynesia Mana node, most scientific knowledge on reef status is relatively recent. The reefs of French Polynesia are the best studied because of many research programmes and institutions (CRIOBE-EPHE, Moorea, University of French Polynesia, and ORSTOM-IRD in Tahiti), and scientific publications continue to increase through the past three decades. There was previously a research station on Fanning Island, Kiribati directed by the University of Hawaii. There are small stations now on Tarawa and Tonga in association with the University of the South Pacific. There is limited information on past events such as cyclones, crown-of-thorns starfish outbreaks, bleaching, major pollution etc. throughout the region. Most of the knowledge continues to be based on surveys by outside scientists, teams or organisations because there is under-developed national capacity even though all members of Polynesian society rely on reefs for direct or indirect food and cash income streams.

Most islands in this area are remote without urban concentrations, so the majority of reefs are in good condition, although this impression is based on spotty information. Compared to other regions of the world, these reefs are amongst the least degraded and under minimal pressures from human and land influences (e.g. few rivers, low terrestrial sedimentation). However, there are increasing anthropogenic pressures on reefs and islands where populations are concentrated, and which will damage the reefs if no management action is undertaken. The major damage is caused by shoreline modification (retaining walls, dredging of channels, mining of coral and sand) which impacts directly on nearby fringing reefs. Irrespective of this localised damage to nearshore reefs, the outer reef slopes are generally healthy except for reduced fish populations due to fishing pressure.

'REEFS AT RISK'

This 1998 report from the World Resources Institute indicated that Pacific reefs were more healthy and under much lower levels of anthropogenic stress than all other regions. The wider Pacific region, (including those bordering the Pacific Ocean: Japan, East Australia, East Pacific coast) contains 60% of reefs considered to be 'low risk', as compared to other coral reefs areas, e.g. South East Asia which as about only about 20% of the reefs at low risk, and 40–50% low risk reefs throughout the Indian Ocean, Caribbean and Middle East. In the Polynesia Mana region, over 50% of the reefs are not threatened by risks in the near future with the exception of impacts from global climate change. The only two 'red' points indicating high estimates of threat to the coral reefs in this region are Tonga and Christmas Islands. The threat on Tonga is due to the urban impacts of Tongatapu, and a similar case could be made for Papeete on Tahiti. Threats for Christmas Island include past nuclear tests, but testing by the United Kingdom ceased in 1958 and by the United States in 1962. At present the only impacts to reefs on Christmas are from small tourist operations (1,712 tourists in 1995).

CORAL REEF FISHERIES

Fish Resources
Reef fish are important in the diets of people of Polynesia and throughout the Pacific, resulting in a strong respect and dependence on reef resources, especially in atoll environments. The annual coral reef fisheries for all of these countries is probably less than 25,000 tons with more than 75% being for subsistence use and less than 25% commercial, especially in Kiribati, Wallis and Futuna and French Polynesia. Most fish are captured by traditional methods (hook and lines, fixed traps near atoll passes and spearguns) by men and/or women according to traditions. The exchange of edible resources from one island to another within a country is unusual, but takes place between the Cook Islands, Kiribati, and French Polynesia. The only significant export industry is the 'live reef fish trade' for mainly grouper (*Serranidae*) and wrasses (*Labridae*), primarily from Kiribati for the Asian market. Fishing for tuna offshore is also an important resource for island peoples, both for the food and licensing royalties received from foreign vessels for the right to harvest within Polynesian country EEZs. For example, about 6,000 tons of tuna are taken annually in French Polynesia, with only 2,300 tons for domestic markets.

Invertebrate and Turtle Resources
These are important for people living on atolls e.g. on Tuamotu in French Polynesia (*Octopus, Tridacna, Turbo*), and on Tonga where 118 tons of molluscs are collected annually (*Anadara, Tridacna*). Many other edible items from the reef are consumed, including algae, worms, sea cucumbers and jellyfish. Choice varies with cultural preferences, but there are other available foods and potential exports not harvested. Sea cucumbers (*Holothuria scabra*) are exported from Wallis-Futuna now, also from Tonga during the 1990s, but export has been banned for 10 years. However, the trade in 'trepang' is not important in any of these countries. Lobsters (*Panulirus*) are over-exploited throughout and populations are now threatened. Although most countries have laws banning exploitation, black coral for export as jewellery has been over-harvested, especially in Kiribati and Tonga when it was once relatively abundant. Turtles (green and hawksbill) have been over-hunted and now are threatened species throughout the Pacific. Some countries try to conserve turtles by protecting nesting sites and banning or regulating exploitation, however enforcement is weak.

Aquarium Trade
This trade has developed recently in Kiribati (with 8 export licences in 1999) and the Cook Islands. The trade in Tonga in 1995-96 consisted of 56% corals, 27% fish (50% damselfish), and 17% invertebrates. A project was launched recently in French Polynesia to collect fish larvae entering the lagoon at night, and after they are sorted, fed and stabilised, they are exported to Europe. These form the broodstock for the aquarium market, with the main ecological advantage being that it is non-destructive of the environment.

Introduced species
Trochus niloticus and *Turbo marmoratus* were introduced to the countries in this region as early as the 1950s in French Polynesia and as recently as Tonga in the 1990s. Today all countries export several tons per year of trochus shell for buttons and paint, in addition to

trochus being an important food and income source for remote areas such as atolls. In Tonga during 1990-91, new stocks of giant clams (*Tridacna gigas* and *Hippopus hippopus*) were introduced to replace previously over-harvested giant clams and mariculture projects for the giant clam, *Tridacna deresa* was established during the 1980s. Four species of *Tridacnidae* occur throughout Kiribati and Tonga, but there is only one species in French Polynesia, *Tridacna maxima*. This species is abundant in some closed lagoons of the Tuamotu Archipelago and is eaten for food, but is not yet over-exploited or exported.

NATURAL DISTURBANCES AND IMPACTS

Volcanic Activity

Several new islands have formed during the last 20 years on the volcanic arc of the Pacific and Australian plates, but most of these have re-submerged. The only recent volcanic activity in the region occurred in Tonga. The last eruptions of Niuafo'ou were in the late 30s and early 40s, and Kao and Tofua are potentially active. The small island of Mehetia (east of Tahiti) in the Society archipelago is close to a hot spot and was probably active a few thousands years ago. While there are corals growing on the island, a fringing reef has not yet formed.

Cyclones

These occur throughout the region, but are more frequent in the west than the east to west and in the north rather than the south e.g. cyclones are relatively frequent in Kiribati, but rare in French Polynesia. The major recent cyclones took place during: 1982, 1995, 1997, 1999, and 2000 in Tonga; 1958, 1960, 1968 and 1979 in Niue; and in 1982-1983, 1991 and 1997 in French Polynesia. The impacts of cyclones are thought to be less damaging to coral reef areas where there is a high prevalence of storms due to adaptation, however it appears that cyclones do more long-term damage to reefs that experience cyclones infrequently. This was the case in French Polynesia which witnessed dramatic destruction of coral reef lagoons and outer slope areas from 6 cyclones that occurred during the El Niño years of 1982-1983

Crown-of-thorns Starfish Infestations

Acanthaster planci outbreaks and elevated numbers were reported in almost all of the Polynesian countries in the late 1970s and early 1980s. But since that time, populations appear to be stable with only sporadic outbreaks in some places e.g. Rarotonga and Aitutaki during 1998.

Coral Bleaching Events

These appear to have been more frequent and severe during the past 20 years, although there are few data for most of these countries before this time. Bleaching was reported for about 80% of the *Acropora* species during March 2000 in Rarotonga, Cook Islands. In Tonga and Fiji bleaching occurred in February, 2000, where up to 90% of the *Goniastrea* and *Platygyra* colonies bleached and coral death on the outer slope was reported to be about 2-5%. Only slight bleaching was reported in Kiribati in 2000. Considerable information on bleaching and mortality has been recorded in French Polynesia where there was major mortality in 1991 (20% of colonies died on the outer slopes of many islands), 1994 (major bleaching, but most colonies recovered), and 1999 (high bleaching and mortality, with great variability at the inter-and intra-island scale).

Global Climate Change

At present there is no definitive evidence that global change effects are impacting more than direct human stresses, however detailed studies are required to detect any future impacts of seawater warming and sea level rise on coral reefs. Politicians and scientists in these countries are aware of the potential impacts from climate scenarios. This was noted in all of the national reports, particularly in relation to the possibility of more frequent El Niño changes. In Kiribati, a Climate Change Working Committee has been launched. Models for Moorea, French Polynesia show that a 1cm a year increase over the next 30 years could destroy all natural shorelines, which would have to be replaced with modified protecting walls. Such construction could have negative impacts on the fringing reefs.

ANTHROPOGENIC THREATS TO CORAL REEFS

Human population densities on all islands are relatively low and all archipelagos are remote from continental influences. Therefore, most of the islands in this region are free of anthropogenic impacts, with only occasional local damage. There has been some degradation of coastal zones, lagoons and coral reefs due to human activities in a few areas near urban centres. Even though there may be some pollution and degradation in lagoons, the important outer slopes of the reef ecosystems remain healthy.

Over-Fishing

This is a major problem in most countries due to changes in governance from traditional and ancestral tenure-based fishing rights and practices to more legislation and policies that are typically common-property based reflecting laws in colonial countries. However, the need for increasing reliance on cash-based income and products (e.g. petrol) is also eroding traditional reef-management practices. For example, Polynesians traditionally collected fish and other food from the lagoons and reefs along their own village 'coastal zone' and collecting was forbidden in other areas. Now individuals and commercial operators can harvest resources from anywhere, and there is over-fishing due to the increased use of power boats, nets and other sophisticated gear. Destructive fishing methods (poisons and dynamite) have only been reported in Tonga, Kiribati (poison) and in Wallis and Futuna (poison and dynamite use). There are no other reports of dynamite fishing.

Coral, Sand and Mangrove Extraction for Construction

Extraction of material from the lagoon and reef areas (e.g. coral sand, rubble) to build houses (Kiribati) or for roads, walls, and shoreline construction is a major problem in all islands with significantly increasing populations. Coral and sand mining have been banned in Kiribati and French Polynesia, but activities continue. Where mangroves exist, they have been cut for firewood and construction, even though it is banned in Kiribati and Tonga.

Groynes and Rock Walls

Construction of walls in conjunction with roads, airports, marinas, wharves, markets and residences to diminish erosion is common everywhere. Often this encroaches on lagoons and results in the disappearance of beaches. This has been a particular problem in Wallis and Futuna, and less than 50% of the coastline remains in a natural unaltered state throughout the Society Islands, French Polynesia. Similar problems have been encountered on Rarotonga, Cook Islands.

Mariculture of Black Pearls

This 'industry' is developing rapidly in French Polynesia e.g. in the Tuamotu atolls there was an increase of 35% in one year to 8 tons in 1999. In the Cook Islands (e.g. Manihiki atoll) the general aquatic ecology of the atoll lagoons has probably been affected. Here the trophic structure of the lagoon food-web has been modified by the large stocks of *Pinctada margaritifera* growing in open waters through the release of large quantities of faecal pellets. This has resulted in eutrophication, algal blooms and significant mortality of pearl oysters, fish and invertebrates.

Sediment, Sewage and Oil Pollution

Poor land management and urbanisation on some of the high islands has resulted in destruction of some reefs and damage to others nearby. However, the impacts are usually localised and sediments are washed into oceanic waters. None of the islands, except for Bora Bora, French Polynesia, have major sewage treatment facilities and most wastes are flushed directly into reef lagoons. This practice is predominately a major problem on the urbanised islands where high densities result in human faecal contamination (Tarawa, Kiribati), as well as pollution from pig breeding areas in coastal margins (e.g. Wallis and Futuna and some islands in French Polynesia). There is evidence of eutrophication from sewage and agriculture fertiliser nutrients in some lagoon waters around Moorea, French Polynesia as well. A sewerage system is planned for Papeete and houses outside the towns are required to have septic tanks. In Aitutaki atoll a decrease in coral cover to 9% and an increase in turf algal of 14% has been reported and associated with localised eutrophication; however, the abundance of fish did not change. There is no oil pollution in the region, except localised damage near the major harbours.

Tourism Activities

These islands are on the outer edge of major tourism development, but the industry is developing in French Polynesia (200,000 tourists per year), Cook Islands (60,000), Tonga (25,000), Kiribati (at least 4,000), and in Niue (about 1,000). Major degradation to reefs has occurred during hotel construction, particularly when these are concentrated on some islands. After construction, there are often conflicts between the different users and interests over lagoon and reef resources, especially between fishermen and operators of tourist aquatic activities. These add to the acknowledged cultural impacts of having too many tourists on islands with low populations.

Nuclear Testing, Military Activities

There were military operations on some atolls during and immediately after World War II (Kiribati and French Polynesia). This also included nuclear weapons tests, but fortunately these no longer occur and the reefs are recovering from the damage.

CORAL REEF PROTECTED AREAS AND COASTAL ZONE MANAGEMENT

The main objective of marine protected areas for coral reefs in this region is to both maintain natural biodiversity as well as provide mechanisms and incentives to encourage sustainable development to maintain natural resources, ecological functioning, economic advancement and cultural heritage values (e.g. fishery recruitment areas, shoreline

STRESS TO REEFS	Kiribati	Wallis and Futuna	Tonga
Over-fishing	3	2	3
Destructive fishing	2	Removing reef rock	Poison, smashing reefs
Extractions	Wall building, now banned	3	2
Embarkment	1	2	Tongatapu
Pearl oyster cultivation	0	0	No
Sedimentation	1	3	1
Sewage	Tarawa	2	1
Eutrophication	2	2	Tongatapu, Nuku'alofa
Tourism	0	0	Tongatapu, Nuku'alofa
Nuclear tests	Christmas, Malden	0	0

Overview of anthrophogenic impacts on reefs in Polynesia Mana node countries

stabilisation areas, and ecotourism areas). While there are only a few coral reef conservation areas in most countries, traditional measures that are enforced by local communities still exist and are often effective at conserving resources. When formal legislation has been declared over coral reef areas, there is frequently a lack of respect for the laws and little enforcement. Community based management or involvement in decision making is a developing concept amongst governments for planning and management of protected areas, recognising that both bottom-up 'community' and top-down 'government' approaches are needed and can be compatible.

INTERNATIONAL CORAL REEF PROTECTED AREAS

Although the Polynesia Mana node encompasses an EEZ area of 12 million km^2 and about 350 islands only a few reefs are formally protected through international designation. For example, the Taiaro atoll in the Tuamotu Archipelago of French Polynesia is a Man and Biosphere Reserve of UNESCO and is currently being proposed that the protected status also include neighbouring atolls and improved zoning provisions. There are no other World Heritage or Ramsar sites that include coral reefs in this region, but similar conservation projects are planned in the Cook Islands, French Polynesia, and Kiribati.

NATIONAL CORAL REEF PROTECTED AREAS

At the national level, each country in this region has some coral reef protected areas, varying in definition and regulation. The Cook Islands has declared 5 coastal areas of Rarotonga (representing 15% of the coastal-marine area) as temporary non-harvesting zones for fishing. These are called 'Ra'ui', which recognises a long-standing traditional Polynesia method utilising rotating zones. These areas receive considerable community support and have encouraging fishery recovery results, but they are not formally gazetted. French Polynesia has two uninhabited atolls denoted as nature reserves (Scilly and Bellinghausen) and also one sand cay on a reef platform in the Marquesas (Motu One). Kiribati has several protected areas as wildlife seabird sanctuaries (e.g. Jarvis and Starbuck islands) but they do not legally address conservation of marine habitats. In Niue, many traditional conservation measures still exist in each village without any legislation, with only formal conservation areas being established in 1988. In Tonga, there have been legal mechanisms since 1946 to establish marine conservation areas. In addition, the Tongan islanders developed 'clam circles' for the protection and rational exploitation of *Tridacna* species. There are no coral reef reserves in Wallis and Futuna. In some places throughout Polynesia 'fishery management areas' (e.g. marine reserves for grouper management) are on the increase, but most are not formally designated as long-term marine conservation areas.

GOVERNMENT POLICIES, LAWS AND LEGISLATION

National policies
Most laws for conservation of the environment and pollution control in this region are recent, with the oldest legislation usually relating to fisheries management goals. However, traditional management regimes relevant to coral reef resources still exist in many countries, but do not in most cases have the backing of formal legislation. Even with such formal designations, government-lead enforcement and compliance is usually limited due to lack of funding, as well as needs for improved knowledge and capacity by government and community leaders alike. Most fishery policies relate to regulating fishing areas, determining fishing periods, allocating fishing quotas, and to species-specific characteristics limiting what can be collected. Most of these regulations are based on traditional knowledge and more recently on scientific investigation in some cases. The objectives are to achieve sustainable use of the marine resources with a particular emphasis on fish and invertebrates that are eaten for subsistence or have commercial values, e.g. giant clams, holothurians and turtles. In French Polynesia environmental impact assessments (EIAs) are required and a bill has recently been drafted for similar assessments in Tonga. The following summarises selected key laws relevant to coral reef conservation measures in this region:

Cook Islands
Marine Resources Act, 1989; Rarotonga Environment Act, 1995; Prevention of Marine Pollution Act, 1998; and an National Environment Bill in preparation;

French Polynesia
various Territorial Assembly decisions and laws; Planning Code, 1995; and Environment Impact Assessment regulation;

Kiribati
Fisheries Ordinance, 1957, prohibiting explosive fishing; National Environment Management Strategy, 1994; Environment Act, 1999; and a National Biodiversity Strategy Action Plan in preparation;

Tonga
Fisheries Act, 1989; Fisheries Regulation Act 1989; Bird and Fish Preservation Act; Parks and Reserves Act;

Wallis and Futuna
Traditional and territorial regulations.

MARINE MANAGEMENT AND STRATEGIC PLANNING

Complementary to, or as part of the marine protected and/or fishery management areas and legislation, there are various types of 'marine management plans' now being established for some areas of French Polynesia. These are typically based on key partners reaching consensus on lagoon activities e.g. numbers of fishermen, sharing of facilities, permitted fishing equipment, and hotel activities. These are essentially a bottom-up decision making process with co-management interventions through government. These marine area plans need to be coordinated with land management plans to realistically form integrated coastal zoning plans. All countries have or are in the process of developing National Environment Management Strategies (NEMS) which highlight involvement of local communities. These have usually been done with assistance from SPREP and other regional governmental and non-governmental assistance organisations.

REGIONAL CONVENTIONS

Several conventions have been developed by UNEP and SPREP that are relevant to coral reef conservation in this node. All countries have ratified the following conventions:
The Apia convention (1976) on conservation of nature in the South Pacific promotes protected areas to preserve examples of natural environments. It came into force in 1990;
The Noumea convention (1996) started in 1986 and came into force in 1990. It focuses on the protection of natural resources and the environment with protocols on dumping at sea and control of pollution emergencies.

INTERNATIONAL CONVENTIONS

- Most countries have ratified the following international conventions that have relevance to coral reef conservation, in particular countries associated with France (French Polynesia, Wallis and Futuna) and New Zealand (Cook Islands, Tokelau);
- Ramsar wetlands convention (1971) concerning the conservation of migrating bird populations. While there are no Ramsar sites in these countries, there is increasing interest in listing selected coral reef areas for wildlife goals;
- Paris convention (1972) or World Heritage of UNESCO; however there are as yet no sites covered by this convention in the region;
- Washington convention (1973) or CITES on endangered species of wild fauna and flora designated to control international trade in species listed in two annexes;
- Bonn convention (1979) on conservation of birds, but including also marine mammals and turtles;
- Rio convention (1992) or UNCED Convention on Biodiversity.

MONITORING

There is a variety of monitoring programmes throughout this Node. In the **Cook Islands** there is a programme to monitor water quality, coral cover and fish and invertebrate abundance in some lagoons. It started in 1996 and developed in 1998 to cover 9 of the 15 islands. In **French Polynesia** there are many monitoring programmes for water quality

(Tahiti and some Society islands), and scientific monitoring of lagoon reefs all around Tahiti, Moorea (e.g. the Tiahura monitoring programme launched in 1991), and on the outer slope of 14 islands throughout the country. This is a long term monitoring programme on the outer reef slopes to detect potential for future impacts, acknowledging that most of these reefs are now healthy, with only natural disturbances (cyclones, bleaching) and no anthropogenic impacts. In **Kiribati** there is no monitoring programme, except one to survey the abundance of the toxic alga, *Gambierdiscus*, which is responsible for ciguatera. In **Niue** and **Tokelau** there is no active monitoring. In **Tonga** there is a current monitoring programme in marine parks and reserves for physical parameters, nutrients, pesticides, metals, and faecal pollution. In **Wallis and Futuna** a monitoring programme started in 1999 focussing on the outer slopes of the 3 islands and the lagoon of Wallis, the main island.

The above survey and monitoring programmes do provide information of percent coral cover on healthy coral reefs, with most outer reef slopes having cover between 16% and 71%, which reflects mostly physical differences and not from anthropogenic impacts. Wallis and Futuna has the lowest coral cover. Tonga and French Polynesia are intermediate, and Kiribati has the highest percent cover. Coral cover in the lagoons is highly variable between locations such that comparisons are of no significance. Few of the results of the above monitoring programmes have been formally published in scientific journals, with French Polynesia being the notable exception. Most monitoring information is predominately in technical reports. Many monitoring endeavours started as surveys, which were occasionally re-surveyed, but did not constitute regular monitoring programmes of sufficient scale to establish degradation or recovery of reefs from year to year (again with the exception of French Polynesia).

PREDICTIONS ON THE FUTURE OF CORAL REEFS

Short, medium and longer term predictions on coral reef health and longevity will vary depending upon the time scales considered and possible synergies between natural and human factors that influence coral reefs within and across the region. In general terms, natural forces will involve either global warming (or cooling), changes in the frequency of El Niño and bleaching events, exceptional outbreaks of coral eating species (e.g. *Acanthaster*), natural diseases etc. The major human factor on the near to long-term horizon will be population increases, evolution of cultures, advances in technology, development on islands, control of pollution etc. Any forecasts must take account of all of these scenarios.

At the time scale of the next decade until the end of this century (3 to 4 human generations), there will definitely be increases in human populations (but not to the extent of Southeast Asia) and movements of Pacific island peoples in and out of the region. Cultural values of Pacific islanders will prevail but probably with more conflicts arising from development activities, western influences and the global economy. Tourism activities and the black pearl industry will continue to grow. Pollution control will be insufficient to reverse degradation around the high populated islands, but there will be gradual progress in the development of natural resource management.

The major unpredictable factors revolve around the rate and impacts of global climate change. With continued warming, it is predicted that there will be more cyclones and bleaching events than during the last 20 years (which already showed increases). Predicted sea level rises of 0.5 or 1cm (up to 2010), will be too small to have a major impact. Predicting outbreaks of species devastating reefs or natural diseases are not possible unless it is established that these are definitely triggered by human stresses and/or deriving from global climate changes. Therefore, significant predictions regarding the status of coral reefs will depend primarily upon the degree of management of human activities.

However, such predictions have to be made with considerable caution. No predictions 40 years ago could have included the growth in tourism or the explosion of the pearl industry, which are both major economic factors in the Cook Islands and French Polynesia. Technology advances are so rapid that predictions on how humans will exploit natural resources are near impossible, but it is certain that exploitation will increase in the short term. Counteracting the demise of coral reef ecosystems is a greater awareness of the need to conserve resources and health of the environment, which increases as economies expand and people become better educated; thus there is a case for both optimism and pessimism.

From a linked anthropogenic and natural context, global climate change is the most worrying concern. There is a strong probability of increases in both the frequency and intensity of bleaching events and cyclones with both having greater impacts on coral reefs. No corrective measures can be proposed other than to reduce the emissions of greenhouse gases that are causing warming (or reinforcing natural warming). Sea level rises of 25-95cm (mean of 50cm) until 2100, will benefit coral reefs, which grew are rates around 60cm per century since the last ice age (10,000 to 14,000 years ago). But the impacts on low-lying coral islands are likely to be severe in the short-term, with seawater invading the fresh water under the sands. In the longer term scenario (second half of this century) any predictions will be too imaginative to have real value, but will rely on whether we believe that humans will take the necessary actions to save themselves and the Earth they depend upon.

CONCLUSIONS AND RECOMMENDATIONS

All countries reported gaps and recommendations for improved management and conservation of their coral reefs and resources in the national reports. The following is a compilation of these findings from the most advanced (French Polynesia) to the least developed:

Political Will
A stronger political will is needed for the conservation of coral reefs and their resources, which is strongly dependent on the availability of information on the ecological, cultural and economic importance of these resources. Large proportions of the population exist in subsistence and rural economies and lack the political perspective and leverage of urban populations. The latter typically drive the agenda for western style economies and receive a disproportionate amount of government services. As an example, politicians are frequently more concerned about tuna exports and the rights of foreign fishermen, than for subsistence fisheries on remote islands. Most governments pay insufficient attention to environmental quality and links to different economic sectors e.g. healthy lagoons and reefs are a requirement for a flourishing tourist industry as well as long-term sustainable livelihoods of Pacific island people.

Funding

A lack of funds for coral reef monitoring, conservation, economic assessment, education and awareness underlies the need for greater understanding and management capacity to maintain healthy coral reefs. This situation is especially true in countries with a low GNP. Some governments are able to allocate funds to manage coral reef resources, whereas others must rely on project funds from international and bi-lateral donor governments and non-governmental organisations. All countries receive assistance through regional organisations, the principal one being SPREP, which organises the input of expertise, meetings, training and education in the region.

Capacity Building and Lack of Knowledge

Most countries lack adequate capacity for reef assessment, research and management. Few have research centres, or tertiary training facilities or even a university campus. While fisheries departments exist in each country, not all have an Environment department that can take a more integrated approach. Even when they do exist, these have little power to coordinate activities with other ministries, which may unknowingly accelerate environmental degradation. Lack of knowledge varies between countries with markedly different levels of resource assessment data and information on exploitation of resources. All countries requested assistance with building capacity to assess resources, particularly at the subsistence level.

Education Programmes

Materials and programmes for education about coral reefs are urgently needed, not only for the formal education sectors, but also for community leaders, stakeholders and decision makers interacting with coral reefs. SPREP has been highly active during the last decade in raising public awareness on coral reefs through radio, TV, posters, leaflets, books etc., but better outreach to remote areas in all countries is needed. Radio and video are ideal mediums for communication, however it is essential that material be presented in languages understood by different population groups.

Marine Protected Areas and Coastal Management Planning

MPAs, Integrated Coastal Management and various zoning initiatives are starting in the more developed countries, but there is still much to do. Environment Impact Assessments, even if mandated in some countries, are not enforced or followed. Most are performed after the event to account for the approval process rather from the design stages. All countries require urgent legislation for sustainable management of activities in the coastal-marine zones. At present there are too few MPAs and poor use of international and regional conventions to effectively conserve the vast and predominately healthy coral reef resources of this Pacific region.

Legislation and Policy

There is a major need to improve legislation and supportive policy contexts to increase the allocation of coral reef protected areas and provide effective management. There is a need for regional cooperation to implement protection under the different categories of national, regional and international conventions and classifications. Few decision-makers and governments understand that protected areas are valuable tools for sustainable development. All countries request assistance in preparing adequate legislation, and especially in the provision of enforcement mechanisms for existing legislation to protect fisheries within protected areas.

Traditional Use and Community Rights
During the 1990s, there have been considerable advances in the recognition in the importance of using traditional forms of management to conserve coral reefs and their resources. The rights over exploitation of resources by communities inhabiting the coastline have been lost in many cases in connection with changes to more open access policies. After the recommendation in Chapter 26 Agenda 21 adopted at the UN Conference on Environment and Development (Rio 1992), Pacific countries developed the South Pacific Biodiversity Conservation Programme (SPBCP) through SPREP to establish community based conservation areas. Using funds from the Global Environment Facility (GEF), they succeeded in launching 17 such areas in 12 Pacific countries, with half having marine components. Governments are now more aware of the situations and problems and some are modifying procedure to empower local communities to manage resources. Establishing a consensus between all stakeholders over coral reefs activities (from fisherman to tourists) is the best way to find solutions. This requires that decision-makers agree to adopt a bottom-up process, which in effect re-establishes traditional marine tenure in all areas instead of open access regimes. At present this is most easily done in lagoon contexts. Tourism provides both opportunities and challenges due to potential investment and returns for different political parties, stakeholders, family and village groups. NGOs play a major role in the decision-making process for coral reef and overall longer-term environmental issues in many countries.

Regional Cooperation and Coordination
In combination with the revitalisation of community based management noted above, there is a complementary need to strengthen regional cooperation and coordination. All these countries have a common language base and culture, which is virtually 'symbiotic' with the sea and coral reefs. Many communities depend on reefs for subsistence, and now the developing industries are tourism and black pearl culture, both of which require healthy coral reef ecosystems. Therefore governments must pay more attention to their respective and highly interlinked natural heritage and cultural resources. The need now is for common approaches and cooperation throughout the Polynesian Mana region to develop their own futures based on their common resources, of which coral reefs are a major component.

ACKNOWLEDGEMENTS AND SUPPORTING DOCUMENTS

Bernard Salvat is the primary compiler of this regional report. He is from both CRIOBE-EPHE marine station on Moorea, French Polynesia and EPHE-CNRS, Université de Perpignan, Avenue de Villeuneuve, 66860, Perpignan Cedex, France, (bsalvat@uni-perp.fr). As noted in the Introduction, the regional report is based on work and information from the following individuals and respective national reports:

Cook Islands by Ben Ponia; Ministry of Marine Resources, PO Box 85, Avarua, Rarotonga, Cook Islands (benponia@hotmail.com);

French Polynesia by Bernard Salvat, Pat Hutchings, Annie Aubanel, Miri Tatarata and Claude Dauphin; contact: Annie Aubanel, Service de l'Urbanisme, B.P. 866, Papeete, Tahiti, French Polynesia, (Annie.Aubanel@services.gov.pf);

Kiribati by Ed Lovell, Taratuu Kirata and Tooti Tekinaiti; contact: Tooti Tekinaiti, Ministry of Natural Resources, Development, Fisheries Division, P.O. Box 276, Bikenibeu, Tarawa, Republic of Kiribati, fax 00.686.28295.

Tonga by Ed Lovell and Asipeli Palaki; contact: Asepeli Palaki, Environmental Planning and Conservation Section, Department of the Environment, Kingdom of Tonga;

Wallis and Futuna by Paino Vanai. Wallis and Futuna by Paino Vanai; contact: Paino Vanai, Service de l'Environnement, Wallis, Wallis et Futuna, (senv@wallis.co.nc)

Kiribati and Tonga reports were produced with the help of IOI-Pacific Islands by Edward R. Lovell, International Ocean Institute, lovell@suva.is.com.fj.

12. Status of Coral Reefs of American Samoa and Micronesia: US-affiliated and Freely Associated Islands of the Pacific

C.E. Birkeland, P. Craig, G. Davis, A. Edward, Y. Golbuu, J. Higgins, J. Gutierrez, N. Idechong, J. Maragos, K. Miller, G. Paulay, R. Richmond, A. Tafileichig and D. Turgeon

Introduction

The US Affiliated and Freely Associated islands of the tropical Pacific include American Samoa and the islands of Micronesia (excluding Kiribati). Most lie north of the equator, except American Samoa, which is considered part of Polynesia, but has the high coral diversity and cultural dependence on coral reefs similar to Micronesia. Micronesia is made up of a group of small tropical islands and atolls in the central and Indo-west pacific, and encompasses an area of approximately 11.6 million km^2, larger than continental United States. The vast majority of this area is ocean, with a land mass just over 3,000km^2. From east to west, the US Affiliated and 'Freely Associated' islands include American Samoa, the Republic of the Marshall Islands, the Federated States of Micronesia (FSM), the Commonwealth of the Northern Mariana Islands (CNMI), Guam (an unincorporated Territory) and the Republic of Belau (Palau). As a region, Micronesia possesses a high diversity of corals and associated organisms and the human population is heavily dependent on coral reefs and related resources both economically and culturally. The coral reefs of American Samoa and Micronesia range in condition from nearly pristine to seriously damaged by anthropogenic disturbance. The human impacts include over-fishing, ship groundings, sedimentation and coastal pollution. In the past, human impacts were largely related to the size of the resident populations, however fishing fleets from other nations have taken their toll on even the most remote islands and atolls. Although nuclear testing in the region stopped in the 1950s, some islands are still used for military exercises and testing of non-nuclear missiles. Micronesia is a highly rated scuba diving destination, and rapid tourism related development, including new roads, hotels, golf courses and personal watercraft, is having a substantial effect on coastal reefs on some islands.

American Samoa

The Territory of American Samoa is a group of 5 volcanic islands and 2 atolls in the central South Pacific Ocean. The islands are small, ranging from Tutuila (142km^2) to the uninhabited and remote Rose Atoll (4km^2). The total reef area is 296km^2, which consist mostly of narrow fringing reefs (85%) growing up against the steep slopes of the main islands, a few offshore banks (12%) and two atolls (3%). The fringing reefs have narrow reef flats (50-500m) and with depths of 1000m within 2-8km of the shore. The coral reefs have a diverse assemblage of 890 fishes, 200+ corals, and 80 algal species and provide an important source of subsistence food and minor income. The reefs also provide shoreline protection, a focus for tourism and small-scale collection for the aquarium trade.

Northern Marianas

The Commonwealth of the Northern Mariana Islands (CNMI) is a chain of 15 islands between 14°-20°5'N, 145°-146°E, divided into two sections, with large variations in coral reef resources between the south and the north. The southern islands (Saipan the capital, Tinian, Agijuan, Rota and Farallon de Medinilla FDM), are mostly raised limestone with well developed fringing coral reefs, whereas the largely uninhabited northern islands (Anantahan, Sariguan, Guguan, Alamagan, Pagan, Agrihan, Ascuncion, Maug, Uracas, and Farallon de Pajaros) are primarily volcanic, including some active volcanos, and much less reef development. The southern islands have gradual sloping coastlines with barrier reefs and well-developed fringing reefs on the western coasts.

FSM

The Federated States of Micronesia (FSM) consists of 4 states: Kosrae, Pohnpei, Chuuk and Yap (all formerly known as the Caroline Islands along with Palau). Each group has its own language, customs, local government, and reef tenure systems, with the traditional Chiefs and community groups being active in governance, in parallel with a western-style democratic government. This dual system provides both opportunities and challenges to reef and marine resource protection. There is a mix of both high islands and atolls, and there is a strong economic and cultural dependence on coral reefs resources. Kosrae (a single island of 109km2; elevation 629m) is surrounded by a fringing reef and has one harbour. Pohnpei is the capital and largest island (345km^2) with a well-developed barrier reef and associated lagoon, and includes 8 nearby smaller island and atolls, to form the State of Pohnpei. Chuuk State (formerly known as Truk) has 15 inhabited islands and atolls, and is famous for the Japanese wrecks sunk in the lagoon during World War II. Yap State has a main island (100km^2) and 15 other islands and atolls, and the people retain most of their traditions, including a highly sophisticated marine tenure and associated marine resource management system.

Guam

This is a US territory and the most southern and largest of the Mariana islands (560km^2; maximum elevation 405m; 13°28' N, 144°45' E). While the island population is 150,000, more than 1 million tourists visit annually. Guam lies close to the centres of coral reef biodiversity, with approximately 270 species of hard corals, 220 species of benthic marine algae, over 1,400 species of molluscs and 1,000 species of reef fishes. Sea surface temperatures are ideal for coral reefs (27-30°C) with higher temperatures over the reef flats and in the lagoons. The northern half of the island is relatively flat and consists of uplifted limestone, whereas the south is primarily volcanic, and more rugged with large areas of erosion prone lateritic soils. The island has fringing, patch, submerged, and a barrier reef, along with offshore banks. The fringing reef flats vary from 10m wide on the windward side to well over 100m elsewhere. The combined areas of reef and lagoon are approximately 60km^2.

Guam is in the middle of the tropical Pacific 'typhoon belt' and experiences about one substantial tropical typhoon each year, therefore the coral reefs play a major role in protecting the land from storm waves. Tourism, primarily from Japan and other Asian countries, is the largest industry. Thus the coral reefs and associated marine recreation are of substantial economic value, contributing over US$70 million per year, specifically associated with scuba diving, snorkelling, submarine tours and beach-related activities.

Coral reef fisheries, including both finfishes and invertebrates, are economically and culturally important. Reef fish have been historically important in the diet of the population, however, 'westernisation' and declining stocks have reduced the role of reef fish whereas, many residents from the other islands in Micronesia continue to include reef fish as a staple part of their diet. Sea cucumbers, a variety of crustaceans, molluscs and marine algae are also eaten locally. In addition to the cash and subsistence value of edible fish and invertebrates, reef-related fisheries are culturally important, as family and group fishing is a common activity in Guam's coastal waters.

Palau

The Republic of Palau is the most western archipelago in Oceania lying 741km east of Mindanao in the southern Philippines and 1,300km southwest of Guam. The islands stretch 700km from Ngaruangel atoll in the north to Helen atoll in the south, with about 20 large and intermediate sized islands and over 500 small islands. The largest island Babeldaob is volcanic, and to the south is Koror and all the southern islands in the chain. Koror is separated from Babeldaob by a deep pass (30-40m), Toachel El Mid, which separates the reefs of Babeldaob from the southern reefs. About 339-599km southwest of the main archipelago lies the Southwest islands. There are a variety of island and reef types; volcanic, Atolls, raised limestone, and low coral islands. There is a well developed barrier reef on the west coast (144km long) which encloses the main cluster of islands from north of Babeldaob to the southern lagoon, merging into the fringing reef with Peleliu, and includes the famous Rock Islands. The barrier reef is not well developed on the east coast of Babeldaob e.g. only Ngchesar and Airai have a protective barrier reef, while Ngerchelong, Ngarard, Melekeok does not.

STATUS OF CORAL REEF BENTHOS AND FISHES

American Samoa

The coral reefs are currently recovering from a series of natural disturbances over the past two decades: a crown-of-thorns starfish invasion (1979); three hurricanes (1986, 1990, 1991); a period of warm water temperatures that caused mass coral bleaching (1994); and chronic human-induced impacts in areas like Pago Pago Harbour. By 1995, coral recovery was evident with an abundance of new recruits, and growth has continued but a full recovery will take more time. The removal of 9 shipwrecks in Pago Pago Harbor, and another at Rose Atoll has assisted recovery. The export of 'live rock' was banned in June 2000. Algal growth is limited and this indicates that nutrient levels are relatively low and grazing herbivore populations are healthy. Encrusting coralline algal cover (*Porolithon*) is high (40-50%), which helps to cement and stabilise the loose surface below.

Water quality is generally good due to the absence of lagoons and efficient flushing by ocean currents, with the exceptions being: 1) sedimentation into coastal waters after heavy rains due to poor land-use practices; 2) nutrient enrichment from human and animal wastes in populated areas; and 3) Pago Pago Harbour, which suffers from two kinds of pollution. Fish and substrates in the harbour are contaminated with heavy metals and other pollutants, and there were increased nutrient loadings from cannery wastes in the inner harbour, resulting in perpetual algal blooms and occasional fish kills due to oxygen depletion. However, nutrient levels were greatly reduced in the early 1990s when the canneries were required to dispose of their wastes beyond the inner harbour.

Northern Marianas
There have been some systematic studies of the coral reefs of Saipan, Tinian and Rota, but only occasional environmental impact assessments of the other islands. These provide useful 'snap-shots' of particular areas, but do not provide long-term monitoring data. The reefs near the southern populated islands of Saipan, Tinian and Rota are of greatest concern as they receive the bulk of anthropogenic impacts from development, population growth, fishing and tourism. The western side of Saipan has a well-developed lagoon with seagrass beds and several pinnacle and patch reefs that are unique to the CNMI. Extensive fringing and apron reefs occur on the northern and eastern sides of the island. Tinian has narrow fringing and apron reefs around most of the western side of the island and at Unai Dankulo on the eastern shore. Small patch reefs occur near Tinian Harbour off Tachogna and Kammer beaches. Well-developed fringing reefs occur along most of the north-west coast of Rota, with narrower fringing reefs found in Sasanhaya Bay and the southern coast. In addition to these, there are a number of submerged sea mounts and shoals in the waters surrounding the CNMI which probably have growing reefs. The coral reef area is estimated at 579km^2 including the offshore banks and shoals.

There are 254 hard coral species, with higher diversity in the Southern Islands where reefs are older and more developed. Coral cover in the inner reef zone and most reef flats on Saipan tends to be patchy with lower species abundance and distribution than on the reef fronts and terraces, probably because these areas are stressed by high variations in temperature and salinity. The detailed surveys on Maug in the northern islands show lower coral diversity, with only 74 species. The southern islands had a crown-of-thorns starfish (*Acanthaster planci*) outbreak in the late 1960s, at the same time as there were damaging outbreaks on Guam. There were smaller starfish outbreaks in the mid 1980s on Saipan, however, the starfish have not been a major problem recently, and most of the reefs appear to be recovering.

FSM
Reef status throughout the FSM is generally good to excellent. Most reefs on the low islands are in excellent condition, with the primary human impacts from fishing and ship groundings. There are no problems in the atolls from sedimentation and erosion or coastal pollution. Reefs in Kosrae have been damaged by coastal development, specifically the construction of the airport on top of a reef. Some dredging and road construction projects have also damaged specific reef areas. The reefs around the Pohnpei vary in condition e.g. following a ship grounding near the Sokehs channel, coral cover was about 20%, and estimates on the barrier reef are of 50% to 70% cover at selected sites. Due to the large annual rainfall, and steep volcanic topography, erosion and sedimentation rates are high, and upland clearing of forests to grow sakau (cava) has resulted in landslides and impacts downstream. The high coral cover in Chuuk indicates generally good water quality, however there is over-fishing by foreign commercial operators, including destructive fishing practices, which are causing localised damage. Surveys of 18 sites around Yap in 1995 and 1997 showed coral cover was similar (28.8% vs 28.7%) even though a typhoon hit the island between the surveys.

Guam
The condition of the reefs is variable, ranging from excellent to poor, depending on adjacent land-use, accessibility, location of ocean outfalls and river discharges, recreational

pressure and water circulation patterns. The northern reefs are generally in better condition because there is limited erosion and sedimentation from the limestone land mass (no surface rivers or streams), but there is some aquifer discharge and associated eutrophication damage to the reefs. Coral cover and diversity are higher on the northeastern (windward) exposures, with a variety of *Acropora* species dominating the reef crest and slope. Reefs on the eastern, central and southern parts are heavily influenced by clay sediments and freshwater runoff during the rainy season (June to November) with these sediments often accumulating on the reef flats and portions of the reef slope. During the early 1990s, a road project in the south resulted in particularly heavy sedimentation on the fringing reefs, with up to 100% mortality of corals. Most of the fringing reefs off the south and southwest shores are in poor to fair condition.

Apra Harbour in the central-west is home to a US Navy Base and the commercial port for the island. The fringing and patch reefs near the harbour mouth are in relatively good condition, whereas those within the harbour have been affected by freshwater runoff, sediment and thermal discharges from the main power generation facilities. Agana Bay and Tumon Bay, north of Apra Harbour, are centres of tourism and recreation activities, with major jet ski operations in Agana Bay. Corals in the inner areas of both bays are in relatively poor condition, influenced by discharges from land as well as the impacts of recreational activities. Improvements in sewerage lines, ocean outfalls and control of runoff are expected to improve water quality in these bays, and some subsequent experimental restoration activities are planned. Coral cover on the good reefs ranges from 35-70%, while the most damaged sites have less than 10% cover, with fleshy algae and sediment dominating. Data compiled in the 1960s generally showed reefs with over 50% live coral cover, but only 7 of the same 113 transects measured in the 1980s - 1990s had over 50% live coral cover, while 88 had less than 25% live coral. Near the Northern District sewer outfall in the early 1990s, coral cover was below 25%, moreover there were very few species, particularly *Porites rus*, indicating a reduction in coral species diversity. Recruitment data also support the observations of an overall decline in coral reef condition. In 1979, 278 coral recruits were found from 525 fouling plates placed on reefs around Guam (0.53 corals/plate), while in 1989 and 1992 there were only 0.004 and 0.009 recruits per plate respectively.

Recent increases in blue-green algae are resulting in overgrowth of corals at some sites. While the crown-of-thorns starfish (*Acanthaster plancii*) occurred in small to moderate numbers over the past few years, many juveniles now are causing concern about the potential for a future outbreak. Coral diseases, the competitive sponge *Terpios*, and coralline algal lethal orange disease (CLOD) have all been observed on Guam's reefs, but none are causing significant damage. Sporadic episodes of coral bleaching have occurred at a number of sites including within Apra Harbor and Piti Bay just to the north, and both hard and soft corals were affected, with a few pockets of high mortality, however, the 1998 mass bleaching did not cause widespread coral mortality in Guam.

Palau
Palau has the most diverse coral fauna in Micronesia; comparable to the highest coral diversity areas of the Philippines, Indonesia and Australia, with 425 coral species. The reefs are generally in good condition with coral cover ranging from 50-70% and sites containing

45-95 coral species per site. Coral cover on the lagoon slopes of the barrier reef is around 60% with about 45 species each time. The Kayangel atoll lagoon has lower coral cover and diversity, whereas the western rim of the channel (Ulach) has healthy coral communities, with 70% coral cover on the outer walls and 50% on the inside in 1992. Coral cover on the ocean slopes averaged 20-25% with the number of coral species varying from 40-50. Kayangel atoll has seagrass beds and at least 126 coral species in the lagoon.

The ocean reef slopes are 62.7km long with cover on the northeast slopes averaging 10% and 35 coral species. The protected area the southern end Ngkesol has a higher coral cover (25%; 45 species) and coral cover on the western facing ocean slopes is from 60-70% with 35 species. Northwest facing reefs have lower coral cover (10-20%) but more species (50). These northern slopes are protected by Ngerael, Ngkesol and Kayangel and have higher coral abundance and diversity, and the western Babeldaob lagoon has over 500 patch reefs with around 50% coral cover, including several rare corals (*Cynarina*, *Zoopilus*, and *Siderastrea*). During 1992, 200 coral species, 170 invertebrates and 277 fish species were found in Ngermeduu Bay which has both estuarine and reef habitats. The channel Toachel Mlengui, north of Ngermeduu Bay entrance, is the most important pass to the open sea with the next pass Sengelokl 85km to the south. There were no changes in coral cover between 1976 and 1991 at a fringing reef at the southern tip of Malakal: reef margin 60.3% compared to 55.7%; reef slope 73.6% compared 82.2%.

STATUS OF CORAL REEF FISHERIES

American Samoa
Despite the on-going recovery of corals, the fish and invertebrate populations are not recovering as well. Giant clams and parrotfish are over-fished, there is heavy fishing pressure on surgeonfish, and fewer or smaller groupers, snappers and jacks are the only remaining large fish. Most village fishermen and elders report that numbers of fish and shellfish have declined, and the fish are now toxic with heavy metals in some areas, particularly Pago Pago Harbour. Hawksbill turtle populations are in serious decline for two major reasons: illegal harvest and loss of nesting habitat. These are listed as 'Endangered' and rapidly approaching extinction in the Pacific.

Only limited information is available for fish and invertebrate catches from American Samoa, but declining subsistence catches were monitored between 1979 and 1991-1995 (there are no data since). The artisanal catch was monitored in 1994, but is only assessed now via market invoices; however, and the data are unreliable because of poor compliance by vendors.

Northern Marianas
The Fisheries Section of the Division of Fish & Wildlife has been collecting data on fish diversity and abundance within the existing and proposed conservation areas on Saipan, Tinian and Rota. The Tinian leadership proposed in 1996 that a marine conservation area be created to cover one-third of the western shoreline of Tinian with harvesting prohibited. No decisions have been made, but baseline surveys are particularly useful. *Trochus* was introduced into the Marianas (Guam, Rota, Tinian, Saipan and Agrihan) in 1938 by the Japanese for commercial harvest (for buttons, paint, and meat). In 1982, the Division of Fish

& Wildlife imposed a moratorium on *Trochus* harvesting after concluding that densities were too low for commercial exploitation. This was reversed in 1996, but two conservation areas were established to house breeding populations (Lighthouse Marine Protected Area, and Tank Beach Conservation Area). Similar evaluations of 9 species of edible sea cucumbers within the Saipan Lagoon in 1996 showed that populations were too low to support a commercial fishery, but a recreational or subsistence fishery may be sustainable. Commercial landings of sea cucumbers on Saipan showed a declined in 1996 surveys.

FSM
The greatest impact on the fisheries has been the commercial export of resources. Yap and Kosrae have limited export largely to personal or family use, such as shipping coolers of fish to relatives in Guam or the CNMI. Chuuk had the largest commercial export in the past, and there was some commercial export of fish and crab from Pohnpei, however, a recent outbreak of cholera shut down the export of fish from Pohnpei, and also affected exports from the other islands. A small sea cucumber fishery operating in Yap was closed in the mid-1990s. Catch and export data are limited, and quantitative assessments of fisheries resources within the FSM are needed.

Guam
Fish populations have declined during the past 15 years, according to statistics collected by the Guam Division of Aquatic and Wildlife Resources. Total finfish harvest (day and night catch) for 1985 was estimated at 151,700kg while in 1999, this dropped to 62,689kg. Surgeonfishes, which are important reef herbivores, accounted for approximately 23% of the total catch in 1999. Catch per unit effort has dropped by more than 50% since 1985, and large reef fish are rare. Seasonal runs of juvenile rabbitfish, a local favourite, were considered poor for the 1997-1999 seasons, but numbers increased in the 2000 season. Regulations and protected reserves are being enforced to deal with this problem. Scuba spearfishing is still allowed on Guam, which is contrary to the conservation ethic. Legislation is now proposed to ban this practice.

Palau
The largest number of reef fish species in Micronesia occur in Palau with 1,278 known species, and the number could potentially rise to 1,449 with more searches. Ulach Channel in Kayangel atoll has high fish abundance and diversity, with high populations on the ocean slopes. Export fisheries are now strictly controlled and most product is sold in the major markets of Koror for local consumption. Reef fish harvests over the last 10 years have been 1,800mt per year. Most fishermen report that fish catches have dropped recently and catch per unit effort has decreased considerably. Some fish spawning aggregations have been seriously depleted and two have been totally lost.

ANTHROPOGENIC THREATS TO CORAL REEF BIODIVERSITY

American Samoa
Human-related impacts are likely to increase due to rapid population growth rate. During the past 10 years, the population increased by 16,300 people; an explosive increase of 35%. The population estimate of 63,000 in 2000 is increasing at 2.5% per year with about

1,600 people added to the population each year. This is unlikely to stop given the high birth rate (4.0 children per female), a high proportion young people (50% of the population is younger than 20), and the high immigration rate. A recent 'Governor's Task Force on Population Growth' recommended a population ceiling of 115,000 people based on limiting resources (particularly drinking water) and diminishing 'quality of life' factors. An 'Action Plan' identified steps to reduce birth and immigration rates.

Recently the American Samoans identified and ranked the following threats to coral reefs:

- HIGH: Over-fishing of reef resources; coastal development and habitat destruction; and oil and hazardous waste spills in Pago Pago harbour.
- MEDIUM: Sedimentation; dumping/improper waste disposal; nutrient loading/eutrophication in Pago Harbour.
- LOW: Nutrient loading/eutrophication other than Pago Harbour; oil and hazardous waste spills other than Pago Harbour; ship groundings; anchor damage; destructive fishing habits; marine debris from marine sources; alien species (e.g. from ballast water); crown-of-thorns starfish predation; coral diseases; collections for aquarium markets; bio-prospecting/natural products.

Northern Marianas
Those coral reefs in proximity to the large population centres in the CNMI receive the greatest impacts from human activities e.g. nearshore reefs of Saipan, Tinian, and Rota, the reefs surrounding Farallon de Medinilla and offshore fishing banks. The primary impacts come from development and land use activities, tourism and recreation, and fishing. In the early 1980s, the economy underwent dramatic growth, with heavy Japan investment in tourism development. The population in the islands increased from 17,000 in the 1980s to 58,900 in 1995 including many foreign workers, and these increases are placing strains on the natural resources. The major concerns are increased sedimentation and nutrient loading in nearshore waters from rapid development on land. Other sources of nutrients on the developed islands include: septic systems and sewage outfalls; fertiliser use on golf courses and agricultural land; and animal waste in runoff. Marine tourism is important in the island economy, with 703,000 visitors in 1996, with many involved in scuba diving, snorkelling, sea-walker programmes, jet ski use, and fishing. Direct damage also comes from net fishermen walking on the reef. The US Navy has leased the entire island of Farallon de Mednilla as a bombing target, and surveys indicate some direct impacts on reefs, as well as sedimentation disturbances on land. Like many other areas in the Pacific, accidental groundings of commercial and recreational vessels has caused some reef damage.

FSM
Road construction and development projects without adequate erosion control have been responsible for reef damage. Dredging associated with airport and harbour construction has destroyed specific reefs, and subsequent increases in freshwater runoff have limited recovery. Increased population size is a rising concern for some islands as they plan new infrastructure, including sewage treatment and sewerage outfalls. Export fisheries have been a problem for Chuuk, and there are numerous reports of destructive fishing practices. Planned tourism development has the potential to damage reefs if not carefully guided. Ship groundings have caused local damage on high and low islands, and there are no funds available for clean up the many foreign long-line fishing boats abandoned on FSM reefs.

Guam
Sedimentation is the major anthropogenic problem for the central and southern reefs of Guam. For the Ugum River Watershed on Guam, the rate of soil erosion was estimated at 290 tons per ha per year; 46% from sloped roads, 34% from 'badlands' – land burnt regularly for hunting. Ugum Watershed erosion rates doubled from 1975 (6,189 T/Y) to 1993 (12,159 T/Y), mainly from road construction and development, and this is accumulating on reefs and reducing coral diversity and abundance. Coastal pollution, eutrophication and sewage have increased on Guam during a development boom in the late 80s and early 90s, as have increased stormwater runoff from the airport expansion, new roads, hotels, shopping centres and golf courses. The salinity of coastal waters can drop below 28°/oo during summer coral spawning events, which results in reproductive failure. Sewage treatment plants divert the wastewater directly into the ocean outfall pipes without treatment during storms, which discharge within 200m of the shore at depths of 20-25m where the corals grow. Extensions of the Northern and Central District outfalls into deeper waters further offshore are planned. A variety of pollutants occur in sediments of Apra Harbour, including PCBs, heavy metals and PAHs. Over-fishing and habitat destruction are continuing problems. Fish populations and catch per unit effort (CPUE) have declined since 1985 through the use of unattended gill nets, bleach, scuba spearfishing and fish traps along with habitat loss due to sedimentation, pollution and physical damage. More than 1 million tourists visit Guam each year, and most have little coral reef conservation ethic or awareness. This leads to damage to reefs by scuba divers and snorkelers, during underwater walking tours using surface-supplied equipment and many watercraft (jet skis). Fishing vessels, recreational watercraft and ships carrying cargo and illegal immigrants often run aground and cause localised damage to reefs, and the Navy has been responsible for reef damage within Apra Harbor. The main power generation facilities on Cabras Island, Apra Harbor dump high temperature seawater and cleaning chemicals, which both result in coral mortality.

Palau
While much has been achieved in guiding tourism to have limited impacts on reef resources, the volume of divers with poor training is increasing, causing damage to some reef sites. The use of mooring buoys, laws preventing coral collecting, and diving tour operator education are all helping to conserve the reefs. The greatest threat is the Compact road project for Babeldaob because of the great potential for erosion and sedimentation, and for opening up of large areas to development, and increased reef use. Dredging is planned for fill materials for the road base, which is more potential for reef damage. Increased population, both local and visitors, will require better sewage treatment. Foreign-based fishing activities are a problem for Palau, with poachers from Indonesia and the Philippines damaging Helen Reef, and moored fishing vessels causing eutrophication in Malakal Harbour.

CURRENT AND POTENTIAL CLIMATE CHANGE IMPACTS

American Samoa
A steady rise in air temperature over the past 20 years suggests future climatic uncertainty and a probable increase in the frequency of hurricanes in the region. The record high temperatures in 1998 El Niño/La Niña caused droughts on land and unusually low tides that caused mortalities among exposed corals. Many reefs experienced extensive bleaching during elevated seawater temperatures in 1994.

Northern Marianas
The coral reefs were spared any damage in the 1998 bleaching event, but some localised bleaching has occurred around Saipan, Tinian and Rota. The CNMI are concerned about alterations in heavy rainfall and typhoon frequency.

FSM
There was some coral bleaching, but there are few data on the extent. There is concern in the low islands and atolls about the potential for sea level rise and inundation, and increases in tropical storms.

Guam
Coral bleaching coincided with elevated water temperatures during El Niño events. Other potential impacts include inundation of low lying coastal areas, and increased sedimentation from heavy rains following drought.

Palau
There was substantial coral bleaching in 1998, with 30% of the reefs strongly affected. Surveys of Ngaruangel, showed live coral cover from 10-30% with an average cover of 5-10%, but 30-50% of the corals were bleached and all table *Acropora* were dead. An estimated 75-85% of soft corals on the west and south side of Ngeruangel were bleached, and on the eastern side, coral cover on the slopes ranged from 1-10%, with about 20% live coral bleached and all staghorn *Acropora* dead. Live corals consisted mainly of massive corals, with dead coral cover averaging 10-25%. The lagoon patch reefs had low live coral cover, with only a few live blue and massive corals. Most of the corals on the southern side of the channel were dead. Moderate coral bleaching was observed at Helen Reef and Tobi in early December 1996, but in 1999 most corals at Helen reef were dead from the 1998 bleaching.

CURRENT MPAS AND MONITORING AND CONSERVATION MANAGEMENT CAPACITY

American Samoa
There are 4MPAs which account for about 6% of the coral reefs, including one 'no-take' area on Rose Atoll. However, regular surveillance and enforcement is generally lacking. There are no 'no-take' MPAs in the main islands where overfishing is occurring. Poaching is on-going in all MPAs, probably accounting for 9% of the local artisanal catch.

Northern Marianas
At present the are two officially designated marine protected areas in the CNMI: Sasanhaya Fish Reserve on Rota, and the Lighthouse Marine Reserve on Saipan. Legislation to create new marine protected areas has passed the CNMI House of Representatives and is awaiting passage by the Senate. This legislation would establish three new marine protected areas on Saipan at Managaha Island, Bird Island, and from Forbidden Island to Tank Beach. The proposed marine protected areas around Bird Island and Forbidden Island are adjacent to existing upland wildlife conservation areas. This would help reduce impacts from land use activities on these proposed reserves. Legislation is also pending to create a substantial marine protected area on the west coast of Tinian.

Within the CNMI, there are three agencies with primary responsibility for conserving and monitoring marine resources. The Division of Fish & Wildlife is primarily responsible for the protection and enhancement of living marine resources. The Division of Environmental Quality (DEQ) has jurisdiction over marine water quality and is responsible for conducting environmental assessments to ensure compliance with local and federal water quality standards. The Coastal Resources Management Office (CRM) is responsible for ensuring protection of coastal resources. The CRM Office is primarily a permitting agency for large scale development projects occurring in the coastal zone, and permitting of various aquatic recreational activities such as jetski operations. CRM is primarily responsible for ensuring that development activities are conducted with limited adverse impact on the marine environment.

Regular monitoring of coral reef resources has been confined to the three populated southern islands (Saipan, Tinian, and Rota) of the Marianas chain, because of inaccessibility of the Northern Islands, and the high cost of conducting marine surveys. Several isolated surveys have been conducted in the northern islands, especially Pagan, however these surveys have been limited in scope and do not provide sufficient information for developing baseline data on the status of reefs in that area.

FSM
All coastal reefs are under state and local control, but there are no 'federal' coral reef reserves. Protection of specific areas is often controlled by chiefs and other traditional leaders. In Yap, the villages own the reefs, and have authority over resource use. A number of the islands have areas set aside for protection and limited resource extraction. The College of Micronesia-FSM and the regulatory agencies (Environmental and Marine Resource) have few staff trained in marine resource assessment and monitoring. Cooperation among regional institutions under the Marine Resources Pacific Consortium (MAREPAC) and funded by the US Dept. of the Interior will increase the local and regional capacity for assessment and monitoring. NGOs are active e.g. The Nature Conservancy, and offer technical and financial assistance for reef programmes. Each State has a Marine Resources Management office and an Environmental Protection Agency office, and Peace Corps volunteers also assist in monitoring programmes.

Guam
Five no-take marine reserves (12% of the coastline) were established in 1998, and are presently under full enforcement. An ocean current assessment is presently underway at the Pati Point Preserve, to determine which areas will be seeded by larvae from this preserve. Fish abundance, size and diversity data are being collected to document the value of the preserves and the success of the program. The University of Guam Marine Laboratory has ongoing coral reef monitoring programmes, in collaboration with the Guam Division of Aquatic and Wildlife Resources (DAWR) and the Guam Environmental Protection Agency (GEPA). The Marine Lab marine database dates back to 1970. The DAWR collects data on fish catch through creel censuses, while GEPA has routine water quality sampling programmes. A joint educational outreach program exists as a collaboration among these three groups as well as the Guam Coastal Management Program.

Palau

The isolated island group known at Ngerukewid or Seventy Islands in the southern lagoon has been a protected area since 1956, and the Rock Islands are now protected by the Koror State. Other conservation areas are being negotiated in partnership with the national government. The Palau Conservation Society is active in conservation, and collaborates on monitoring and assessment programmes with the Coral Reef Research Foundation, the Palau Community College, the Environmental Quality Protection Board and the Marine Resources Division. The Nature Conservancy also has an office on Palau, and works with the other organisations on coral reef conservation. Palau has substantial expertise, but not financial resources.

GOVERNMENT POLICIES, LAWS AND LEGISLATION

American Samoa

Legislation is in place for water quality standards, land use regulations, waste disposal, fishery management, habitat protection, endangered species, protected areas, ship pollution and other environmental issues. Environmental violations are more frequently detected and prosecuted, but enforcement of these regulations is not widespread and many problems persist. Local environmental agencies have also undertaken aggressive education programmes to increase community understanding of environmental issues. This effort is commendable, but it is difficult to keep pace with the territory's rapidly growing population and concurrent development pressures.

Northern Marianas

Coral reefs are protected through a variety of regulations and requirements. The permit requirements established by the Coastal Resources Management Agency contain provisions to ensure minimal impact to marine resources. Likewise, the water quality standards and requirements were established by the Division of Environmental Quality to protect marine waters and thereby limit the damage to reefs from marine pollution, sedimentation and nutrient loading. The Division of Fish & Wildlife is the agency with primary authority for developing and implementing regulations regarding protection of living marine resources.

Guam

The natural resource laws which protect coral reef resources are presented in two statutory laws, which are relatively old and have not been revised for over 20 years e.g. coral is considered under the definition of a fish, because at that time, the only people likely to have an impact were fishermen. Now the statutory laws regulate coral collection and have penalties for damage when fishing. Coral collecting requires a permit, but no permits have been issued since 1982. This same law (5 GCA chapter 63) also regulates fish net mesh sizes and illegal chemical and explosive use in coastal waters. Recently the Division of Aquatic and Wildlife Resources amended and expanded the existing fishing regulations (Title 16) to contain size restrictions and seasons for aquatic fauna, and proper definitions of corals etc. A major component of these new regulations are the rules for the 5 new permanent marine reserves. These areas account for nearly 12% of Guam's coastline as no-take fisheries areas, with violations being treated as a petty misdemeanour. The Guam Environmental Protection Agency has local water quality standards, which they enforce with fines of up to US$10,000 a day. The Seashore Protection Commission has authority over construction projects within the area from 10m of the mean high tide mark out to a depth of 20m. A panel reviews proposals and provides recommendations to a commission of appointed members of the public for consideration for approval or rejection. This commission has not been very successful in upholding their responsibilities. Guam is still in need of laws to address ship groundings and levy fines for the amount of coral reef damage.

Palau

The National Government environmental policies are contained in the Palau National Master Development Plan (SAGRIC, 1996). The plan recommends several actions and steps necessary to protect the environment of Palau including policy process, institution strengthening, education and research, protected areas and wildlife, waste management and pollution, and legislation. The Marine Protection Act of 1994 regulates the taking of certain species of marine organisms, prohibiting or limiting certain fishing methods, and authorising the Minister of Resources and Development to develop regulations regulating the collection of marine animals for aquaria or research. The Environmental Quality Protection Act was enacted in 1981, and Division 2 (Chapter 10) deals with wildlife protection. Protected sea life includes turtles, sponges, mother-of-pearl, dugong, trochus and clams. Chapter 13 deals with illegal fishing methods including the use of explosives, poisons or chemicals. Division 3 deals with the protected areas of Ngerukewid and Ngerumkaol.

The Rock Island Management and Preservation Act designated certain areas of the Rock Islands as reserve areas and others as tourist activity areas. Koror State Public Law K6-101-99 established the Ngerukewid Islands Wildlife Preserve and prohibited fishing in Ngerumkaol spawning areas. The Ngatpang Conservation Act of 1999 established the Ngatpang Reserve which includes three areas and Ngatpang's portion of Ngermeduu Bay. The Ngiwal State Conservation Act of 1997 established the Ngemai Conservation Area.

RESTORING CORAL REEFS IN PAGO PAGO HARBOR, AMERICAN SAMOA

During a cyclone in 1991, 9 Taiwanese longline fishing boats were driven up onto the coral reefs of Pago Pago Harbor, American Samoa. Not only were these an eyesore, they were also leaking fuel and preventing recovery of the once beautiful reefs on the reef flat. So the National Oceanic and Atmospheric Administration worked with the US Coast Guard, US Department of the Interior and the Government of Samoa to remove them and start the restoration efforts. The first job was to completely remove the grounded vessels and associated debris to allow natural recovery of about 3,000m2 of bottom that had been smashed. Then they transplanted back the corals that had been taken from the sites soon after the incident to save them from further damage, so that the injured areas would recover 'naturally'. Finally they re-established the Aua transect in the area to provide very long-term monitoring data on Pacific coral ecosystems.

GAPS IN CURRENT MONITORING AND CONSERVATION CAPACITY

American Samoa
On-island expertise in environmental protection has increased in recent years, and considerable research, monitoring and environmental compliance efforts are conducted by several groups: Department of Marine and Wildlife Resources; American Samoa Environmental Protection Agency; Department of Commerce and Fagatele Bay National Marine Sanctuary; American Samoa Community College; and National Park of American Samoa. However, these programmes are generally small and would benefit from increased on-island capacity. There is also little on-island expertise to identify the 200 coral species and 890 fishes on local reefs, consequently, local staff need to focus on indicator species and monitor key reef resources at appropriate time intervals.

Northern Marianas
There are a variety of monitoring activities to obtain information on the status and health of coral reef resources, however, there are significant gaps in the data and a need for greater coordination among agencies to ensure that data are compatible and that monitoring activities are complementary. Of paramount importance is the need for CNMI agencies to identify the objectives of marine monitoring. Some monitoring has not had clear objectives, which resulted in data that could not be used to answer management questions, such as the relationship between marine water quality and coral health or the relationship between observed coral damage and activities contributing to coral decline.

FSM
The expertise in the FSM has been increasing, with a number of highly trained individuals dispersed among the College of Micronesia-FSM, the regulatory agencies and local institutions such as the Pohnpei Environmental Research Institute. Funds are a limiting factor, especially as financial support from the Compacts of Free Association is decreasing. More emphasis needs to be placed on development of local talent and less dependence on expatriate technical staff.

Guam
Guam is fortunate in having considerable technical expertise in coral reef ecology, management and policy. Three government agencies have staff dedicated to dealing with coral reef and coastal resource monitoring and protection: the Guam Department of Agriculture's Division of Aquatic and Wildlife Resources; the Guam Coastal Management Program within the Bureau of Planning; and the Guam Environmental Protection Agency. In addition, there are two research units at the University of Guam that focus on coral reef and coastal environmental issues: The University of Guam Marine Laboratory; and the Water and Environmental Research Institute. Collaboration and cooperation is high among these agencies and institutions, however, funds specifically earmarked for monitoring are limited, especially for water quality monitoring, and more stakeholder involvement is needed.

Palau
Training in coral reef monitoring and assessment is required so that changes in conservation areas can be monitored. Only with regular monitoring program can resource managers effectively implement management plans. Currently, the few people in Palau with the

ability to perform monitoring and assessment are employed on other tasks. Thus additional people should be trained to ensure adequate monitoring of coral reefs. Many of the training exercises are short and focus on techniques that are easy to use and not time consuming, and coral identification is usually not covered. Coral surveys using life forms would not give an accurate measure of diversity e.g. at Malakal sewer outfall. There were no significant difference in percent coral cover between 1976 and 1991, but species diversity has decreased dramatically since 1973. Without taxonomic evaluation, this evidence would have been missed. Periodically, funds are needed to bring many people together to perform assessment of the whole of Palau from Ngeruangel to the Southwest Islands. Reef studies tend to be limited to Koror and selected places in Babeldaob, while unique areas far from Koror are usually neglected. The last comprehensive work on Palauan reefs was in 1992 during the REA, and many of the places have not been studied since.

Conclusions and Recommendations for Coral Reef Conservation

Overall, the reefs of American Samoa and Micronesia are very diverse and in good condition, but development pressures and increasing population demands are resulting in increases in sedimentation, commercial fisheries, coastal pollution, ship groundings and recreational activities. While reefs in the atolls and less populated areas are in good to excellent condition, many of those adjacent to population centres, particularly on high islands, are declining to fair and poor categories, with noticeable decreases in coral cover, fish abundance and resilience to natural disturbances. Coral reefs in the small islands have a distinct carry capacity, which is clearly being exceeded in some instances.

Pro-active steps in preserving coral reefs are evident in all areas presented in this report. American Samoa deserves recognition for establishing a task force on human population. While all acknowledge this is a sensitive issue, it must also be recognised that human population growth is the key element in increasing levels of anthropogenic disturbance. A regional consortium to deal with marine resource issues was recently established with support from the US Department of the Interior, Office of Insular Affairs, that includes American Samoa, the CNMI, the FSM, Guam, Palau and the Republic of the Marshall Islands. The Marine Resources Pacific Consortium (MAREPAC) has developed specific objectives, including:

- To develop the capabilities of the regional resource agencies, institutions of higher education, and community-based organisations within Micronesia and American Samoa to deal with issues surrounding sustainable use of marine resources of cultural, economic and scientific value.
- To foster cooperation and collaboration among the local and federal resource agencies, research facilities, community-based organisations, educational institutions and the private sector in meeting their mandates, goals and community needs.
- To collect, synthesise and disseminate adequate and accurate information in support of sound policy development on marine resource use, addressing present needs as well as the concerns for future generations.

The 16-member governing board includes 2 members each from regional resource agencies, institutions of higher education and community-based organisations.

Jurisdictions pertinent to coral reefs are as follows:

American Samoa

Coral reefs in American Samoa have been resilient so far to a series of natural perturbations, although a full recovery may take another decade. Although the message for managers is that the reefs will recover if we 'do no harm', we are faced with several examples where serious harm has been done, particularly over-fishing of reef resources, reduced water quality in populated areas, and loss of turtle nesting beaches due to coastal construction. Additionally, we are faced with climate change with uncertain consequences for local reefs.

Several areas which require further work for coral reef conservation are:

- Expand coral reef monitoring efforts and focus the objectives. Most monitoring in American Samoa can be characterised as 'ecological monitoring' that tracks changes in the ecosystem over time. These studies have provided valuable insight, highlighting the dynamic nature of coral reef systems. From another perspective, however, these studies were not designed to address questions commonly faced by reef managers: Is over-fishing occurring? Is sediment from poor land-use practices harming the reef? etc. Consequently, it is important at the outset of any monitoring program to clearly identify (a) the intended users of the monitoring program, and (b) the parameters that should be measured to provide the information that they need. In addition to this management driven approach, a monitoring program in American Samoa should be: (a) achievable with local staff, although we recognise that off-island scientific expertise may be needed to address some issues, (b) stable to rotations of technical staff who are typically hired on 2-year contracts, (c) comparable to other programmes to the degree possible, and (d) open to community input and management.
- Restart monitoring of coral reef fisheries. A workshop in American Samoa identified over-fishing as the major problem hindering recovery of local reefs (ASCRTF 1999), so it is essential that basic harvest data be collected to monitor trends in total catch, catch-per-unit-effort, etc.
- Create a territorial network of 'no-take' MPAs. There are two issues here: existing Marine Protected Areas are not yet adequately enforced; and 'no-take' MPAs are needed in the main islands where over-fishing is occurring.
- Improve land-use practices that reduce water quality. Despite welcome improvements in water quality in Pago Pago Harbour, there has not been a full recovery of coral reefs, swimming is not safe, nor are fish safe to eat. A step-wise recovery plan should be implemented to build on progress, and harbour fish and sediments need to be tested for toxicity at regular intervals. Improvements in land-use practices and waste disposal are needed to reduce sedimentation and pollution impacts on coral reefs.

Northern Marianas

The reefs in the CNMI outside of the main population and development centres are in relatively good condition, but more surveys and quantitative data are needed. Greater effort has been made in the last few years to establish a long-term marine monitoring program, with ongoing monitoring at a number of sites on Saipan, Tinian and Rota, and opportunistic monitoring in the northern islands. Data from these surveys will be important in assessing changes in these sites over time. The CNMI should improve cooperation between individual agencies that collect and analyse monitoring data. Monitoring programmes should start with a plan that identifies the objectives and establishes a repeatable protocol for data collection, storage and analysis. Additional funding is needed to address the budget gaps of specific agencies, but more specifically to address priority needs identified by the interagency marine monitoring group. Basic training in data collection, data analysis and quality control should be developed with specialised training in identification of marine flora and fauna. This additional training would greatly improve the quality of the data being collected and would empower local staff in the protection of reef resources. In the long run, funding for enforcement remains the most critical need. While there are regulations and measures to protect coral reefs, the CNMI is unable to enforce these provisions, thus protection exists only on paper.

FSM
The reefs are in relatively good condition, however, land-use practices on the high islands are causing concern. Development projects and agriculture have already been responsible for reef damage, and this is expected to increase in magnitude and effect. Thus, integrated watershed management programmes need to be developed. Reef fisheries on some islands have been over-exploited, therefore improved coordination of management activities among the states is recommended. Community education and stakeholder involvement programmes need to be expanded. Ship groundings need to be addressed at the state and national level, with consideration for having vessels post bonds. Additional support for the resource agencies is necessary if they are to meet their mandates.

Guam
In general, the condition of the coastal reefs continues to decline, primarily as a result of land based activities. Data on coral recruitment indicate significant reductions over the past 2 decades. Many areas impacted by natural disturbances, including typhoons, crown-of-thorns starfish outbreaks and earthquakes are not recovering. Community education initiatives are increasing awareness, and the political will to address reef decline is improving, however far more is needed. Over-fishing is still a concern, however the establishment of 5 Marine Reserves which formally received enforcement protection in June, 2000, is a positive step forward. Enforcement of existing laws and environmental regulations is still a major problem on Guam. More improvements in infrastructure and erosion control programmes are needed to reduce the land-based stressors on coastal reefs.

Palau
Before the 1998 bleaching event in Palau, the remote reefs were generally healthy and in good condition. The reefs closer to population centres or near development are less healthy with distinct signs of degradation. After the bleaching event, most reefs in Palau were severely affected. Crown-of-thorns starfish are also a problem and in many areas they are targeting the few remaining *Acropora* that survived the bleaching. The fish populations around the main islands of Palau show signs of over-fishing compared to the Southwest Islands, where there are fewer fishing pressures. Highly desired species of fish are present in low numbers or absent around the main islands of Palau.

Regular assessment and monitoring of Palau coral reefs is required for coral reef management to be effective. Regular monitoring programmes can detect problems early so that corrective strategies can be developed. Reports and publications from the monitoring program are critically important.

Education is an important aspect of reef conservation. School curricula from elementary to secondary and post secondary should incorporate environmental issues and concerns. Education should not be limited to schools, but extend to all levels of the communities. Efforts should focus on raising environmental awareness among policy makers, traditional and political leaders and villagers. Effective management of marine resources requires an informed and supportive public.

Catch levels and trends for reef fisheries should be monitored closely and accurately so that effective management of resources can be implemented. Currently, only two fish markets in Koror provide landings information to Marine Resources Division. To have accurate market data, Marine Resources need to have landings and catch data from all fish markets. Other information such as type of fishing gear used and the number of hours spent fishing would help to determine the level of exploitation.

All agencies and organisations in Palau involved with coral reef monitoring and management need to combine for strategic planning to set priorities and determine areas of focus for each group and problems that can be better addressed through collaborative work. With limited resources (time, money, human resources) and a big coral reef area, all groups need to work together, and avoid duplication of efforts and competition among groups. With more coordinated efforts of all groups in Palau, coral reef conservation will no doubt improve.

SUPPORTING DOCUMENTS AND AUTHORS

Birkeland, C.E. 1997. Status of Coral Reefs in the Marianas. Pp. 91-100 in R.W. Grigg and C.E. Birkeland, eds., Status of Coral Reefs in the Pacific. UNIHI-SEAGRANT-CP-98-01.

Green, A. 1997. Is There A Need For Federal Management Of Coral Reef Resources In the Western Pacific Region? Prepared for the Western Pacific Regional Fisheries Management Council under Award No. NA67AC0940. Draft. 249 pp.

Golbuu, Y., G. Mereb, D. Uehara, A. Bauman and J. Umang. 1999. Biological Survey at Ngerumkaol, Koror State, Republic of Palau. PCC-CRE Publication 17/99.

Johannes, R.E., L. Squire, T. Graham, Y. Sadovy, and H. Renguul (1999). Spawning Aggregations of Groupers (Serranidae) in Palau. Marine Conservation Research Series Publ. #1. The Nature Conservancy.

Maragos, J.E. and many others (1994). Marine and Coastal Areas Survey of the Main Palau Islands: Part 1Natural and Cultural Resources Survey of the Southwest Islands of Palau and 2. Rapid Ecological Assessment Synthesis Report. Prepared by CORIAL and The Nature Conservancy, Pacific Region for Republic of Palau, Ministry of Resources and Development.

Palau Conservation Society. 1999. Resource Surveys of Ngemai Reef, Ngiwal, 1997-1998. Report Prepared for Ngiwal State. PCS Report 99-01.

Richmond, R.H. 1996. Reproduction and Recruitment in Corals: Critical Links in the Persistence of Reefs. Pp. 175-197 In C.E. Birkeland, (ed.), Life and Death of Coral Reefs. Chapman and Hall, Publishers. N.Y.

NATIONAL AUTHORS AND CONTACTS

American Samoa: Peter Craig, National Park of American Samoa; peter_craig@nps.gov
CNMI: Katharine Miller, Division of Fish and Wildlife, Dept. of Lands and Natural Resources; kath@itecnmi.com

FSM: Asher Edward, College of Micronesia, FSM; aedward@mail.fm
Guam: Robert Richmond, Marine Laboratory, University of Guam; richmond@uog9.uog.edu
Palau: Yimnang Golbuu, Palau Community College; ygolbuu@yahoo.com

13. Status of Coral Reefs in the Hawaiian Archipelago

David Gulko, James Maragos, Alan Friedlander, Cynthia Hunter and Russell Brainard

Abstract

Approximately 85% of all US coral reefs occur within the Hawaiian Archipelago. The reefs that make up this region stretch over 2,000km and contain a majority of the reef types seen across the Pacific. They are characterised by their isolation from other Pacific reefs and high endemism across most phyla. The Archipelago consists of two distinct regions: the Main Hawaiian Islands (MHI) made up of populated, high, volcanic islands with non-structural reef communities and fringing reefs abutting the shore, and the Northwestern Hawaiian Islands (NWHI) consisting of mostly uninhabited atolls and banks accounting for the majority (69%) of US reefs. The differences in anthropogenic impacts on these two regions are striking. The MHI are urbanised with extensive coastal development and associated runoff; overfishing for food and marine ornamentals, alien species invasions, and marine debris are each of increasing concern. The NWHI coral reefs are affected primarily by marine debris and the impacts from lobster and bottom fisheries. Concerns exist over current and proposed ecotourism activities and new fisheries. Recent coral reef management initiatives include new forms of MPAs related to marine ornamental collection and tourism, as well as assessment and long-term monitoring programmes involving active partnerships between scientists, community groups and management agencies.

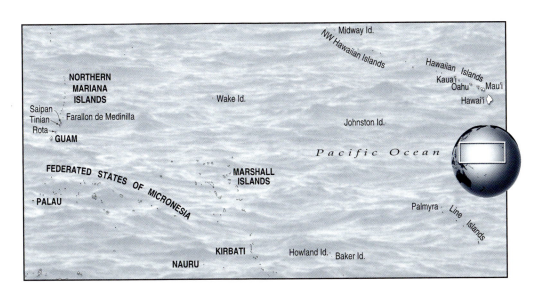

No. islands, islets & coral cays		Land Area (km²)	Reef Area (km²)	Reef Type
Hawaii Island	(4)	10,433	253	Fringing, reef communities
Maui*	(18)	3,037	1421	Fringing, banks, reef communities
O'ahu	(15)	1547	504	Barrier, fringing, platform
Kaua'l#	(5)	1612	348	Barrier, fringing, reef communities
NWHI	(29)	14	8,318	Atolls, fringing, platform, banks

Basic geographic details of the islands covered by this report
* Maui includes the high islands of Maui, Lana'i, Moloka'i & Kaho'olawe
 # Kaua'i includes the high islands of Kaua'i and Ni'ihau.

INTRODUCTION

The Hawaiian Archipelago stretches over 2,400km from 19–28°N latitude and is one of the most isolated, yet populated areas on earth. Therefore regional and global impacts seen in many other areas are of less concern than the increasing impacts of urban-related activities. The geographic isolation has lead to many endemic species in many marine phyla, including corals. There are 8 large islands and 124 small islands, reefs and shoals in 2 distinct regions: the Northwestern Hawaiian Islands (NWHI) consisting of mainly uninhabited atolls and banks containing 60% of US reefs; and the populated, high, volcanic Main Hawaiian Islands (MHI) surrounded by reef communities, fringing reefs and two barrier reefs

MHI

These consist of high volcanic islands which range from 7 million years old (Kaua'i) to active lava flows on the east side of the 'Big Island' of Hawaii. The Polynesian peoples who first settled these islands had strong ties to coral reefs; reef resources provided food, medicines, and building materials, and were a major part of cultural and social customs in ancient Hawaii.

Broad-scale degradation of Hawaiian coral reefs began 100-200 years ago with the expansion of Western influence bringing livestock and agriculture and contributing to erosion and sediment flow onto the reefs. Dredging and reclamation of nearshore reefs for military, commercial, and residential use by over 1.2 million inhabitants has resulted in more damage. Streams have been channelised and increased paving of land area has increased the rate of water runoff near the urban centres. Despite these changes, most of the nearshore reefs remain in fair to very good condition. Coral reefs are a critical component of the $800 million per year marine tourism industry, providing white sandy beaches, surfing and diving sites, and recreational, subsistence, and commercial fishing.

	Population (Distribution)	Population per km²	GDP in US$ (per capita)	Tourists/yr
Hawaii Island	138,400 (27% Hilo)	13	$2,400 million ($17,341)	1,255,480
Maui, Kaho'olawe, Lana'i & Moloka'i	100,504 (17% Kahului)	33	$2,800 million ($30,648)	2,376,330
O'ahu	871,800 (43% Honolulu)	564	$26,800 million ($30,741)	5,002,530
Kaua'i & Ni'ihau	56,665 (14% Kapa'a)	35	$1,200 million ($21,177)	992,780
NWHI	164 (99% Midway)	11	Not applicable	5,000

Basic demographic and economic data for islands of the Hawaiian Archipelago.

NWHI

These extend for more than 1600km northwest of Kaua'i, starting with Nihoa and Necker (7 and 10 million years old) to Midway and Kure atolls (28 million years old). Most of the atolls are uninhabited, although Midway, Kure and French Frigate Shoals have all been military bases. The US Fish and Wildlife Service (USFWS) manages the Midway National Wildlife Refuge and a relatively large ecotourism operation has started on Midway. The remainder of the NWHI are part of the State of Hawaii and the majority are within the Hawaiian Islands National Wildlife Refuge, established by President Roosevelt in 1909, and currently administered by the USFWS. The NWHI reefs are in very good to excellent condition, but are subject to species-specific over-fishing.

STATUS OF CORAL REEFS

MHI

The coral reefs of the Hawaiian Archipelago are considered to be a separate eco-region because of their isolation from other Pacific reefs. There are 55 species of hard (scleractinian) corals in Hawaii, with over 25% of these being endemic. The highest diversity of corals in the MHI occurs around the islands of O'ahu and Hawaii. Live coral cover ranged from 4-50% at the 25 sites between 2-3m and 10m depths surveyed by the Hawaii Coral Reef Assessment and Monitoring Program (CRAMP) in 1999-2000. Marine invertebrate biodiversity and endemism are relatively high in Hawaii, with over 100 species of sponges, 1071 species of marine molluscs, 884 species of crustaceans, and 278 species of echinoderms. The Hawaiian Islands are thought to support over 400 algal species, many of which are endemic, as well as one endemic seagrass, *Halophila hawaiiana*. There are 557 species of reef and shore fishes in Hawaii; very low compared to other Indo-West Pacific reefs (e.g. >2,500 fish species in the Philippines). Hawaii has the highest percentage of endemics (24.3%) in the world, and these endemics are often dominant members of Hawaiian reef communities. The low number of species and even families is due to the geographic and oceanographic isolation from other reefs far to the west and south.

NWHI

There are 22 hard coral species and about 200 algal species known at present, but more complete surveys are expected to increase these numbers substantially. The *Acropora* species found here do not occur in the MHI. French Frigate Shoals, Pearl and Hermes Reef, and Maro Reef have the highest coral biodiversity and cover, but coral cover is generally low throughout the NWHI. Growth rates for corals in the northern portion of the chain are significantly slower than farther south raising concern about potential recovery from human impacts. The endemic seagrass *Halophila hawaiiensis* has been found at Midway Atoll and may occur elsewhere in the NWHI. There have been no recent surveys of other invertebrates, but rapid ecological assessments of the NWHI are scheduled for October 2000.

Reef and shore fish diversity is lower in the NWHI, with 266 species recorded from Midway Atoll. The cooler waters, lack of critical habitats (and lower sampling effort) may be the reasons for these low numbers. The fish community in the NWHI is different from the MHI with reduced abundance of herbivores (mostly surgeonfishes) and more predators (mostly jacks, sharks, goatfishes, scorpionfishes, and bigeyes). The fish assemblage in the NWHI is dominated both in numbers and biomass by carnivores.

Bleaching Events

Hawaiian reefs escaped the widespread bleaching that devastated many Pacific reefs during the last decade, although Hawaiian corals are susceptible and local rises in water temperatures (or increased UV penetration) have caused occasional bleaching. Following three days of exceptionally calm and warm conditions in September 1996, many shallow corals in Kane'ohe Bay and Kailua Bay on O'ahu, showed partial or complete bleaching that resulted in limited coral mortality.

Disease

Coral diseases and tumours have been found on most major coral species of the MHI (*Porites lobata, P. compressa, Montipora capitata, M. patula*, and *Pocillopora meandrina*). Surveys from 1991 to 1998 showed a mean frequency of 22% of diseased corals over a range of depths and habitats. At some sites, up to 60% of colonies were affected with disease covering up to 32% of the surface of the dominant *P. lobata*. Like other areas, the incidence of coral disease in Hawaii does not appear to be related to anthropogenic stress e.g. pollution, proximity to urban centres, and no major disease outbreaks have yet occurred in Hawaii.

Turtle tumours (fibropapillomas) were rarely seen on Hawaiian green sea turtles (*Chelonia mydas*) prior to 1985, but the incidence has increased drastically, and this disease now commonly afflicts turtles on Kaua'i, O'ahu, Moloka'i, and Maui; up to 60% of turtles in Kane'ohe Bay have tumours.

STATUS OF REMOTE CORAL REEFS IN THE US INSULAR PACIFIC

Jim Maragos and Rusty Brainard

The US has jurisdiction over 9 remote islands in the tropical Pacific: Howland and Baker Islands in the Phoenix Islands; Jarvis Island, Palmyra Atoll, Johnston Atoll, and Kingman Reef in the northern Line Islands; and Wake Atoll north of the Marshall Islands. Midway Atoll and Rose Atoll are off Hawaii and American Samoa respectively and are also covered elsewhere. Wake, Midway, Kingman, Palmyra, Rose, and Johnston are coral atolls, while Jarvis, Howland, and Baker are low coral islands surrounded by narrow fringing reefs. These islands were uninhabited until 200 years ago, and now there are people only on Wake, Johnston and Midway. The Nature Conservancy is establishing an ecotourism facility at Palmyra Atoll, and the USFWS authorises compatible ecotourism in the Midway National Wildlife Refuge (NWR). Midway, Howland, Baker, Palmyra, Kingman, and Jarvis were former US military bases, and Johnston and Wake are still active militarily, although there are plans to close them. Johnston, Rose, Howland, Baker, and Jarvis are NWRs, and Kingman and Palmyra are being proposed as NWRs. Rose NWR is jointly administered by American Samoa and the USFWS. All of these reefs have been surveyed since 1995, and they are all in near pristine condition, as there are no anthropogenic impacts, except for occasional ship groundings, and illegal fishing.

Rose Atoll
Dissolved iron concentrations, algae, giant clams, sea cucumbers, and fish have been monitored several times following the grounding of a long-line fishing vessel at Rose Atoll NWR in late 1993. Additional monitoring sites for giant clams and corals were established during the 1999-2000 ship debris removal from the atoll. Monitoring and observations documented the impacts of the fuel spill and iron enrichment on reef habitats and helped to justify the funds to remove the wreckage. About 80% of the ship debris have been removed, leading to recovery of giant clams (corals and coralline algae, and reduction of invading blue-green algae stimulated by iron enrichment.

Johnston and Wake Atolls
The US military has sponsored ecological monitoring programmes at Johnston Atoll during the past 20 years, including fish populations and seabirds. In 1997, coral reef monitoring activities were extended to Wake Atoll (and in 2000 to Johnston Atoll where permanent coral monitoring transects were established during installation of mooring buoys). Both atolls suffered extensive damage from dredging and filling during World War II, but reefs have recovered and are now in healthy condition.

Howland, Baker Jarvis, Kingman, and Palmyra
These were surveyed in early 2000 and found to be in relatively pristine condition except for some over-fishing. The disposal of military debris off the west side of Baker and the west lagoon of Palmyra may have stimuled some invasive algal growths. Only small sharks were observed at Howland and Baker, which indicates recent shark fishing. Sharks and most other large edible fish were relatively rare at Kingman, and sharks are

still depleted at Palmyra, although more abundant than observed in 1998. Recent storm damage was also evident off the north coast of Jarvis and Palmyra.

Midway Atoll
Fish populations have been monitored during surveys of the NW Hawaiian Islands, and the ocean reefs are in excellent condition, but the lagoon reefs are not as flourishing, possibly due to restricted circulation. Reef fish are abundant and healthy, due to fishing restrictions as a NWR.

Fish stocks are in excellent to pristine condition at most of the islands, apart from evidence of recent shark finning and some targeting of large reef fish on the remote reefs, despite being protected as a NWR. Military or USFWS presence at Johnston, Wake, Midway, and Palmyra is discouraging illegal poaching, and giant clam populations are healthy with spectacular densities on Kingman. The only anthropogenic stresses other than fishing impacts (shark finning and depletion of large fish), are vessel groundings, as these low coral islands or atolls are difficult to spot at a distance, and are frequently hit by ships (with associated fuel spills). Marine debris and abandoned fish nets are a major problem at Midway and the other NW Hawaiian Islands, where they become easily snagged, damaging reefs and corals and entangling seabirds, marine mammals, sea turtles and fish. Observations at Palmyra, Kingman, Baker, and Howland in 1998 to 2000 suggest a major bleaching event at these islands in 1997 or early 1998. A major bleaching event was observed at Rose Atoll in April 1994 and at Johnston Atoll in September 1996. Although Howland, Baker, Jarvis, and Rose are no-take NWR, unauthorised fishing probably occurs because they are so remote. Johnston and Midway are limited take NWRs, with on-site management, therefore fish stocks are better managed and abundant. Kingman and Palmyra are proposed as MPAs, but a lack of easy enforcement and surveillance will always put them under threat from over-fishing. Wake Atoll is well protected by the US military. From: Jim Maragos, U. S. Fish & Wildlife Service, Hawaii (jim_maragos@fws.gov), Rusty Brainard, National Marine Fisheries Service, Honolulu (Rusty.Brainard@noaa.gov) and Phillip Lobel Boston University and the Woods Hole Marine Biology Laboratory (pLobel@bu.edu).

Crown-of-thorns Starfish Outbreaks
The last major outbreak of the crown-of-thorns starfish (*Acanthaster planci*) was off the island of Moloka'i in the late 1960s and early 1970s. Lack of *Acropora*, dominance of *Porites* and the island's overall isolation may help to explain the lack of major outbreaks.

STATUS OF CORAL REEF FISHERIES

MHI
There has been a steady decline in abundance of fishes over the past century, with over-fishing listed as the major cause of this decline. Fishing pressure near the heavily populated areas now exceeds the renewal capacity of these resources and the abundance of reef fishes in unprotected areas is much lower than in protected areas. Fisheries catch statistics are unreliable owing to under-reporting by commercial fishers and the large resident

recreational and subsistence fishing population. The recreational catch on nearshore reefs is likely equal to or greater than commercial fisheries catch, and these recreational fishers take more species using a wider range of fishing gear. Hawaii is one of the few coastal states that does not require a saltwater recreational fishing license. The Hawaii Department of Land and Natural Resources is currently expanding its creel program to sample the recreational catch and is also revising existing fishing regulations.

Subsistence fishing is culturally and economically important to many rural communities. In previous times, Hawaiians developed a management system that provided for sustainable harvest of their natural resources. This management put limits on the times and places for fishing based on knowledge of natural cycles and habitats. The Hawaii State Legislature created a process in 1994 to designate community-based subsistence fishing areas and there are efforts currently underway to create such areas on Moloka'i, Kaua'i and Hawaii. Poaching is a recurring problem with the catching of undersized fish and invertebrates and fishing out-of-season contributing to the depletion of coastal resources. For example, in Hanalei Bay, Kaua'i, less than 30% of the omilu (a highly prized jack species) harvested were legal size and only 3% were sexually mature. Currently there are numerous long, cheap gill nets being set deeper and in locations not previously fished (e.g. officials seized 8km of illegal nets from inshore areas over a 6 month period in 1998). Enforcement is poor and fines are minimal, with the few prosecutions not acting as a deterrent to violators of fisheries regulations.

Hawaii provides most of the ornamental fish and invertebrates taken in the USA, because quality is high and the rare endemic species are highly prized. The annual harvest of aquarium fishes rose from 90,000 in 1973 to 422,823 in 1995, with most coming from the island of Hawaiii. Commercial permits increased by 39% between 1995 and 1998 and the trade is now having a major impact on both dominant and rare aquarium species. There are no regulations limiting the size, number and collecting season for most species and the full impacts may not be felt yet. Conflicts are common between marine ornamental collectors and subsistence fishers, commercial fishers, environmentalists and the tourism industry. In response, the Hawaii State legislature established the West Hawaii Regional Fishery Management Area in 1998 to improve reef management and declared that 30% of the West Hawaii coastline would be Fish Replenishment Areas (FRAs), where aquarium fish collecting is prohibited.

NWHI
Only commercial fishers with vessels larger than 20m can regularly fish in these remote and exposed waters. Fisheries for bottom fish and lobsters to 100m depth are managed by the NMFS through an advisory body: the Western Pacific Regional Fisheries Management Council (WPRFMC). The abundance of lobsters has declined since the 1980s and the fishery is currently closed. One concern is the impact on the endangered Hawaiian monk seal that consumes lobster in their diet. Resource fish species tend to be larger and more abundant in the NWHI, compared to the MHI. The mean biomass of fish on shallow reefs at French Frigate Shoals and Midway Atoll is almost twice that found in the MHI, probably reflecting lower fishing pressure at these sites. There is minimal collection of corals and aquarium fishes in the NWHI, but there is concern about rare and highly valued species like the masked angelfish (*Genicanthus personatus*), dragon eel (*Enchelycore pardalis*) and the Hawaiian lionfish (*Pterois sphex*).

ANTHROPOGENIC THREATS TO CORAL REEF BIODIVERSITY

Anthropogenic pressures are high on coral reefs of the MHI because most reefs are close to urban centres, but impacts are variable:

Alien Species
At least 19 species of macroalgae have been introduced to O'ahu since 1950 with 4 of these being highly successful and now prevalent throughout the MHI. One alien species, *Kappaphycus alvarezii*, is overgrowing and killing corals in Kane'ohe Bay, some invertebrates have also been introduced, but there are no reports of damage to coral reefs. At least 13 species of marine fishes have been introduced, mostly target species of fishers (6 groupers, 4 snappers, and 1 emperor), with the blueline (*Lutjanus kasmira*) and blacktail (*Lutjanus fulvus*) snapper, and the bluespotted grouper (*Cephalopholis argus*) being the most successful and damaging fish introductions. Other introductions include sardines, herrings, mullet, tilapia and goatfish, which are displacing local species.

Destructive Fishing Practices
Dynamite and cyanide fishing are not problems in Hawaii, but there have been recent cases of chlorine being used to capture lobster or fish. This is highly damaging as it destroys the habitat for all species. Long gill nets are used extensively throughout the Hawaiian Islands and cause localised depletion of fish stocks, including endangered species as bycatch (sea turtles, Hawaiian monk seals), and continual damage to fragile coral colonies through nets being lost, abandoned, or improperly used. State laws require that nets be checked every 2 hours and removed after 4 hours.

Eutrophication
In the MHI, high nutrient levels are known to encourage algal blooms, which can out compete and overgrow living corals. Algal blooms have been a recurring problem on reef flats off the southern and western coasts of Maui for almost 10 years; *Hypnea, Sargassum, Dictyota* and *Cladophora* have all dominated reef flat areas at various times, presumably due to leaching of nutrients from cesspools, injection wells or other non-point sources. The Federal Clean Water Act requires a National Pollutant Discharge Elimination (NPDES) permit for any pollutant discharged into nearshore waters, and the Hawaii Coastal Zone Management Program and the Hawaii Department of Health are developing a pollution control plan to address discharges not covered under NPDES permits.

Marine Debris
Marine debris from beach goers, storm drains, industrial facilities and waste disposal sites is common on reefs and shores throughout the MHI. In the last 10 years, the amount of derelict fishing gear washing ashore has increased e.g. in 1998, a 3 day community clean up removed over 3,000kg of nets and debris from Kane'ohe Bay and Wai'anae waters.

The major marine debris problem in the NWHI is from derelict fishing gear, including drift nets, trawls, traps and lines. These result in dislodging and breaking of coral colonies, and entangling and killing of seabirds, monk seals, turtles and fish. Drifting marine debris may also serve as a vector for alien species introductions. In 1998 and 1999, NMFS lead multi-agency project teams which removed 39 tons of marine debris from French Frigate Shoals,

Stress and (score)		Hawaii	Maui	O'ahu	Kauai	NWHI
Overfishing	(18)	4	4	5	3	2
Coastal Alteration	(16)	4	4	5	3	0
Tourism	(15)	3	4	4	2	2
Alien Species	(13)	2	4	5	1	1
Urbanisation	(14)	2	4	5	3	0
Marine Ornamental Trade	(14)	4	3	4	2	1
Sedimentation	(12)	1	4	4	3	0
Pollution	(12)	2	3	4	2	1
Destructive Fishing	(10)	2	2	3	2	1
Hurricanes	(10)	2	2	2	3	1
Oil Spills	(9)	1	1	3	2	2
Climate Change	(5)	1	1	1	1	1
Coral Bleaching	(5)	1	1	2	1	0
Acanthaster planci	(1)	0	1	0	0	0
TOTAL IMPACT RATINGS	33	42	52	31	14	12

This is a summary to the major causes of coral reef degradation in the US NE Pacific with 0 = no threat; 1 = low threat; 2 = localised low threat; 3 = average threat; 4 = localised major threat; 5 = general major threat.

Lisianski, Pearl & Hermes and Midway. It is estimated that over 4,000 tons of marine debris still remain in the NWHI.

Military Impacts

The island of Kaho'olawe was used as a military firing and bombing range, which contributed to extensive sedimentation and debris on nearby reef areas. The island is now managed by the Kaho'olawe Island Reserve Commission (KIRC) which is assessing damage and cleaning up debris. Kaula Rock, a small islet located off the southern tip of the island of Ni'ihau, is currently used as a live-firing and bombing target by the US military. Military amphibious vessels have run aground on reefs off the windward side of O'ahu, and last year (1999) a large military troop vessel smashed into a patch reef in Kane'ohe Bay. As with most ship groundings in the MHI, no penalties were assessed or paid. Installation of a US Navy-supported low frequency transmitter near Kaua'i is raising concerns about possible behavioural disruptions to reef fish, turtles, monk seals and humpback whales.

Oil and Chemical Spills

Hawaii has been increasingly reliant on imported crude oil (52 billion barrels/year) for electricity generation and transportation. Large tankers use MHI ports that are immediately adjacent to coral reefs, and there was a 200% increase in the number of oil spills between 1980 to 1990. While 40% of reported spills are small, larger spills have resulted from ship groundings or offloading accidents. In a recent case, an oil spill near O'ahu caused reef damage when it washed ashore 75 miles northwest on Kaua'i. Recent oil spills in the NWHI are almost entirely due to groundings of fishing vessels on the isolated atolls. The October 1998 grounding of a 25m fishing vessel at Kure Atoll released over half of its 11,000 gallons of diesel onto the shallow reef environment.

Given the extensive urbanisation of coastal areas in Hawaii, the number of motor vehicles used on these small islands, and the channelisation of streams and storm drains into nearshore reef environments, the introduction of toxic chemicals (heavy metals, oils, PCBs, PAHs, tributyltin, pesticides and herbicides) in nearshore environments has the potential to affect biodiversity and fisheries stocks. High concentrations of dieldrin and chlordane were found in oyster tissues sampled near stream mouths in Kane'ohe Bay in 1991, five years after their use was banned in Hawaii. In the NWHI, lead and PCBs have been detected in monk seals, possibly resulting from decaying seawalls at French Frigate Shoals and from dumps established by the U. S. Coast Guard in the past. As a response to increasing shore-based chemical spills (e.g. sulphuric acid, PCBs, and refrigerants), the Hawaii Coral Reef Emergency Response Team (HCRERT) is being created to help assess impacts of these events to the environment.

Sediment and Runoff

Sedimentation from runoff continues to be a chronic problem throughout the MHI as a result of poor land-use practices upslope of reefs. Sediment runoff is estimated at more than 1 million tons per year from agricultural, ranching, urban and industrial activities. 'Slash and burn' agriculture (sugar and pineapple) has decreased, and alternatives such as coffee, macadamia, cocoa and fruit trees may also result in a long-term decrease in sedimentation. This decrease, along with the relatively rapid removal of sediments along exposed coasts into deep water by strong current and wave action, will improve conditions for reef growth in the coming decades. Stream channeling, and paving of coastal and upland areas has also contributed to urban impacts on reefs. Two major freshwater kills of corals occurred in Kane'ohe Bay, O'ahu in 1965 and 1987 from '100 year storms'. Salinity within 1-2m of the surface was reduced to less than half (15ppt) for 2-3 days, causing mass mortality of corals and invertebrates on shallow reefs.

Ship Groundings

With over 16,000 commercial and recreational vessels registered in Hawaii, and numerous others passing through the islands, groundings are a persistent problem. A long-line fishing vessel ran aground on a fringing reef of Kaua'i in 1999 and portions of the vessel still remain; no restitution for damage was made. Large inter-island cruise ships make over 400 port calls a year; such traffic is expected to triple in the next 4 years and with it the potential for groundings and spills. Although longline fishing vessels are not allowed to operate within a 50-mile zone around the NWHI, two have run aground on these pristine reefs in the past couple years.

Tourism and Leisure

These constitute the largest industry, employer and revenue generator in Hawaii, with marine tourism bringing in over US$800 million per year and employing over 7,000 people in over 1000 small businesses. Rodale's Scuba Diving Magazine rated Hawaii as one of the 5 most popular diving destinations in the world, with the small islet of Molokini (and its 1km^2 coral reef) being rated as the third best dive site in the world for the year 2000.

Damage from tourism activities includes: boating impacts; coastal developments; and overuse of reef resources by commercial dive and snorkel tours. These concerns have triggered limitations on carrying capacity in Hanauma Bay, and restrictions on mooring permits for Molokini e.g. 40 commercial tour boats may be moored at one time. Anchor damage has been reduced at major dive sites through the installation of day-use mooring buoys by the government and dive industry. New technologies such as tourist submarines, underwater propulsion units, 'seawalkers' (a type of surface-supplied helmet rig), and rebreathers are being introduced without assessment of their potential impacts. Currently there is insufficient monitoring to assess damage and the marine parks do not have resident rangers or managers.

In contrast, tourism activity is minimal in the NWHI. However, the fragile nature of these small reef areas is evidenced by observations that populations of large jacks appear to have been reduced compared to unfished reefs, although the number of visitors are limited and a catch-and-release fishery is encouraged at Midway Atoll.

Urbanisation
Most (85%) of the State's growing resident population resides on O'ahu and Maui. Many segments of coastlines have been extensively altered by filling of reef flats for coastal development and airstrips, dredging of ship channels and harbours, and construction of seawalls. This has resulted in a loss of coastal wetlands and estuaries, which were important fish nurseries. Modifications have increased erosion rates and sedimentation. In contrast, the only recent construction in the NWHI was a Naval Air Station on Midway Atoll and other modifications for military use.

CURRENT AND POTENTIAL CLIMATE CHANGE IMPACTS

Because the location of the Hawaiian chain is favoured by oceanic gyres and deep water surrounding all the islands, these are predicted to be among the last reefs to experience major bleaching events. There were declines in seabirds, monk seals, reef fishes, and phytoplankton in the NWHI in the 1980s, which may have resulted from regional decreases in oceanic productivity caused by climate shifts or decadal oscillations. Early indications suggest that NWHI populations may now be in recovery.

CURRENT MPAs, MONITORING AND CONSERVATION MANAGEMENT CAPACITY

Current MPAs
Hawaiians traditionally closed areas to fishing (*kapu*) to ensure catches for special events or as replenishment areas for regular fishing grounds. Several types of protected status exist today for Hawaiian reefs, including: Marine Life Conservation Districts (MLCDs); Fisheries Management Areas (FMAs); Marine Laboratory Refuge; Natural Area Reserves (NARs); and National Wildlife Refuges. Several no-take MLCDs have proven to increase fish standing stocks, and some less restrictive MLCDs have provided limited protection from fishing, but fish populations have not increased significantly. Other MLCDs are popular tourist destinations featuring fish-feeding and large-scale commercial activities. A FMA on the Kona Coast restricts collecting for the ornamental trade and has increasing stocks of targeted fish species compared to nearby unprotected areas.

MHI
Functional no-take areas account for only 0.3% of the reefs in the MHI. On the island of O'ahu, there are 3 MLCDs and one FMA. Hanauma Bay (41ha) was the first MLCD in 1967 with taking of all marine life prohibited. Pupukea MLCD (10ha) in northern O'ahu, was established in 1983 and permits pole-and-line fishing, spearfishing (without scuba) and collecting of seaweeds, along with the use of gillnets in the northern portion. The Waikiki MLCD (1988; 31ha) is at one end of Waikiki Beach containing a reef flat that has been greatly altered by shoreline construction, beach replenishment, terrestrial inputs and proximity to a large urban population. The Waikiki-Diamond Head Shoreline Fisheries Management Area (FMA) is a regulated fishing area adjacent to the Waikiki MLCD where fishing (but not at night or with gillnets) is permitted only during even-numbered years. Not withstanding differences in location, size, and habitat type, the completely protected Hanauma Bay MLCD has an order of magnitude higher standing stock of fishes compared with the other areas that had either limited protection or poor quality habitat. Other no-take MLCDs of Hawaii have also been effective in increasing fish stocks. Kealakekua Bay MLCD on the island Hawaii was established in 1969 and Honolua-Mokuleia Bay MLCD on Maui in 1978. All fishing is prohibited in specific sub-zones of these MLCDs and fish biomass has increased steadily since their inception. These examples clearly demonstrate that no-take MPAs with good habitat diversity and complexity increase fish standing stocks.

NWHI
There are several species-specific, limited-take MPAs. The majority of the area is a Critical Habitat for the endangered Hawaiian Monk Seal (*Monachus schauinslandi*). A 50-mile protected species zone is supposed to restrict longline fishing and seasonal area closures prohibited taking lobster until the entire fishery was closed in 2000. The Hawaii Board of Land and Natural Resources will soon hold State-wide public hearings to establish Fishery Management Areas in the NWHI to improve coral reef resource management, including the use of permits to regulate access and manage all activities in coral reef areas in State waters. Recreational and commercial fishing activities are prohibited in the 10-20 fathom isobath of most islands northwest of Kaua'i owing to their status as a National Wildlife Refuge managed by the US Fish and Wildlife Service.

Hawaii Coral Reef Initiative - Research Program (HCRI-RP)
The US government provided funds, through the National Ocean Service (NOS) of NOAA to establish the HCRI-RP as a partnership between the University of Hawaii and the State of Hawaii Department of Land and Natural Resources, Division of Aquatic Resources. The goals are to:

- Assess major threats to coral reef ecosystems, and provide information for more effective management;
- Advance the understanding of the biological and physical processes that affect the health of coral reefs and build management capacity;
- Develop a database to store and access data;
- Conduct public awareness programmes on the threats to coral reefs; and
- Implement education and training for coral reef scientists and managers.

The HCRI-RP has awarded US$700,000 for research projects in the 3rd year including:

- Continued CRAMP monitoring of coral reef health at sites around the main Hawaiian Islands;
- Continued monitoring of impacts of aquarium fish harvesting on West Hawaiian coral reefs;
- Algal identification and the development of a quantitative sampling method that supports coral reef monitoring;
- Real-time water quality monitoring of some coral reefs and the impact of runoff, using macroalgae as an indicator of pollution;
- Development of a rapid assessment method for describing coral reef resources of the Northwestern Hawaiian Islands (NOW-RAMP); and
- Assessment of effectiveness of marine protected areas to conserve fishery resources and the impact of fishing in a management area.

Past research projects funded by HCRI-RP include:

- Coral Reef Assessment and Monitoring Program (CRAMP);
- Kane'ohe Bay Decision Support System;
- Fine-scale Processes on Hawaiian Reefs involving Fish, Algae & Corals;
- Genetic Variation and Status in Hawaiian Coral Species;
- Impacts of Aquarium Fish Collecting Assessment;
- Effectiveness of Marine Protected Areas Assessment;
- Macroalgal Ecology and Taxonomy Support for CRAMP; and
- Ecological Success of Alien/Invasive Algae Assessment.

Under a separate grant from NOAA-NOS, the State will purchase satellite imagery to support the assessment of coral reef resources in the NWHI discussed above. The grant also provides for the establishment of permanent ecological monitoring sites on Kure Atoll and other NWHI resource areas in subsequent years.

Monitoring Programmes

In recognition of the major contribution of coral reefs to the Hawaiian economy and the need to provide better information for managers, the establishment of regional monitoring programmes was the focus of a workshop in June 1998. The Hawaii Coral Reef Assessment and Monitoring Program (CRAMP) was developed to describe the spatial and temporal variation in coral reef communities due to natural and man-made disturbances. The partners are the University of Hawaii, the Hawaii State Department of Land and Natural Resources, federal agencies, and NGOs. CRAMP has initiated monitoring at 25 sites on Kaua'i, O'ahu, Maui, Moloka'i and Hawaii, providing a cross section of different reef types across the MHI, and includes impacted and pristine sites. Initial assessments suggest that Hawaiian coral reefs are in better condition than reefs in many other regions. Nine new Fish Replenishment Areas (roughly 35% of the West Hawaii coastline) prohibit aquarium fish collecting and monitoring of reef fish stocks is undertaken every 2 months aims to document the impact of collecting, and the effect of closure on the sites.

> ## PALMYRA ATOLL: JEWEL OF AMERICA'S PACIFIC CORAL REEFS
>
> Palmyra is the second largest US atoll as well as the only undeveloped and unpopulated 'wet' one left in the Pacific. Because of its abundant water, remote location and development potential, Palmyra was threatened with development as a tourism resort, a base for commercial fishing, a refuelling stop for aircraft, a storage site for spent nuclear fuel rods, and a satellite and rocket launching base. So when the chance came to purchase the whole atoll and its incredibly rich biodiversity in early 2000, The Nature Conservancy (TNC) did just that to establish a wildlife sanctuary with some compatible ecotourism. Furthermore, the US Fish and Wildlife Service proposed establishing Palmyra as a National Wildlife Refuge, ideally in cooperation with TNC. Palmyra is a circular string of 50 emerald islets with white sand beaches scattered around aquamarine lagoons with 6,500ha of shallow and deep reefs. It is at the northern end of the Line Islands with the nearest neighbours: Kingman Reef, 50km northwest; and Hawaii 1,700km to the north. It has virtually never been populated, except for visits by voyaging Polynesians and 20 years straddling World War II, therefore the fish and wildlife resources are undisturbed as they have been for thousands of years. That could have changed if 'development' had come with rats, weeds, and shark and cyanide fishing. Palmyra Atoll is in a great location for coral reef biodiversity as it has three times the coral species (135) of Hawaii (and three times more than the entire Caribbean) because it lies in the Inter-tropical Convergence Zone (the doldrums), and receives larvae from all directions, either via the Pacific Equatorial Current, or the Pacific Equatorial Countercurrent. This way it also serves as a natural stepping stone for the marine and terrestrial species that island and reef hop across the Pacific. Major pelagic and migratory fish stocks pass, along with large populations of seabirds, marine mammals, and sea turtles. For example, there are 29 bird species, including some of the largest nesting colonies of red-footed boobies and black noddies in the world, and large nesting beaches for the threatened green turtle. There is also the world's largest land invertebrate, the rare coconut crab, amongst groves of coconut trees. The reefs have amazing populations of fishes that are being over-fished almost everywhere else in the world, as well as rare giant clams, black-lipped pearl shells, and pen shells. Protecting this atoll will conserve all these valuable resources while allowing a small-scale ecotourism operation to ensure that some people can enjoy this gem in the Pacific.
> From: J.E. Maragos, US Fish and Wildlife Service, Honolulu
> (Jim_Maragos@r1.fws.gov)

Populations of the endangered Hawaiian Monk seals and sea turtles have been monitored by the NMFS off French Frigate Shoals, Laysan, Lisianski, Pearl and Hermes, Midway and Kure in the NWHI since the early 1980s. Annual reef fish surveys are conducted at French Frigate Shoals and Midway Atoll, and the reefs are being habitat mapped using towed divers, video cameras and submersibles.

There are many volunteer monitoring programmes including periodic Reef Check assessments on several islands, and impact-specific monitoring by community groups in the MHI. The DLNR is seeking to coordinate these activities to ensure that data and data-collecting impacts are managed. Part of this effort includes producing the following:

- Best Practices Guidelines: a pamphlet on best management practices for various activities on Hawaiian coral reefs, for direct distribution to the marine tourism industry;
- Visual Impact Cards: use photographs to train divers to identify specific types of coral reef impacts e.g. to distinguish coral disease from fish bites;
- Volunteer Monitoring Techniques Training Manuals.
- NOAA Coral Reef Managers Habitat Classification Workshop: NOS sponsored a Hawaii workshop in 2000 for national coral reef managers to define parameters used to describe reef types and habitats for detailed reef mapping.
- Marine Ecosystems Global Informational Systems (MEGIS) Group: the USFWS formed a multi-agency partnership to create a GIS computer database to share habitat maps and management data for Hawaii and the rest of US Pacific.
- Hawaii Coral Reef Emergency Response Team: the DLNR, with other agency partners is forming a rapid response team to provide resource damage assessment and advice during events e.g. chemical spills, boat groundings, disease outbreaks, etc. The National Fish and Wildlife Foundation has helped equip the team.
- Hawaiian Candidate Species for ESA: the DLNR held a workshop to help develop a proposed list of Hawaiian marine species to submit for Candidate listing under the Endangered Species Act (ESA). Candidate listing helps alert the public, user groups, managers and policy makers of concerns about these species to facilitate voluntary conservation and possible legal listing in the future. The criteria chosen for listing were: restricted range; threats throughout range; limited dispersal; limited reproduction; prolonged time to reach maturity; biological dependency (obligate associations with other organisms); life history characteristics; depleted food-prey; over-fished (includes food, aquarium, research and bioprospecting); and competitive exclusion. Examples of species include: endemic corals with limited Hawaiian range (*Montipora dilatata*, *Porites pukoensis*, *Porites duerdeni* etc.); fish and invertebrates over-fished for food (*Cellana talcosa*, *Epinephalus quernus*, *Epinephalus lanceolatus*, *Scarus perspicillatus* etc.); targets of the marine ornamental trade (*Heteractis malu*, *Centropyge loriculus* etc.); or research (*Lingula reevii*, *Euprymna scolopes* etc.); and habitat-forming organisms of limited range threatened by human activities (*Halophila hawaiiana* etc.).

GOVERNMENT POLICIES, LAWS AND LEGISLATION

The majority of the coral reefs around the MHI are under the jurisdiction of the State of Hawaii (primarily the Department of Land and Natural Resources). In June, 1998, US President Clinton signed Executive Order 13089 for Coral Reef Protection which mandated that: 'All Federal agencies whose actions may affect U. S. coral reef ecosystems shall:

- identify their actions that may affect U. S. coral reef ecosystems;
- utilize their programmes and authorities to protect and enhance the conditions of such ecosystems; and
- to the extent permitted by law, ensure that any actions they authorise, fund, or carry out will not degrade the conditions of such ecosystems.'

This Executive Order requested that various federal agencies assist State and Territorial resource trustees to protect national coral reef resources. The President also created the US Coral Reef Task Force with cabinet-level appointees charged with implementing the Executive Order.

Recently the US President also directed the Departments of Commerce and the Interior, to work directly with the State of Hawaii (and consult the WPRFMC) to develop recommendations for coordinated management of coral reef resources of the NWHI, and resolve questions of jurisdiction. The State of Hawaii holds trusteeship of all reef resources out to 3 nautical miles from land, and the USFWS administers a National Wildlife Refuge throughout the Islands (with the exception of Kure Atoll). NOAA has authority over most reef resources outside 3 nautical miles primarily through NMFS, with the WPRFMC serving in an advisory role.

Hawaii State Laws and Regulations
The State has laws and regulations on uses and impacts on corals and coral reefs e.g. sand, rubble, live rock and coral are protected from harvest or destruction in State waters, and many Hawaiian stony corals cannot be sold. The State Constitution notes that 'the State and its political subdivisions shall conserve and protect Hawaii's natural beauty and all natural resources, including land, air, mineral and energy sources, and shall promote the development and utilisation of these resources in a manner consistent with their conservation and in furtherance of the self-sufficiency of the State.' In addition, the Constitution states that 'each person has the right to a clean and healthful environment, as defined by laws relating to environmental quality, including control of pollution and conservation, protection and enhancement of natural resources. Any person may enforce this right against any party, public or private, through appropriate legal proceedings, subject to reasonable limitations and regulations as provided by law.' These can be applied to conserve coral reefs. In 2000, a law was enacted banning the harvest of shark fins or export through the State.

Federal Laws and Regulations
A number of Federal laws and regulations apply directly to coral reefs:

- Clean Water Act (CWA) - regulates the discharge of dredged or fill materials into marine and fresh waters; contains provisions governing the filling or draining of wetlands, including seagrass beds, mangroves and salt marshes;
- Coastal Zone Management Act (CZMA) - provides for controlled management and development of shoreline areas through cooperation between federal and state agencies;
- Endangered Species Act (ESA) - provides for monitoring, conservation, and recovery of species listed as either Endangered or Threatened as it relates to the use of federal monies or institutions; and
- National Environmental Policy Act (NEPA) - requires Environmental Impact Assessments (EIAs) or Statements (EISs) for proposed projects that involve federal sponsorship or approval, including coral reef-related projects.

Public Education and Outreach for Coral Reef Protection
The DLNR is producing annual State-of-the-Reefs Reports for distribution to the public, policy-makers and government agencies, and distributes pamphlets on coral reef MPAs, fishing regulations, coral laws and regulations and basic natural history information. There are also public education and outreach projects designed to facilitate coral reef management, including community-based monitoring initiatives, a coral reef awareness-raising campaign, community-based marine debris removal, the installation of day-use mooring buoys at Molokini MLCD, and assessments of marine tourism use of MPAs. The Hawaii Coastal Zone Management Program is supporting research and education on Hawaiian coral reefs and has just produced a new booklet to educate visitors on limiting their damaging impacts.

GAPS IN CURRENT MONITORING AND CONSERVATION CAPACITY

MHI
A primary problem is lack of enforcement of existing laws and regulations, compounded by the lack of specificity in many State and Federal laws and regulations about impacts on coral reefs. Often existing laws are out of date with no mechanisms to deal with new technologies, new uses of natural resources, or over-use of these ecosystems. The Presidential Executive Order to protect coral reefs has yet to make a significant impact in Hawaii; of particular concern is continued government support of economic activities, such as offshore aquaculture, bioprospecting, marine ornamental aquaculture, underwater sensor technology, and shoreline or harbour modification. Current fisheries monitoring provides insufficient information on the large recreational and subsistence fisheries sectors for management. Overuse of MPAs by the tourist industry and a lack of fully protected (no-take) reserves means that there are no coral reef areas in the MHI that are pristine and not exploited. There is a need for a series of fully protected coral reef reserves, however, increasing populations and the proliferation of user groups all demanding access to reef resources, mean that the creation of such new types of MPA are unlikely in the near future.

Ongoing Progress

- The National Ocean Service of NOAA started in mid 2000 to map all shallow (< 20 m) coral reef habitats of the MHI using a specially-outfitted aircraft and will spend the next year producing maps and analyses. These are critical for determining management strategies for these resources;
- The USFWS, the State of Hawaii and the academic community are planning a workshop on alien species for resource managers in mid-2001;
- DLNR is producing a training manual for community volunteer groups on monitoring Hawaii's coral reefs;
- The State, using money from NOAA, is studying the social and economic uses of coral reefs;
- The City and County of Honolulu is constructing a major environmental education centre focused on coral reefs at the county park directly adjacent to the heavily visited Hanauma Bay MLCD; and
- The Waikiki Aquarium and DLNR are attempting to restore damaged coral habitats smashed by large marine tourism vessel moorings in Kealakekua Bay MLCD on the island of Hawaii.

NWHI

There are concerns about the effectiveness of multiple agency jurisdiction over one of the last major reef ecosystems in the world that has not been heavily impacted by human activities. Recent Presidential initiatives may be successful in creating the collaboration necessary to protect these resources. A major challenge is patrolling and enforcing regulations over 1600km of the island chain to ensure compliance by fishing, research and ecotourism users. There is a need to create an automated Vessel Monitoring System (VMS), with all vessels carrying transmitters that automatically notify the ship and enforcement authorities of its location in relation to protected areas. Current management is fragmented among agencies with differing missions. Reliance on some agencies to inform others on the management of fisheries resources has had limited success and Federal court action has been required to force the NMFS to comply with fisheries management and endangered species laws. The proposed WPRFMC management plan aims to manage the coral reef ecosystems, yet does not include other fisheries management plans operating in these ecosystems in the proposed measures. Since many coral reef species are listed under existing management plans, the ecosystem approach is subordinate to single-species decision-making. Successful mechanisms like automated VMS, active zoning etc. have not yet been incorporated into proposed regulations.

Ongoing Progress

- The USFWS, DLNR, NMFS, NOS and the academic community will undertake a large-scale Rapid Ecological Assessment (REA) of the coral reefs of the NWHI during September and October, 2000 with 3 teams on 2 research ships. They will assess benthic and fish coral reef resources, map habitat complexity and assess impacts to the ecosystem at all reefs along the 1600km of the NWHI. These results will be used to allow for better management of the coral reef resources;

- DLNR is planning the establishment of a Fisheries Management Area (FMA) for all State waters in the NWHI to protect these unique resources. A State-managed FMA, overlapping and cooperating with the 2 USFWS managed National Wildlife Refuges, will provide better protection for the majority of coral-dominated NWHI coral reefs.

CONCLUSIONS

Coral reefs have always been important for the Hawaiian people, providing food and shoreline protection. The nearshore reefs once provided most of the animal protein and subsistence activities continue, along with commercial, and recreational exploitation and non-extractive activities. Despite this importance, coral reefs are now being degraded in parallel with continued population growth, urbanisation and development. Ocean outfalls, urban and recreational coastal development (hotels, golf courses, etc.) are focal points for coral reef degradation, and new technologies for extraction, offshore aquaculture, and bioprospecting raise concerns about the ability of management agencies to keep ahead of damage to coral reef resources. Short-term economic pressures that ignore the detrimental impacts of development will have severe consequences for Hawaiian coral reefs, which protect shorelines and support the growing marine tourism industry, which is a major component of the economy.

Over-exploitation of most target food fish and invertebrates, and key species for the marine aquarium trade is evident in the MHI. The extent is certainly an underestimate as there is serious under-reporting of current levels of exploitation, further compounding problems for resource management.

The Hawaiian Archipelago reefs constitute the vast majority of US reef area, and Hawaiian reefs contain a high proportion of different reef habitats and endemic species. But over-exploitation, alien species introduction and marine ornamental collection threaten these resources more than in other US coral reef areas.

Hawaii has a wide variety of MPAs which are partially effective in protecting some coral reef habitats, however very few of the reefs (0.3%) around the MHI are fully protected as 'no-take MPAs', where extraction is prohibited. Thus it is going to be extremely difficult to meet the target set by the US Coral Reef Task Force to establish a minimum of 20% of coral reef habitat as 'no-take MPAs' by 2010. Even the few existing MHI 'no-take MPAs' are exposed to heavy human use for recreation and marine tourism, potentially undermining their effectiveness as representing 'natural' coral reef ecosystems. The situation is more promising in the NWHI where anthropogenic pressures are minimal, mainly due to the isolation of these reefs from human populations; however, because corals grow at far slower rates in the NWHI than the MHI, any reef damage in the NWHI is of particular concern.

While new partnerships have been formed by management agencies, academia, NGOs and user communities, there is a need for more financial and political support for the existing and proposed efforts at conserving the exceptionally wide variety of coral reef habitats and resources in the Hawaiian Archipelago.

AUTHOR CONTACTS

Dave Gulko is with the Hawaii Department of Land & Natural Resources david_a_gulko@exec.state.hi.us, Jim Maragos works for the US Fish & Wildlife Service in Hawaii- jim_maragos@fws.gov, Alan Friedlander is with the Oceanic Institute, Waimanalo - afriedlander@oceanicinstitute.org, Cynthia Hunter is Curator at the Waikiki Aquarium Honolulu - cindyh@hawaii.edu, and Rusty Brainard is with the National Marine Fisheries Service in Honolulu - Rusty.Brainard@noaa.gov

14. Status of Coral Reefs in the US Caribbean and Gulf of Mexico: Florida, Texas, Puerto Rico, U.S Virgin Islands and Navassa

Billy Causey, Joanne Delaney, Ernesto Diaz, Dick Dodge, Jorge R. Garcia, Jamie Higgins, Walter Jaap, Cruz A. Matos, George P. Schmahl, Caroline Rogers, Margaret W. Miller and Donna D. Turgeon

Abstract

United States Federal funding for regional mapping, monitoring, and management of coral reef ecosystems of Florida, Texas, Puerto Rico, US Virgin Islands and Navassa increased significantly in 2000. The US Coral Reef Task Force outlined a National Action Plan, which the National Ocean Service (NOS) of NOAA (National Oceanic and Atmospheric Administration) funded some of the activities relevant to the US Caribbean and the Gulf of Mexico. Besides the NOS cooperative grants of over US$0.6 million to assess and monitor coral reef ecosystems and to complete several projects outlined in the All Islands Initiative 'Green Book', NOS has spent several months mapping coral reefs off the US Virgin Islands and Puerto Rico. Additionally, NOS provided a further US$1million of FY2000 funding cooperative grants for coral reef research and monitoring to the National Coral Reef Initiative (NCRI), and the Puerto Rico Department of Natural and Environmental Resources. NCRI grant projects include mapping, monitoring, and research conducted on Florida East Coast coral reef ecosystems. Puerto Rico will initiate assessments and set up at least 16 long-term monitoring sites at selected reefs around Puerto Rico, and a cooperative grant will result in a baseline characterisation of coral reef and seagrass communities off Vieques Island. The US Virgin Island grant will support a multi-agency effort among the Virgin Island Department of Planning and Natural Resources, the US Geological Survey, the University of the Virgin Islands, and the National Park Service to establish baseline conditions prior to the anticipated establishment of marine reserves so that their effectiveness in promoting recovery of fish assemblages and marine habitats can be evaluated.

The hope is that such heightened attention to coral reef issues in this region will continue into the near future, the following recommendations are proposed for regional scale action:

- Identify hydrological and ecological ecosystem linkages at regional scales;
- Establish procedures to reduce regional water quality stresses to coral reefs;
- Determine sources and sinks of marine resources on a regional scale;
- Use satellite resources to map and track watershed influences and aid in characterising the oceanographic patterns both upstream and downstream of coral reef areas;
- Establish cross-boundary and cross-jurisdictional agreements to manage reef areas using an ecosystem approach;

- Establish domestic and international agreements within a region to address coral reef issues, including protocols for marine resource management and protection;
- Establish and implement consistent research and monitoring procedures; and
- Share information that will facilitate cross-jurisdictional management of marine resources, such as when spawning aggregations of species within another jurisdiction feed larvae into the nursery areas elsewhere.

INTRODUCTION

The following report on the status of US Caribbean coral reef ecosystems has been summarised from more extensive reports submitted to the US Coral Reef Task Force (USCRTF) working group that implemented in 2000 'A National Program to Assess, Inventory, and Monitor US Coral Reef Ecosystems'. The more-lengthy reports are also the basis for the biennial-issued document, 'Status and Trends of US Coral Reef Ecosystems'. Each author is a recognised technical expert with responsibility for monitoring and/or managing aspects of their respective coral reef ecosystems.

Florida East Coast

The reefs from northern Monroe County to Vero Beach are a series of 3 discontinuous reef lines parallel to the shore: First Reef to 3–5m of the surface with very low profile cover of algae and small octocorals; Second Reef 6–8m has more relief including dissecting channels and conspicuous octocorals often in high density; and Third Reef 15-22m has the most diversity and abundant stony corals including *Diploria clivosa, Dichocoenia stokesii, Montastraea cavernosa*, and *Solenastrea bournoni*. There has been strong recruitment of *Acropora cervicornis* in the past 3 years and clusters of staghorn coral are common, especially in Broward County, along with large barrel sponges (*Xestospongia muta*). Stony corals are larger on the Third than the Second Reef and moderate sized colonies of *Montastraea annularis* complex are common. Active growth of *Acropora palmata* ceased more than 5,000 years ago on these reefs.

Most reefs have been mapped and coral cover estimated for the first two rows, but less so for the deeper Third Reef. These reefs are protected from some impacts by Florida State statutes and regulations e.g. fishing regulations, dredging permits, prohibition against harvest, sale, or destruction of corals etc., in addition, mooring buoys are being established.

Florida Keys

The Florida Keys coral reefs extend from just south of Miami to the Dry Tortugas and include the only emergent reefs off the continental United States. The Florida Keys National Marine Sanctuary was designated in 1990 to protect and conserve the nationally significant biological and cultural marine resources of the area, including critical coral reef habitats. The Sanctuary covers 9600km^2 with 325km^2 of coral reef (44% in Florida State territorial waters; 66% in Federal waters i.e. more than 3 nautical miles offshore). The reefs comprise a bank reef system of almost continuous reef communities in lines that run parallel to each another. There are several distinct habitats represented including offshore patch reefs, seagrass beds, back reefs and reef flats, bank or transitional reefs, intermediate reefs, deep reefs, outlier reefs, and sand and softbottom areas.

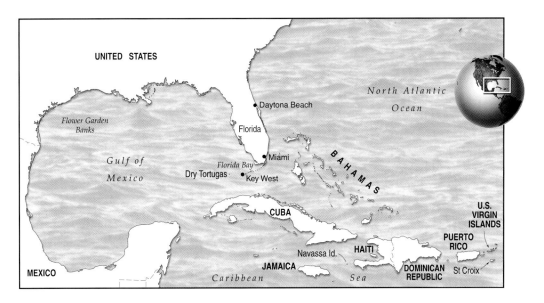

Texas Flower Garden Banks

These are two prominent geological features on the edge of the continental shelf in the northwest Gulf of Mexico, approximately 190km southeast of Galveston, Texas. The Banks are uplifted Jurassic salt domes, rising from 100m depth to within 17m of the surface and have about 1.4km² of luxurious bank reefs on the shallowest portions of the East and West Flower Garden Banks. These are the most northerly coral reefs on the continental shelf of North America (27°52' to 27°56' North) and also some of the most isolated reefs of the Caribbean, being over 690km from the nearest reefs of Campeche Bank off Yucatan, Mexico.

The East Flower Garden Bank (27°54'N; 93°36'W) has about 70% of the coral area, with the rest on West Flower Garden Bank (27°52'N; 93°49'W) about 22km away. These reefs have relatively low diversity of reef-building corals (21 species), probably because their isolation. The dominant coral species are *Montastrea annularis* complex, *Diploria strigosa*, *Montastrea cavernosa*, *Porites astreoides* and *Colpolphyllia natans*, but there are no branching *Acropora palmata* and *A. cervicornis*, and no sea whips or sea fans (gorgonians), which are common elsewhere in the Caribbean. Coral cover is high, ranging between 45 and 52% with crustose, coralline and calcareous green algae also common. The Flower Garden Banks are composed of large, closely spaced coral heads up to 3m in diameter and height, which are hollow in places due to bioerosion and separated by sand patches and channels. These corals grow from the top near 17m down to about 50m.

The reefs were designated as the Flower Garden Banks National Marine Sanctuary (FGBNMS) in 1992, and Stetson Bank was added in 1996. The Sanctuary covers 143km² and includes all the coral reef areas. Regulations are designed to protect the corals with prohibitions on: oil and gas exploration; anchoring or mooring of vessels over 30m; anchoring of smaller vessels near mooring buoys; injuring or taking coral and other marine organism; use of fishing gear other than hook and line; discharging or depositing any substances or materials; altering the seabed; building or abandoning any structures; and using explosives or electrical charges.

Puerto Rico

Along with the main island, there are two inhabited small islands off the east coast (Culebra and Vieques), and three uninhabited islands (Mona, Monito, Desecheo) off the west coast. Most coral reefs occur in the east, south and west coasts, with fringing reefs being the most common type. The western two-thirds of the north coast consists of mainly hard ground and reef rock with low to very low coral cover and some small, sparse, low coral colonies. Coral reefs cover approximately 3,370km^2 within 3 nautical miles of the coasts, which are about 3% of the total coral reef area under US jurisdiction (Hawaiian Islands are first with 85%). The main islands of Puerto Rico, including Culebra and Vieques, are almost completely ringed by reefs, although coral reef abundance is highly variable and dependent of local conditions.

US Virgin Islands

Coral reefs occur around all the major islands; St. Croix, St. John, and St. Thomas, as well as the offshore cays. Fringing reefs, deep reefs (wall and shelf-edge), patch reefs, and spur and groove formations are present, although only St. Croix has barrier reefs. Bank reefs and scattered patch reefs with high coral diversity occur deeper offshore. The US Department of Interior, the US Department of Commerce, and the Virgin Islands Government have jurisdiction over submerged lands with coral reefs within the USVI.

Navassa

This tiny (5km^2) uninhabited US protectorate, located between Jamaica and Haiti, is under the jurisdiction of the US Fish and Wildlife Service. Knowledge about the status of Navassa's reefs is extremely limited, being based primarily on several expeditions over the past 2 years sponsored by the Center for Marine Conservation.

STATUS OF THE CORAL REEFS

Florida East Coast

Reef Health

There have been no comprehensive surveys of the reefs in this area. Their general condition is similar to reefs in the Keys; relatively healthy coral populations with comparable bleaching. The only monitoring is on environmental impacts to specific sites (dredging, ship groundings, pipeline and cable deployments, and beach renourishment), and these have lasted a few years. Recent monitoring has expanded to include environmental conditions (physical and chemical oceanography), coral, sponge, and fish abundance and cover.

Florida Keys

Coral Health

Damage to reefs from human activities has been noticed for decades in the Florida Keys, including boat groundings, propeller scouring of seagrass, accumulation of debris, and improper anchoring practices, which have damaged large areas. In addition to these direct impacts, there are many indirect stresses that are affecting corals and other reef organisms. This is seen as a decrease in coral cover and species diversity and an alarming increase in coral diseases and coral bleaching along the Florida Keys. The Coral Reef/Hardbottom Monitoring Project (CRMP) is monitoring the status and trends of coral communities throughout the Florida Keys National Marine Sanctuary at 43 permanent sites. Since 1996,

73% of sites show losses in hard coral diversity with a maximum of 7 species lost, and most sites have lost coral cover, with an overall loss of 16.4% at 9 sites. Coral diseases are becoming an increasing threat to the overall health of the reef system with over 10 coral diseases observed. Most of these are due to unknown causes, indicating an urgent research need. More and more sites contain diseased coral, and the number of species affected by disease is increasing. This increase in diseases may be directly linked to increases in seawater temperatures.

Coral Bleaching
There has been an increase in frequency and duration of coral bleaching events in the Florida Keys over the past 20 years. In June 1980, millions of reef fish died following 6 weeks of calm, hot weather, and soon after there was slight coral bleaching along the lower Keys reef tract. In June-July 1983, many corals were bleached severely from Big Pine Key to Key West during similar calm weather with warmer water temperatures. This coincided with the massive die-off of long-spined sea urchins. There were large outbreaks of black-band disease on corals in May-September 1986 throughout the Florida Keys, concentrated at the Looe Key National Marine Sanctuary, but coral bleaching was minor. Similar doldrum-like weather in June 1987 triggered severe coral bleaching which lasted 3 weeks. Slight bleaching was seen in 1989, before the advent of more calm weather in July 1990 provoked massive coral bleaching including coral communities in the near-shore waters for the first time. These corals were considered to be acclimatised to wide temperature variations, but the extremes of the summer of 1990 pushed them beyond their previous limits. Sanctuary biologists and managers documented 65% losses of *Millepora* corals at Looe Key Reef. Prior to this only a few percent of corals had died after bleaching episodes.

The next major coral bleaching was in 1997 during the same calm weather patterns with widespread bleaching of inshore coral communities. The following winter was mild and also the spring of 1998, so water temperatures remained high and bleaching virtually continued into the summer of 1998, when it was again severe. Then Hurricane Georges and Tropical Storm Mitch hit the Florida Keys while the corals were still stressed and large amounts of living coral cover was lost on some shallow reefs as a result of these combined impacts. These losses of corals have highlighted the importance of continued monitoring to improve management strategies to maintain coral health.

Research & Monitoring Programmes
The Sanctuary implemented a comprehensive monitoring programme in the Florida Keys reef. The largest component is the Water Quality Protection Program (WQPP) which began in 1994 with 3 components: water quality; corals and hardground communities; and seagrasses. The status of reef fishes, spiny lobster, queen conch, benthic cover, and algal blooms are also monitored at fixed stations. The Zone Monitoring Program monitors the effects of no-take management, which began in 1997 at 23 discrete marine reserves. The goal is to determine by 2002 whether the no-take zones are protecting marine biodiversity and enhancing human values related to the Sanctuary. Measures of effectiveness include the abundance and size of fish, invertebrates, and algae; and economic and aesthetic values of Sanctuary users and their compliance with regulations. This is a three-level program which includes monitoring changes in ecosystem structure (size and number of invertebrates, fish, corals, and other organisms) and function (such as coral recruitment, herbivory, predation). Human uses of zoned areas are also being tracked.

Water Quality
The Water Quality Protection Program of the Sanctuary has assessed the status and trends in a range of water quality parameters at 154 sites in the Florida Keys, Florida Bay, and western Florida shelf since 1995. These show higher nutrient concentrations in the Middle and Lower Keys than in the Upper Keys and Dry Tortugas regions, and declining trends inshore to offshore for nitrite, nitrate, and ammonium. There have also been significant increases in phosphorous.

Algae, Seagrasses, and Benthic Organisms
Rapid, large scale sampling of benthic communities along the Florida Keys National Marine Sanctuary began in 1998 at 80 sites, with 34 in no-take zones compared to others in fished reference areas. Algae were the dominant bottom cover at all sites (average cover above 75%) with fine and thick turf algae, brown fleshy algae and green calcareous algae dominating deep water sites, and calcareous algae such as *Halimeda* spp. dominant in shallower sites. Sponges and soft corals were abundant, but variable at all deep and shallow sites due to the impacts of Hurricanes Georges and Floyd in 1998 and Hurricane Irene in 1999. The cover and abundance of seagrasses has also been shown to vary.

Texas Flower Garden Banks
Reef Health
These are probably the least disturbed coral reefs in the Caribbean and western Atlantic i.e. in nearly pristine condition as they sit in clean oceanic waters with exceptional visibility of around 30m, and stable temperatures (minimums in February 19°-20°C; maximums in August 29°-30°C). Since the early 1970s, the coral communities have not changed and appear in excellent health. There is isolated damage from anchoring of vessels, laying of tow cables and seismic arrays and the impact of illegal fishing gear. Some of the targeted fish populations may have been reduced as anecdotal reports from the 1970s mention large populations of snapper, jewfish and other grouper. Like other reefs, these experienced significant increases in algal cover within 2 years of the die-off of the long-spined sea urchin (*Diadema antillarum*) in 1983, but unlike other reefs, the algal populations returned to earlier levels with apparent increases in other herbivores. There is periodic coral bleaching, but it is usually minor with little mortality. Bleaching during the summer of 1998 was slightly higher, but was still less than 5% of the entire reef, with mortality of less than 1%. Predation of living coral by parrotfish is commonly observed with no apparent long-term consequences.

In 1989, the National Oceanic and Atmospheric Administration (NOAA) and the Minerals Management Service (MMS) established long-term monitoring of coral cover, relative abundance, diversity, and coral growth rates and health. This enhanced another long-term study of the benthic community started in 1972 that assessed the potential impact of oil and gas development in the area. Other studies have included reef fish counts, coral spawning, recruitment and genetic observations. Coral health and cover has been stable since 1972 at an average of 47% with higher cover of 52% (54% on East Bank; 50% on West Bank) on the reef cap. Other components are bare rock (45%), algae (3%), sponges (2%) and sand (<0.1% -outside sand channel areas). *Montastrea annularis* is the dominant species (29%), followed by *Diploria strigosa* (9%), *Porites astreoides* (5%), and *Montastrea cavernosa* (4%). No bleaching was observed in 1996, and in 1997 less than 2% bleaching was seen on both banks, and a very low incidence of coral disease. Bleaching in 1998 was less than 5% for both banks.

Puerto Rico
Coral Health
The dominant bottom cover of the shallow reefs (1-5m) was algae, ranging from 31.8% to 82.1% (mean 65%) and live corals varied from 48.9% to 3.7% (mean 15.5%). The dominant bottom cover at 6-12m was algae (28.2-98%) and live coral (0.6 - 49.1%). On deeper reefs (15-25m), coral cover varied from 0% to 44%. A total of 228 coral species, including 117 scleractinian (stony) corals, 99 soft corals and gorgonians, 13 corallimorphs, 3 fire corals, and 5 hydrocorals have been reported, such that the coral reefs of Puerto Rico are the best developed in the US. Caribbean. Generalisations about reef health are difficult because there are few long-term monitoring data, but it is obvious that reefs near the main island have been damaged because of their close proximity to coastal developments. Coral reef research at 52 reefs during the past 10 years has focused on community studies and monitoring programmes to select areas for marine reserves and long term data sets are available for only a few reefs from 1984 to 1999. Several monitoring programmes are underway to document changes in the community structure.

Coral Diseases
Mass mortalities of corals and other reef organisms have been seen in Puerto Rico e.g. white and black band disease destroyed large areas of corals in 1979 and mass mortalities of sea urchins in 1983, but the urchins have since reappeared. Whereas the brain corals were affected by an outbreak of the White Plague in 1996 less than a month after Hurricane Hortense.

Coral Bleaching
There was massive coral bleaching in the late 1980s with considerable mortality, and there was also major bleaching in 1998, followed by major recovery with very little mortality (see Chapter 1).

US Virgin Islands
Coral Health
An array of stresses has degraded the coral reefs and other marine ecosystems, as well as the fishery resources. Among the natural stresses are 8 hurricanes that caused conspicuous damage to USVI reefs in the last 20 years, and coral diseases that have caused coral losses on all 3 islands, especially in 1997 off St. John. Anchoring and ship groundings on coral reefs and seagrass beds are examples of acute stresses with immediate, and sometimes long-term effects. Dredging, sand extraction, groyne construction and sewage effluent have affected reefs, especially around St. Thomas and St. Croix. Throughout the islands, chronic stresses like over-fishing (commercial, hand-line, trap fishing, spear fishing, net, long-line, trolling, driftnet), point and non-point sources of pollution, and sedimentation from accelerated runoff after land clearing are harder to quantify and track, but may do the most damage. Moreover, many of these stresses can combine with natural disturbances to accelerate damage to reefs or slow their rate of recovery. Over-fishing has markedly reduced resources, including those within Virgin Islands National Park (VINP) and Buck Island Reef National Monument (BUIS). Reports from 20 years ago suggested that fishing was already changing the reef fish populations, even before developments on land caused

extensive loss of habitat as well coral diseases, hurricanes, and other stresses. Fisheries are close to collapse, with even areas within the boundaries of 'marine protected areas' are deteriorating. Existing zoning, erosion control, and fishing regulations are not providing sufficient protection against this unprecedented combination of natural and human stresses.

Coral Diseases
These have caused extensive mortality on reefs around St. John and St. Croix. Corals around Buck Island experienced less disease than those around St. John, except for white band disease. Recent measures show the incidence of disease on corals to be 5.4% (St. Thomas), 5.6% (St. John) and 2.0% (St. Croix). The branching *Acropora palmata, A. cervicornis* are the most vulnerable to storm damage and are also susceptible to white band disease e.g. the impressive cover of *A. palmata* on Buck Island fell from 85% in 1976 to 5% in 1988. Only 6 living *A. palmata* colonies were found on Haulover Bay, St. John in 1999, where previously there had been impressive stands. Small patch reefs of *Porites porites* died in many bays, around St. John, St. Thomas, and St. Croix, possibly from an undetected disease, and some have been dead for over 12 years. Gorgonians have been killed by a fungal disease (*Aspergillis syndowii*), which coincided with strains of the fungus isolated from African dust carried over the Caribbean.

Coral Bleaching
There was extensive coral bleaching in 1998, but mortality was relatively minor. Bleaching on Newfound and Lameshur Reefs, St. John was 43% and 47%, and 41% on Caret Bay reef, St. Thomas during the hottest summer seawater temperatures on record. All corals that bleached at Buck Island recovered, but only 50% of bleached colonies of *Montastraea* on Saba Island, St. Thomas had fully recovered their pigmentation 6 months later. There was previous bleaching in 1987 and 1990 around St. John.

Navassa
Reef Health
The shallow reefs of Navassa have high live coral cover and a high degree of architectural complexity, which is particularly valuable as reef fish habitat. There is also a high abundance of small, newly recruiting corals and extremely low incidence of coral diseases, which suggests that the benthic reef communities are 'healthy. At least 36 hard coral species are reported from 4 sites along the west (lee) coast. Octocorals were less varied with 12 species. Mean percent cover of live coral ranged from 20 to 26.1% for 4 sites along the west coast. Other major community components (in terms of space occupation) were sponges (7-27%), fleshy brown algae (10-24%), and crustose coralline algae (5-16%). Finally, the keystone grazing urchin, *Diadema antillarum*, was moderately abundant in all sites e.g. 2.9 urchins per 30 m^2 belt transect.

STATUS OF CORAL REEF FISHES AND INVERTEBRATES

Florida East Coast
Reef Fish
Recent observations indicate that some of these reefs are in relatively good condition based on corals and fish populations. The Florida Current (Gulf Stream) often influences these reefs (particularly the reefs off Palm Beach County). The current moderates winter

temperature and on occasion brings algal blooms over the reefs. During cold-water upwelling events there are also reef fish kills. In summary, these reefs are in similar condition as those in the Keys.

Florida Keys
Reef Fish
Over 500 fish species have been identified from the Florida Keys, with 180 being reef fish. Fish monitoring of 263 species began in 1979, and monitoring results indicate that over 59% of all fish come from 10 species. The bulk of the biomass was from a few species, namely bluehead wrasse, bicolor damselfish, tomtate, sergeant major, striped parrotfish, yellowtail snapper, bluestriped grunt, white grunt, masked goby, and French grunt. Other species of large fish seen are tarpon, barracuda, gray snapper, Bermuda chub, stoplight parrotfish, smallmouth grunt, and yellow goatfish. The most significant observations have been that there are relatively few individual fish of legal, harvestable size, confirming other data indicating that reef fish are over-exploited. Data show that 13 of 16 groupers, 7 of 13 snappers, one wrasse, and 2 of 5 grunts are over-fished in the Florida Keys and such non-sustainable fishing practices are changing fish trophic structures on the reefs and leading to reduced reproductive capacity.

But the news is not all bad. Fish numbers in the no-take areas are increasing and the numbers of target reef fish per are higher in no-take Sanctuary Preservation Areas than in fished reference sites. There has been an overall increase in abundance of 4 species of snapper (*Lutjanidae*) and hogfish (*Lachnolaimus maximus*) at 7 of 9 areas after the establishment of the no-take zones in 1997. Commercial fishing is one of the largest industries in the Florida Keys and is heavily regulated through annual catch quotas, closed seasons, gear restrictions, and guidelines which set minimum catch sizes. State and Federal fisheries management councils have developed laws and the State collects catch information on 400 species of fishes, invertebrates and plants in order to follow trends.

Invertebrates
Populations of spiny lobster (*Panulirus argus*) are now more abundant and larger in protected areas than in similar sites outside. However, queen conch populations (*Strombus gigas*) continue to decline in both no-take areas and reference sites despite a cessation of harvesting by commercial and recreational fishermen in the mid-1980s. Attempts are underway to supplement wild populations with laboratory-reared stock. Populations of the black-spined urchin (*Diadema antillarum*) have shown poor recovery since the die-off in 1983. So far 62 hard coral species and 90 species of algae have been identified in the Florida Keys, as well as 38 species of sponges, 33 species of polychaete worms, 82 species of echinoderms, two species of fire coral, and 42 species of octocoral.

Texas Flower Garden Banks
Fish Health
Fish diversity on the very isolated Flower Garden Banks is low compared to other Caribbean reefs (approximately 260 fish species), but abundances are high. The fish population includes both resident tropical species and migratory pelagic species. Some of the most abundant species at all three banks are butterflyfish (*Chaetodon sedentarius*), Spanish hogfish (*Bodianus rufus*), bluehead wrasse (*Thalassoma bifasciatum*), brown chromis

(*Chromis multilineata*), bicolor damselfish (*Stegastes partitus*), creolefish (*Paranthias furcifer*), and sharpnose puffer (*Canthigaster rostrata*). The banks are a year-round habitats for manta rays and whale sharks, and serve as a winter habitat for several species of sharks and juvenile loggerhead sea turtles. Anecdotal information suggests that populations of the major target fish (snapper and grouper) have declined.

Puerto Rico
Fish Health
There is no large-scale commercial fishing, but there are modest artisanal fisheries with clear evidence of over-fishing (reduced total landings, declining catch per unit effort, shifts in catch to smaller sized individuals and recruitment failures). Reef fisheries have plummeted with a reduction of 69% in fish landings from 1979 to 1990. The catch in 1996 was 1,640mt of fish and shellfish, valued at US$7.06 million for the artisanal fisheries. There is an absence of large fish predators, and parrotfishes, which has stimulated a proliferation of damselfishes (e.g. *Stegastes planifrons*) which farm algae and inhibit coral growth. Persistent fishing pressure on spiny lobster has reduced their abundance, with a parallel increase in their favourite prey, coral eating molluscs, which are now affecting *Acropora palmata* populations in the southwest of Puerto Rico. There are approximately 800 species of coral reef fishes, but there are no endemic species.

US Virgin Islands
Fish Health
It is difficult to separate out the effects of mangrove, seagrass bed and reef loss from the effects of over-fishing. Degradation of these habitats has undoubtedly contributed to the significant changes in reef fish populations, but there is also clear evidence of heavy fishing pressure with reduced stocks even within the national parks. Existing regulations have failed to protect reef fishes or return populations of large groupers and snappers to natural levels. Enforcement is poor e.g. over 50% of traps did not have the mandated biodegradable panels to allow fish to escape if traps were lost. However, it is unlikely that full compliance with existing regulations will reverse these alarming trends. Queen conchs used to be very abundant around St. John, but populations are decreasing and density of conchs inside the Virgin Islands National Park are no higher than outside. Normally conchs live in seagrass beds, but this habitat has been reduced greatly as a result of hurricane and anchor damage. Similar estimates show decreases in the average size of lobsters since 1970.

Navassa
Fish Health
Despite its remoteness, an active artisinal fishery (primarily using traps and hook and line) by Haitians is the primary human stress to Navassa reefs. Subsistence fishing is allowed under Wildlife Refuge regulations, nevertheless, shallow reef fish communities have high density (range 97-140 fish/60m^2) with healthy populations of large snapper, grouper, and herbivores, which are largely absent in nearby Caribbean locations with high fishing pressure.

ANTHROPOGENIC THREATS TO CORAL REEF BIODIVERSITY

Florida East Coast
There are varied and chronic stresses from this extremely urbanised coast. Dredging for beach renourishment, channel deepening and maintenance have significantly reduced water quality, smothering corals and other invertebrates and lowering productivity e.g. Boca Raton and Sunny Isles. Recreational usage can be extreme especially in warmer months, with clear evidence of fishing gear impact and anchor damage. Shipping from the large ports (Miami, Port Everglades, and Palm Beach) means that ships frequently run aground or anchor on reefs. Ocean outfalls pour large volumes of secondary treated sewage into the coastal waters.

Florida Keys
The major threats to the coral reefs stem from over 4 million annual visitors and 100,000 residents in the Florida Keys. The population of Monroe County has grown by 160% during the past 40 years, and this trend is expected to continue, with estimated increases of 65% over the next 20 years. Similarly, the number of registered private boats has increased over 6 times since 1965. Most visible damage in the last 20 years has been from direct human impacts such as grounding boats in coral, seagrasses, or hardbottom areas, breaking and damaging corals with ship anchors, destructive fishing methods, and divers and snorkelers standing on corals. Boat propellers have permanently damaged over 12,000ha of seagrasses, and over 500 small boat groundings are reported annually in the Florida Keys National Marine Sanctuary. Large ships have been responsible for damaging or destroying over 10ha of coral reef habitat.

Indirect human impacts are significantly affecting the coral reefs with euthrophication in nearshore waters a major documented problem. Wastewater and stormwater treatment, and solid waste disposal facilities are inadequate in the Keys. In Florida Bay, reduced freshwater flow has resulted in plankton bloom increases, sponge and seagrass die-offs, fish kills, and the loss of critical nursery and juvenile habitat for reef species, which affects populations on the offshore coral reefs. Other indirect pressures on reef resources include serial over-fishing that has dramatically altered fish and other animal populations. It is critical to reduce these impacts if the Florida Keys economy, largely based on the coral reefs, is to be sustained. Tourism is the number one industry, with commercial and recreational fishing also being important with annual contributions of US$57 million and US$500 million respectively.

Texas Flower Garden Banks
Physical damage from vessel anchoring, potential water quality degradation, impacts of fishing and fishing related activities, and impacts from oil and gas exploration and development are the primary anthropogenic threats to the coral reefs. Anchors from large ships have devastating local impacts with some clear examples, including foreign-flagged cargo vessels unaware of anchoring restrictions. This continues to be a problem, as the National Marine Sanctuary regulations are not included on foreign navigational charts. Potential pollution sources include coastal runoff, river discharges and effluent discharges from offshore activities such as oil and gas development and marine transportation. Oxygen-depleted or 'hypoxic' bottom waters occur in large areas of the northern Gulf ($16,500 km^2$) from the Mississippi delta to the Texas coast. Though not currently in proximity to the Flower Gardens, these could potentially influence resources on the Banks.

Puerto Rico
The present status of Puerto Rican coral reefs is amongst the most critical in the Caribbean, due to accelerated urban and industrial coastal development during the last 40 years combined with a lack of effective management of these resources. Massive clearing of mangroves, dredging of rivers for sand and harbours, runoff from large scale agricultural developments, deforestation in large watersheds, raw sewage disposal and power plants are all important factors stressing the coral reefs. Other major anthropogenic activities include oil spills, anchoring of large oil cargo vessels, over-fishing, uncontrolled recreational activities, eutrophication, and military bombing activities (at Vieques and Culebra Islands). The coastal waters are monitored and evaluated for direct human and indirect human health problems.

US Virgin Islands
Destruction from boat anchors and from boats running aground on reefs has been severe on St. John's reefs. The worst was the cruise ship Windspirit, which destroyed 283m^2 of reef in 1988, with no recovery 10 years later. Small boats frequently run aground on shallow reefs, destroying corals, particularly elkhorn coral making them more susceptible to storm damage and white band disease. Critical habitats have been damaged by anchoring gear, and deployment of fish traps on coral reef areas. Most coral breakage occurs during major hurricanes, however chronic coral damage occurs at areas of high recreational use by snorkelers and divers. On St. Croix, many popular snorkel and dive sites experience heavy visitor use (100-200 visitors/site) on days when cruise ships are in port e.g. Cane Bay, Davis Bay, Buck Island Reef NM, Carambola, Protestant Cay and Frederiksted beaches. The intensive use of the underwater trail at Buck Island Reef National Monument shows damage from snorkelers.

Navassa
Fishing is the only anthropogenic threat to Navassa reefs and this is unlikely to change.

CURRENT AND POTENTIAL CLIMATE CHANGE IMPACTS

Florida
The principal natural environmental controls in this area are hurricanes, severe storms, winter cold fronts, cold-water upwelling, and ground water effects. The assumed climate change scenario is for warmer waters, rising sea levels, more frequent and stronger hurricanes. This will probably cause significant changes to the reefs, including more bleaching. Rising sea levels will flood coastal areas and introduce water quality problems. Therefore, management strategies must focus on alleviating the controllable, anthropogenic impacts while working toward legislation and policy that will address global emissions in the long-term.

Texas Flower Garden Banks
There are no anticipated problems, as the location and depth of these reefs buffer them from the short-term effects of global warming and climate change. However, if summer water temperatures approach or exceed 30°C on a more consistent basis, the current minor incidences of bleaching will probably increase in severity.

Puerto Rico
Current levels of natural factors (hurricanes, coral bleaching, coral diseases) are resulting in considerable coral reef degradation which may mask any signals from climate change.

US Virgin Islands
Hurricanes David (1979) and Hugo (1989) caused severe destruction on the reefs in the USVI, and recovery has been very slow due to subsequent hurricanes in 1995, 1998 and in late 1999. Any increases in these hurricane events (as predicted by many climate models) will inhibit the recovery of elkhorn coral at some places around St. John, St. Thomas and, St. Croix.

Navassa
There is very little ecological information on Navassa reefs, and hence no basis for assessing trends in current and potential climate change impacts.

CURRENT MPAs AND MANAGEMENT CAPACITY

Florida East Coast
There are no MPAs in this area; however, the Biscayne National Park and the Florida Keys National Marine Sanctuary are immediately south. There is an *Oculina* MPA in the far north, established to protect the coral from dredging, trawling and long-line fishing gear damage. A suggested goal of the US Coral Reef Task Force is to declare 20% of all US coral reef ecosystems as MPAs as a wide swath from the intertidal to the state boundary, protecting all species from harvest.

Florida Keys
With the designation of the Florida Keys National Marine Sanctuary in 1990, the entire coral reef tract of the Florida Keys was afforded some level of protection, with oil exploration, mining, and large shipping traffic being excluded. Anchoring on or touching corals in shallow water is prohibited, as is collecting living or dead coral and harvesting 'live rock' for the aquarium trade. Potential pollution sources from outside the Sanctuary that have impacts within can be controlled. After 6 years, a management plan was implemented with strategies for conserving, protecting and managing the significant natural and cultural resources of the Florida Keys marine environment based on an overall ecosystem approach. There are several marine zones to protect specific reef areas more intensely e.g. 23 no-take zones, which cover less than 1% of the Sanctuary but protect 65% of shallow coral reef habitats and were implemented in 1997. Most of the smaller zones (Sanctuary Preservation Areas) are on the offshore reef tract in heavily used spur and groove coral formations. The 31km^2 Western Sambo Ecological Reserve protects offshore reefs and mangrove fringes, seagrasses, productive hardbottom communities and patch reefs. The Sanctuary is planning to create a new ecological reserve in the Tortugas (far west Florida Keys), which will increase the total protection of coral reefs within the Sanctuary to 10%.

Texas Flower Garden Banks
The Sanctuary protects the fragile ecosystem from threats posed by anchoring, oil and gas development and destructive fishing techniques. Sanctuary staff direct resource protection, education, research, and enforcement efforts. There is also an ongoing long-term

monitoring programme. Additional protection is provided by the Minerals Management Service through requirements imposed on industry operators such as the 'Topographic Features Stipulation' for the Flower Garden Banks.

Puerto Rico

The Department of Natural and Environmental Resources (DNER) has designated 8 Special Planning Areas (including all mangroves) and 23 coastal and marine natural reserves. Management plans for these have been developed to contribute to coral reef protection and management. Guidelines and funding under Section 6217 of the Coastal Zone Management Act will be provided in late 2000 to control non-point sources of pollution. A Natural Protected Areas Strategy has been prepared and includes a MPA Sub-system with guidelines for important coastal area and resources identification, management and protection. The new Action Plan maintains the original objective of addressing the lack of information and adequate management of coral reefs. Puerto Rico has completed coral reef assessments for Jobos Bay, Caja de Muerto, Guanica, Tourmaline, and Fajardo with 15 permanent monitoring transects established per site. A joint community and government initiative aims to undertake long-term monitoring of these sites. The Puerto Rico Coastal Zone Management Program has established an inter-agency Coral Reef Committee and it is compiling historic and new information on the coral reefs using a centralised data management system to facilitate the exchange of information. The first Natural Reserve in Culebra now incorporates 'No Take Zones'. A 'Marine Reserve' around Desecheo Island was added to the 23 existing Coastal Natural Reserves. New and revised laws and regulations for the protection of coral reefs, fisheries, and related habitats have been approved and the coral reefs in Puerto Rico are now being thoroughly mapped.

US Virgin Islands

- Hind Bank Marine Conservation District - a seasonal federal closure was enacted at the Red Hind spawning site off St. Thomas in 1990, and improvements in the fishery were documented. In November 1999, the closed area was designated a marine reserve with all fishing and anchoring prohibited;
- Buck Island Reef National Monument (BUIS) was designated in 1961 to protect 280ha around Buck Island, including the reef system. The eastern 66% is a no-take zone including most of the barrier reef, and limited fishing is allowed in the rest. No spearfishing is allowed anywhere in the Monument, but illegal trap fishing occurs throughout, and all types of fishing occur immediately outside the boundary of the Marine Garden. Due to inadequate enforcement and the lack of a buffer area, there has been no effective protection of fish populations;
- Virgin Islands National Park (VINP) occupies 56% of the 48km^2 island of St. John and 2,286ha of the surrounding waters. Traditional fishing with traps is allowed in the park, although illegal commercial fishing is occurring. No fishing, including spearfishing, is allowed in Trunk Bay, the site of an underwater trail;
- Salt River Bay National Historical Park and Ecological Preserve, this park and reserve has 160ha of land and 245ha of water to the 100m depth, including the marine resources of the Salt River Bay, Triton and Sugar Bays. (Omnibus Insular Areas Act of 1992. 16 USC 410tt).

> ### THE TORTUGAS ECOLOGICAL RESERVE: PROTECTING OCEAN WILDERNESS
>
> The Tortugas reefs boast the healthiest coral and highest water quality in the Florida Keys region. These reefs, which lie 120km west of Key West, also contain a diversity of fishes and other organisms that are not seen elsewhere in the Keys. Some populations may contribute a major source of fish and lobster larvae for the rest of the Florida Keys. The Florida Keys National Marine Sanctuary (FKNMS) is working with the State of Florida, the Gulf of Mexico Fishery Management Council, and the National Marine Fisheries Service to preserve the richness of species and health of fish stocks in the Tortugas. The threats to these resources include commercial and recreational fishing, anchoring by freighters, and high visitor levels. A 400km2 'no-take' ecological reserve was proposed for the remote western part of the sanctuary that would protect important spawning areas for snapper and grouper and deepwater habitat for other commercial species.
>
> The Sanctuary convened a 25-member Working Group of commercial and recreational fishers, divers, conservationists, scientists, concerned citizens, and government agencies in 1998 to assist in designing the reserve. They used a series of public meetings and socioeconomic and resource evaluations to examine the whole ecosystem, and then recommended alternatives based on this information. In May 1999, the Working Group reached a consensus on proposed boundaries and regulations for the reserve. In June 1999, the Sanctuary Advisory Council unanimously approved their proposal. A Draft Supplemental Environmental Impact Statement (SEIS) for the proposed reserve was then released which detailed the alternatives considered. Over 4000 comments from locals, residents elsewhere in the USA and from around the world were received on the Draft SEIS, and the vast majority of strongly endorsed the concept of an ecological reserve in the Tortugas. This support was a good demonstration of value of having diverse stakeholder involvement in the planning process and the strong sense of ownership for the proposed plan. Currently, responses are being prepared to address public comments received on the Draft SEIS.

The USVI Government has also designated Marine Reserves and Wildlife Sanctuaries (Salt River, Cas Cay/Mangrove Lagoon and St. James) where fishing is allowed only with handlines or for baitfish with a permit (St. James).

Navassa
This island is part of the US Fish and Wildlife Service's Caribbean Islands National Wildlife Refuge.

GOVERNMENT POLICIES AND LEGISLATION

East Coast Florida
Policies on environmental impacts of dredging, fresh-water management, and nutrient input should receive attention. Vessel anchorages off Miami, Port Everglades and Palm Beach should be reviewed and changed to provide maximum protection for the reef system.

Texas Flower Garden Banks
Regulations governing the FGBNMS under the National Marine Sanctuaries Act, as amended, 16 U.S.C. 1431 are contained within the Code of Federal Regulations and can be viewed on the web at: http://www.sanctuaries.nos.noaa.gov/oms/pdfs/FlowerGardensRegs.pdf.

US Virgin Islands
The US Department of Interior, the US Department of Commerce (including NOAA and the Caribbean Fishery Management Council), and the USVI Government all have policies, laws and legislation relating to coral reefs in the USVI. The Code of Federal Regulations Title 36 and the enabling legislation for Virgin Islands National Park and Buck Island Reef National Monument relate to reefs in the national parks. The Caribbean Fishery Management Council has Reef Fish and Coral Reef Management Plans with regulations pertaining to federal waters. Title 12 of the Virgin Islands Code presents environmental laws and regulations of the Virgin Islands. Several specific Acts relate to regulations on corals, fishing, etc.

Navassa
A 12-mile fringe of marine habitat around Navassa (estimated at 134,000ha) is under US Fish and Wildlife management. Refuge policies allow subsistence fishing.

GAPS IN CURRENT MONITORING AND CONSERVATION CAPACITY

Florida East Coast
There is no comprehensive and systematic monitoring programme for these reefs, yet one is needed to provide a baseline assessment. Site selection should ensure that representative habitats and unique sites are mapped and monitored. This will require that a selection committee of academic, county, state, conservation and fishing groups be constituted, and decisions rapidly disseminated for public discussion.

Florida Keys
Current monitoring in the Sanctuary has focused largely on detecting changes in designated no-take zones and establishing the status and trends in corals, seagrasses, and water quality. Such monitoring must continue in the short-term until solid baseline data are obtained. This baseline will assist in detecting possible long-term changes in communities that may result from management practices (e.g. zoning) or from massive restoration efforts soon to be implemented in south Florida's Everglades.

Texas Flower Garden Banks
Recent observations of increased algal abundance highlight the need to improve water quality monitoring and assess currents and water circulation. The monitoring should include studies on algal populations, coral diseases, and extend to deeper coral reef communities. The great distance of the Sanctuary offshore makes surveillance and enforcement more difficult. Currently, the Sanctuary does not own a boat and relies on charter vessels to get to the area. Recent indications that the Banks may be important spawning areas for several grouper species highlight the need to create a marine reserve to protect the biodiversity of this area.

US Virgin Islands

Some of the longest data sets on coral reefs in the Caribbean come from a diverse array of ongoing monitoring activities. However, intensive, long-term monitoring has only been conducted at a few sites around St. Croix and St. John, with less information for St. Thomas. Coral reef monitoring needs to be extended to include a wider variety of coral habitats and more sites to provide managers with critical information to enable further protection and preservation of key reef areas. Very little is known of the deeper reefs around the USVI especially in the critical grouper and snapper spawning aggregation sites along the shelf edge. Some of these reefs have exceptionally high coral cover. Little is known about the interactions among reefs, mangroves and seagrass beds and how the deterioration of mangroves and seagrass beds contributes to the degradation of coral reefs. All agencies involved in coral reef monitoring suffer from a shortage of staff, and enforcement of regulations has been limited. The Code of Federal Regulations states that commercial fishing is prohibited 'except where specifically authorised by Federal Statutory law'. However, commercial fishing is occurring in the waters of Virgin Islands National Park and Buck Island Reef National Monument.

Navassa

There is no monitoring program ongoing nor even planned for Navassa reefs, nor the artisinal fisheries. This presents an important opportunity to assess the impacts of artisanal reef fisheries in the absence of other direct anthropogenic effects.

CONCLUSIONS

Florida East Coast and Keys

These coral reefs are in a rapid and unprecedented state of declining health, signalling that habitats in the ocean are responding to high human stresses over 4 decades. Corrective actions are required at all levels of national governments; this further emphasises the importance of the US Coral Reef Task Force. National and State leaders in the United States are paying closer attention to the problems confronting coral reefs, but it will take continued commitment and dedication for coral reefs to be protected and conserved for future generations. While there is a fine line between being an alarmist and a strong advocate for coral reef protection and conservation, decisions by people at the local level will have the greatest influence on the survival of coral reefs. It is not too late to take action and it is imperative that people continue to seek solutions to the problems which affect coral reefs. Coral colonies fragmented during Hurricane Georges in 1998 are re-establishing in the Florida Keys National Marine Sanctuary, indicating that the strong regeneration capacity of the reefs will result in recovery of these damaged reefs, provided that the major human stresses (pollution, sedimentation, over-fishing, and physical impact damage) are controlled. 2000 has been a good year for recruitment of staghorn coral (*Acropora cervicornis*), which indicates that management interventions can help coral reefs survive.

The following are recommendations for action in the Florida Keys that if implemented along with other local recommendations could have significant positive impacts on a regional scale:

- Map the benthic habitats of the coral reef community;
- Establish minimal water quality standards that exceed those of existing legislation;
- Eliminate sources of nutrients entering nearshore waters, and sources of heavy metal pollution near coral reefs;
- Protect and conserve all marine habitats;
- Establish 'no take' ecological reserves in strategically well-sited areas;
- Determine sources and sinks of marine larvae at local scales;
- Implement research and monitoring programmes to detect change and sources of change;
- Detect 'hot spot' areas of coral diseases at a local scale;
- Implement education and outreach programmes that focus on problems and solutions;
- Implement an enforcement program or determine alternative means of achieving compliance;
- Learn and monitor the socioeconomic aspects of the area and use these data to support management actions; and
- Form knowledgeable Advisory Groups of local resource users, conservation groups, scientists, and educators to provide information and personal experiences and observations to managers.

US Virgin Islands

There is compelling evidence that more marine protected areas need to be established in the USVI, specifically a 'marine reserve' that prohibits recreational and commercial uses that cause damage and restricts development of adjacent shorelines. This is essential to allow the recovery and replenishment of the fishery and benthic resources. There is irrefutable evidence that additional regulations and enhanced enforcement of existing regulations are necessary to reverse serious decline and degradation in the marine resources. Resource managers from the local government and National Park Service have expressed an interest in establishing marine reserves to protect functional reef ecosystems, to allow their recovery where damage has occurred; to allow recovery of fish assemblages.

- The USVI need management plans for all designated Areas of Particular Concern (APCs);
- There is an urgent need to designate 'no-take' areas (e.g. Lang Bank, Salt River, Buck Island ecosystem);
- The USVI Government needs to become more effective at enforcing existing environmental regulations;
- More stringent environmental regulations need to be created (e.g. no gill-netting, no spearfishing); and
- Environmental education for residents and visitors needs to be improved and extended.

In conclusion, we recommend the establishment of marine reserves in the USVI as soon as possible to reverse the alarming declines and degradation in fishery and benthic resources. Establishment of such recovery zones will be a start in implementing the National Action Plan to Conserve Coral Reefs that was officially adopted in March 2000.

Navassa
The presence of a relatively intact Caribbean reef could provide a unique opportunity for research on the ecological function of Caribbean reefs that could aid in understanding and effective management and restoration of such reefs in other areas of the Caribbean.

AUTHORS AND KEY RESOURCE MATERIAL

Florida East Coast
Walter Jaap, Florida Marine Research Institute (Jim@wahoo.fmri.usf.edu)

Florida Keys
Billy Causey, Superintendent, Florida Keys National Marine Sanctuary, (billy.causey@noaa.gov)

Texas
George P. Schmahl, Sanctuary Manager, Texas Flower Garden Banks National Marine Sanctuary

Puerto Rico
Ernesto Diaz, Puerto Rico Department of Natural and Environmental Resources (eldiaz@caribe.net)

US Virgin Islands
Dr. Caroline Rogers, US Geological Survey, US Virgin Islands (caroline_rogers@usgs.gov)

Navassa
Margaret W. Miller, National Marine Fisheries Service Southeast Fisheries Center (margaret.w.miller@noaa.gov)

Florida East Coast
Ettinger, B.D., D.S. Gilliam, L.K.B. Jordan, R.L. Sherman and R.E. Spieler. The coral reef fishes off Broward County Florida, species and abundance: a work in progress. 52nd Annual Meeting of the Proc. Gulf Caribb. Fish. Instit. November 1-5, 1999. Key West, Florida. USA

Ettinger, B.D. 2000. Coral Reef Fishes of Broward County, Florida: Species and Abundance. Nova Southeastern University, Unpublished M. S. Thesis, 100 pp.

Florida Keys
Bohnsack, J.A., McClellan, D.B., et al. (1999). Baseline data for evaluating reef fish populations in the Florida Keys, 1979 – 1998. NOAA Technical Memorandum NMFS-SEFSC-427. 61 pp.

Richardson, L.L. (1999). The Florida Keys National Marine Sanctuary and State of Florida Zone Performance Review: First Year Report 1998. 17 pp. & (2000). The Florida Keys National Marine Sanctuary and State of Florida Zone Performance Review: Second Year Report 1999. 23 pp.

The Flower Garden Banks

Deslarzes, K.J.P., ed. (1998). The Flower Garden Banks (Northwest Gulf of Mexico): Environmental Characteristics and Human Interaction. OCS Report MMS 98-0010. US Dept. of the Interior, Minerals Management Service, Gulf of Mexico OCS Region, New Orleans, La. 100pp.

Pattengill, C.V., Semmens, B.X. and Gittings S.R. (1997). Reef fish trophic structure of the Flower Gardens and Stetson Bank, NW Gulf of Mexico. Proc. 8th Int. Coral Reef Symposium 1:1023-1028.

Puerto Rico

Department of Natural and Environmental Resources, San Juan, P.R. (1999). Puerto Rico and the Sea, 1999 Report.

Environmental Quality Board, San Juan, P.R. Goals and Progress of Statewide Water Quality Management Planning for Puerto Rico. 2000 305(b) Report.

US Virgin Islands

Garrison, V.H., Rogers, C.S., Beets, J. 1998. Of reef fishes, overfishing and in situ observations of fish traps in St. John, US Virgin Islands. Rev. Biol. Trop. 46 Supl. 5: 41-59.

Rogers, C., Miller, J. (in press). Coral bleaching, hurricane damage, and benthic cover on coral reefs in St. John, US Virgin Islands: a comparison of surveys with the chain transect method and videography. Proceedings of National Coral Reef Institute Conference. April 1999.

Navassa

Atkinson, M.J. (1999) Topographical relief as a proxy for the friction factors of reefs: estimates of nutrient uptake into coral reef benthos. IN Maragos JE and Grober-Dunsmore R (eds.) Proceedings of the Hawaii Coral Reef Monitoring Workshop. Hawaii Divs. of Land and Natural Resources, Honolulu, HI. pp 99-103.

Littler, M.M., Littler, D.S., Brooks, B.L. (1999) The first oceanographic expedition to Navassa Island, USA: Status of marine plant and animal communities. Reef Encounter, 25 July 1999, pp. 26-30.

THE US CORAL REEF TASK FORCE

US President Bill Clinton issued Executive Order 13089 in 1998 'to conserve and protect US coral reef ecosystems', along with a proposal to form the US Coral Reef Task Force (USCRTF) to implement the presidential edict. One of the first actions of the USCRTF was to form Working Groups to review the current position and propose activities within the themes of coastal uses, ecosystem science and conservation, mapping and information synthesis, water and air quality, international dimensions, and education and outreach. Working Group membership included representatives from Federal, State, Commonwealth, and Territorial agencies, non-governmental organisations, and the All Islands Coral Reef Initiative Committee. The USCRTF adopted the 1999 'US All Islands Coral Reef Initiative Strategy', which is the 'Green Book' for action. At the 4th meeting of the USCRTF, representatives from the US associated Pacific Island nations were invited to join the USCRTF. The Task Force issued a National Action Plan in mid 2000, which should generate considerable new coral reef work to start in 2000 and continue for the next few years.

The US Congress allocated US$6 million for coral reef mapping and management activities to the National Oceanic and Atmospheric Administration's National Ocean Service (NOAA NOS), and to implement many of the critical activities identified in the National Plan. Among those actions, US$1 million was devoted to mapping the coral reefs of the US Virgin Islands, Puerto Rico, and the Hawaiian Archipelago. Along with funds from the US Department of the Interior, another US$1M was provided to support the All Island Green Book projects, and another US$1M to help implement a multi-agency, comprehensive 'National Program to Assess, Inventory, and Monitor US Coral Reef Ecosystems'. Within this program, NOS gave support for 6 cooperative grants to start 3-year projects beginning in 2000 in Hawaii, Guam, Puerto Rico, American Samoa, and the Commonwealth of the Northern Mariana Islands. These projects will help build island capacity for assessing their coral reef resources, and filling gaps in monitoring coverage of US coral reef ecosystems.

The prognosis for further support for US States and Territories, and US affiliated Island coral reef ecosystems is reasonably good. Both the Departments of the Interior and Commerce have indications from congress of funding specified for US coral reef activities in FY2001. There are also several bills currently being debated on the floor of the US Senate and the House of Representatives to provide substantial funding for the mapping, monitoring, research, management, and restoration of US coral reefs. Information on the USCRTF can be obtained on http://coralreef.gov/

15. Status of Coral Reefs in the Northern Caribbean and Western Atlantic

Jeremy Woodley, Pedro Alcolado, Timothy Austin, John Barnes, Rodolfo Claro-Madruga, Gina Ebanks-Petrie, Reynaldo Estrada, Francisco Geraldes, Anne Glasspool, Floyd Homer, Brian Luckhurst, Eleanor Phillips, David Shim, Robbie Smith, Kathleen Sullivan Sealey, Mónica Vega, Jack Ward and Jean Wiener

Abstract

All countries have reported a deterioration of reef resources. The most widespread direct human impact is over-fishing, particularly acute in Jamaica, Haiti and the Dominican Republic, where narrow fringing reefs are easily accessible. Reef fish stocks dispersed over broad shelves are less depleted as in Cuba and, especially, in the Bahamas and the Turks and Caicos Islands. Higher standards of living in Bermuda and Cayman Islands have resulted in lower fishing pressures. Impacts from the mass mortality of the sea-urchin *Diadema antillarum* in 1983 are still apparent as excessive growth of macroalgae, at least where over-fishing had depleted herbivores. White-band disease in the *Acropora* spp. has led to catastrophic declines in coral cover, particularly in Jamaica (where there has been some recent recovery). In 1998, coral bleaching was severe in the Caymans and Cuba, but mortality appears to be low. Sediment runoff and nutrient pollution are especially prevalent in the three high islands. Careless coastal development for tourism has damaged reefs in most countries, while increasing pressures of diving tourism are apparent in the Cayman and Turks and Caicos Islands. There is increasing local awareness of the need for coastal conservation, and all countries have declared Marine Protected Areas, except Haiti. There is generally little enforcement of conservation laws and most agencies need more resources, trained personnel and political support.

Introduction

This report covers reefs from the Pedro Bank in the centre of the Caribbean Sea (17°N) to the Bermuda platform in the western Atlantic at 32°N. It includes: Cuba, Hispaniola (with Haiti and the Dominican Republic) and Jamaica of the Greater Antilles with their offshore banks and islands, and 3 clusters of smaller, low islands; the Cayman Islands, within the Caribbean Sea; the Bahamian Archipelago (shared between the Bahamas and the Turks and Caicos Islands) just north of the Caribbean; and Bermuda. Coral reefs are well developed everywhere, most obviously as fringing reefs, but also as patch and bank reefs, especially on broad shallow banks and island shelves. They are of great economic importance in all 8 countries for coastal protection, reef fisheries and, especially, tourism.

Geographical Reef Coverage and Extent

The major physical factors affecting the reef development include: freshwater runoff from the land, which is greater in the high islands; routine wave exposure, especially from the north-east and open to the Atlantic; and sea temperatures which may be modified by cool upwellings from adjacent deep water or by the warm Gulf Stream. Hurricanes are prevalent everywhere, but irregular in frequency. Hard coral diversity is high in most of the region, except for the northern Bahamas and, especially, Bermuda.

Bahamas
The islands lie on a limestone platform in several sections with carbonate banks. The largest is the Great Bahama Bank with Andros Island, while smaller ones form a chain extending from the Straits of Florida to the Caicos Islands. The total area of the banks is nearly 260,000km^2, much greater than the 1,300 islands (12,000km^2). The islands are low-lying and mostly very porous limestone so there is no surface water. Therefore the corals can grow close to the shore. Coral reefs occur mostly fringing the bank margins, with some small patch reefs on the bank in areas with high tidal circulation, and a few bank-barrier reefs. The most important reef regions, with their approximate areas are, from the North: Little Bahama Bank (323km^2), Bimini (90km^2), Berry Islands/Andros (182km^2), New Providence (30km^2), Eleuthera/Cat islands (200km^2), San Salvador/Rum Cay and Conception Islands (132km^2), Exuma Cays/Ragged Islands (386km^2) Samana Cays (50km^2), Plana Cays (31km^2), Mayaguana (72km^2), the Inaguas (164km^2), Hogsty Reef (23km^2), Cay Sal Bank (153km^2) and Crooked/Acklin Islands (151km^2).

Bermuda
These are the most northerly coral reefs in the Atlantic and survive because of warm-water eddies from the nearby Gulf Stream. Cool winters are probably responsible for the low diversity of stony corals (only 20 species, with *Acropora* notably absent). Nonetheless, the reefs are well developed on the Bermuda platform in an atoll-like form (750km^2), with patch reefs in the central lagoon.

Cayman Islands
The 3 low islands (Grand Cayman, Cayman Brac and Little Cayman, with a total land area of 259km^2) have well-developed fringing reefs on narrow coastal shelves, with a total area of 241km^2.

Cuba
This is a large (110,000km^2) high island, of which 98% of the 3,966km long shelf edge has coral reefs, and over 50% of these are separated from the mainland by cays or by broad shallow lagoons with many patch reefs. This separation has provided protection for the outer reefs from anthropogenic influences, except for fishing and, in some places, tourist diving. Important reef areas, clockwise from the north-west, include the Archipielago de los Colorados, the Archipielago de Sabana and the Archipielago de Camaguey on the north coast; the Golfo de Guacanayabo, the Golfo de Ana Maria, the Archipielago de los Jardines de la Reina, the Archipielago de los Canarreos and the Isla de Juventad on the south.

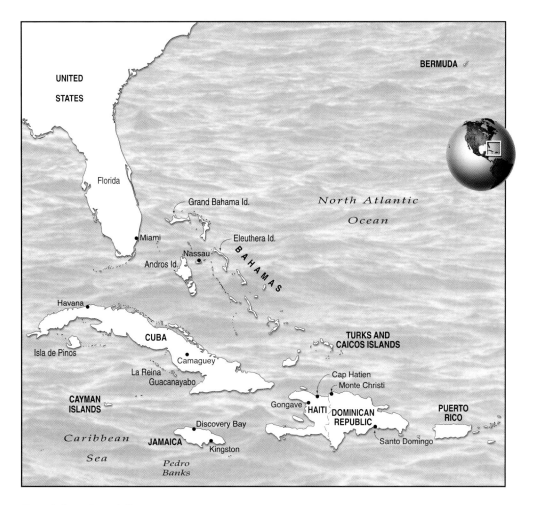

Dominican Republic

The eastern part (48,500km²) of the high island of Hispaniola is mountainous, and large rivers drain extensive watersheds, limiting reef growth with freshwater and sediments. Only 27% of the 1,400km shore (average shelf width 8km) is fringed by mangroves and only 12% by coral reefs. Important reef areas on the north (Atlantic) coast include the Montecristi barrier reef in the north-west (where the shelf is widest), narrow high-energy reefs in the central region and the Bávaro-El Macao-Punta Cana barrier reef system at the eastern end. North of that, Samaná Bay receives many rivers and is the largest estuary of the insular Caribbean: reefs in the vicinity are poorly developed, but the Navidad Shoals and Silver Banks reef systems are about 100km to the north. To the south, on the Caribbean coast, are the well-studied reefs of Parque Nacional del Este and the adjacent Isla Saona. Westward past Isla Catalina to beyond Santo Domingo are uplifted carbonate terraces with reefs growing on narrow platforms e.g. Boca Chica and the Parque Nacional Submarino de Caleta. Conditions are not good for reefs in the south-west, except on the shallow sheltered shelf east of Cabo Beata at Parque Nacional Jaragua.

Haiti
This is the western part (27,600km^2) of Hispaniola. Reefs occur: in the south, near Ile a Vache; all around Ile de la Gonave in the central bay of Port-au-Prince; on the Rochelois Bank and at Les Iles Cayemites, off the northern coast of the southern peninsula; and in the north, from the border with the Dominican Republic in the east to the Baie de l'Acul just west of Cap Haitien.

Jamaica
This high island (10,800km^2) has major fringing reefs along the narrow submarine shelves of the north and east coasts while, on the south, where the shelf extends for up to 20km, shallow reef formations are patchy and interrupted by rivers and sediment slopes. Reefs also occur on 9 offshore banks including the Pedro Bank (two-thirds the size of the island of Jamaica) 70km south, and the Morant Bank, 50km south-west, both of which have coral cays.

Turks and Caicos Islands
There are 8 coral islands and 40 small cays, a total land area of about 425km^2 on two shallow banks. The four largest islands (Providenciales, North Caicos, Middle Caicos and East Caicos) have fringing reefs at about 1 to 2.5km offshore, along the entire northern coasts, generally with a wall drop off starting at 14-18m. Providenciales, West Caicos and Grand Turk have fringing reefs along their western coasts. Shallow patch reefs are common around all of the islands and cays.

BIODIVERSITY ON THE REEFS

The diversity of reef organisms in the Caribbean and Atlantic is much less than in the Indo-Pacific. There are about 65 species in 25 genera of reef-building stony corals in the Greater Antilles. The number of species is reduced by the cooler climate in the northern Bahamas, while there are about 20 species in Bermuda, which is highly seasonal. Fishes found in reefs and associated habitats show a similar pattern, with about 500 species in the Caribbean.

STATUS OF THE CORAL REEFS

Reef communities were seriously damaged over the last 20 years by a series of apparently natural disturbances. Mass mortality of the sea-urchin *Diadema antillarum* in 1983 resulted in macroalgal over-growth, at least where herbivorous fish had been depleted by fishing. White-band disease in the *Acropora* spp, led to major declines in coral cover, especially in Jamaica where T.F. Goreau had ranked the reefs among the best in the world 20 years earlier. Large-scale coral bleaching associated with unusually high sea-temperatures was unknown in this region before 1987, but has occurred many times since, including 1998 and 1999. Other diseases of stony corals and gorgonians have been reported with increasing frequency.

Continuing growth in human populations and economic development is increasing the stresses on coral reefs. Deforestation, especially on the mountainsides of the high islands is resulting in increased runoff of suspended sediment, which is damaging coral reefs near river mouths, most notably in Haiti, Cuba, Jamaica and the Dominican Republic. In the last two countries, as well as in Cayman and the Bahamas, careless coastal tourism development has also caused sediment damage. Other forms of land-based pollution, from agriculture industry, tourism, and human waste disposal, are causing even more concern. Sewage pollution is particularly severe near large coastal cities such as Kingston, Havana, Port-au-Prince and Santo Domingo, but negative impacts are also detectable from smaller coastal communities. These problems, and those from over-fishing, are less severe in countries with relatively low population density.

The previous report in 1998 stressed the considerable reef degradation from the combined effects of over-fishing that removes important predators and herbivores, and the unrelated loss of the major grazing sea-urchin *Diadema antillarum* back in 1983. These initiated blooms of macroalgae that are still evident in many countries but not in Bermuda, Cayman or other places where herbivorous fish remain relatively abundant. Over-fishing has also been blamed for increased abundance of the coral-eating snail, *Coralliophila abbreviatum*, and of territorial damselfishes, *Stegastes* spp. that farm algal mats.

Bahamas
Coral reefs are in relatively good condition but are no longer pristine. *Acropora* and *Diadema* populations were decimated by disease. There was extensive bleaching in 1998 and some mortality in the Exuma Cays. CARICOMP monitoring at San Salvador (where coastal development is expanding) showed hard coral cover declined from 9.6% in 1994 to 6.1% in 1998, with algae rising from 17.5% to 39.3%. There have been extensive blooms of the green alga *Microdictyon marinum* on some islands. The best reefs are in remote locations like Cay Sal Bank. Hard coral cover was higher on Exuma patch reefs and highest nearshore where the reefs are especially vulnerable to changes in land use. Algal cover is higher on patch reefs in Montagu Bay. Comparison of aerial photographs of the north coast of New Providence island showed that 60% of the coral reef habitat has been lost from dredge and fill, construction of the cruise ship port, and sedimentation since 1943.

Bermuda
The reefs are generally in excellent condition, with an average coral cover of 30-35%; as high as 50% on the outer terrace, declining to about 9% inshore. Although coral bleaching occurred in 1998, there is no evidence of any significant mortality.

Cayman Islands
Coral reefs are in generally good condition, with an average coral cover of about 20%, ranging between 15% and 52%. *Acropora* species were severely impacted by white-band disease, although isolated healthy stands exist. Black band disease has been locally significant and most other coral diseases have been reported. Coral bleaching in 1998 was as severe as in 1995, when 10% mortality was measured in affected *Montastraea annularis* colonies. The sea-urchin *Diadema antillarum* is making a significant recovery from the mass mortality of 1983.

Cuba
Less than 3% of the shelf edge, including the vicinity of Havana, has been affected by severe pollution but extensive areas are affected by proliferation of macroalgae and blue green (cyanobacterial) mats, perhaps due to nutrification and the continuing low

populations of *Diadema*. Acroporid reefs are in decline, presumably because of white-band disease; other hard corals are affected by white plague and other diseases, and sea-fans by aspergillosis. Exceptionally intense coral bleaching occurred on both coasts in 1998, but there was widespread recovery. Low carnivorous and herbivorous fish populations and biomass occur in several reefs, which may partly explain the abundance of coral predators and scrapers, and macroalgae. Reefs at Herradura (31% coral cover), west of Havana, and in the Archiepelago de los Canarreos are reported to be in good condition.

Dominican Republic
Numerous surveys since 1992 all document the progressive degradation of Dominican coral reefs and other coastal environments due to human and natural impacts. High coral cover is found only at some deep or offshore reefs e.g. hard coral cover can be as high as 50% (average 35%) on the outer part of the Montecristi barrier reef. At the Silver Banks, mean coral cover is 40%, with turf algae at 51%. On inaccessible steep slopes below high cliffs at Samaná, cover by corals is about 40%, and by tube sponges 28%. Leeward (west) of the Parque Nacional del Este and south of Isla Saona are spur and groove reefs, far from the coast, at the northern part of which coral cover reaches 34% (with 25 species) with algal cover at 35%. Coral cover is generally less than 10% on most fore-reef slopes, although a back-reef at 3m depth had 25% cover, with 14 species. dominated by *Porites furcata*. Other examples of high coral cover (33%) were found on deep spurs beyond the highly impacted reefs of the south coast, exceeding algal cover (26%).

Most nearshore reefs have been damaged by terrestrial runoff. Some watersheds have been deforested for centuries but sedimentation and other agricultural and industrial pollution continue today, especially from the many rivers. Shallow reefs near rivers on the central and south coast consist of mostly dead colonies, covered by turf algae, detritus and garbage. On deeper spurs at Boca Chica, coral cover is 20%, and that by algae 56%. At Parque Nacional Submarino La Caleta, the corresponding figures are 28% and 41%, while sponges and octocorals cover 13% each. On shallow, less disturbed coasts away from rivers, recent development for tourism has destroyed mangroves and wetlands, thus removing natural sediment traps and bringing nutrient pollution. Recent studies on shallow (3-10m) reefs of the north coast (Puerto Plata and Las Terrenas) show 80% coral mortality and 92% cover by algae.

The poor status of reefs is partly due to the diseases affecting *Acropora, Diadema* (the absence of which enhances the spread of algae) and *Gorgonia* in the last 20 years. Still erect stands of dead *A. palmata* were seen from Bávaro and Boca Chica, with signs of white band disease in *A. cervicornis* at Bahía de Las Aguilas in the south-west, but there are also some recent recruits and healthy stands of both species.

Haiti
While it is presumed that reefs have suffered from the recent Caribbean-wide mortalities in *Acropora* and *Diadema*, as well as the effects of extreme deforestation, over-fishing and local pollution, few data are available. Urban runoff from Port-au-Prince has caused obvious impacts.

Jamaica
There have been dramatic changes in Jamaican coral reefs in the last 20 years: once dominated by corals, they are now dominated by algae. Major causes have been white-

band disease in *Acropora* spp., hurricanes Allen (1980) and Gilbert (1988), and the loss of herbivores through extreme over-fishing and the mortality of *Diadema*. Sewage pollution has been implicated at many sites. The thickets of elkhorn and staghorn that dominated the north coast reefs are gone. On fringing reefs around the island, mean coral cover at 10m depth fell from 52% in the late 1970s to 3% in the early 1990s, while mean algal cover rose from 4% to 92%. Since then, studies at 27 sites along 10km of the north coast around Discovery Bay showed that coral cover has increased slightly. In 1997, at 5m it was 15% (algae 35%), at 10m it was 16% (algae 56%), and at 15m it was 11%, up from 2%, although algal cover was 63%. This increase was mainly due to recruitment by opportunistic species such as *Porites astreoides, P. porites* and *Agaricia agaricites*, rather than by the original frame-builders such as the *Acropora* spp. and massive corals. An AGRRA survey in August 2000 from the south-west, along the north coast to the east reported that 10 sites above 5m (mean depth 2m) had live coral cover of 5%, and at 47 sites below 5m (mean 9m) coral cover was 12%. Deeper reefs, especially on the southern shelf edge, are in relatively good condition with *Acropora palmata* and *A. cervicornis* growing actively near Port Royal and with no current signs of white-band disease, in contrast to most of the north coast. But, in the last 10 years, other coral diseases have appeared on Jamaican reefs, with black-band and yellow-band (first seen in *M. annularis* in 1996) being locally destructive. The most severe occurrence of mass coral bleaching was in 1995 and although there was bleaching again in 1998 and 1999 there was little mortality.

Turks and Caicos Islands

Reefs are generally in good condition, with some observed pollution damage along the north and east coasts of Providenciales and along the west coast of Grand Turk, some diver impact at heavily used sites, and much *Acropora palmata* apparently lost to disease. At 5 dive sites on Grand Turk in 1994, coral cover ranged from 26% to 42%, sponges from 0.1% to 16%, while macroalgal abundance was always low. The most frequently dived site showed signs of diver impacts. AGRRA surveys in 1999 at Grand Turk, Providenciales, West Caicos, South Caicos, Ambergris Cay and the Mouchoir Bank found that coral mortality was low (<1%), diversity was high (37 coral species), and coral cover as high as 30% at several locations. Almost no macroalgae were found except in the Mouchoir Bank, Ambergris Cay and in the shallow *Acropora palmata* zone. On the east-facing banks, dead *A. palmata* stands were more abundant than live ones and *A. cervicornis* was rare. The level of active coral diseases was low but many different diseases were seen, especially on the north side of Providenciales where tourism activities are intense, and at other heavily dived sites.

In 2000, the Coastal Resources Management Project (CRMP) assessed reefs around Providenciales and West Caicos. At Northwest Point and West Caicos, coral cover on the flat at the top of the wall was relatively low (<20%) and algae (*Dictyota* and *Lobophora* spp) were abundant. On the reef face at 15-25m hard coral cover ranged from 20-50%; generally with 30%-60% algal cover (*Lobophora*). Deeper on the reef wall there was higher hard coral cover of 30-60%, and lower *Lobophora* (20-50%). In the most popular near shore patch reef in Providenciales (1-3m depth on Bight Reef), there is repeated damage by snorkelers trampling and breaking coral, especially at low tide.

STATUS OF CORAL REEF FISH

Bahamas
Fisheries resources are abundant throughout the islands, because there are large areas of shallow banks compared to the land area with relatively few fishers. Groupers and snappers still dominate the catch but their abundance and size have diminished.

Bermuda
Towards the end of the 20th century, coral reef fishes were being over-exploited, but fish-traps were banned in 1990 and herbivorous fish numbers have increased, although large groupers remain rare.

Cayman Islands
The diversity and abundance of reef fish remains high, although larger individuals are becoming scarcer, presumably because of fishing.

Cuba
Reef fish populations are declining, due to habitat deterioration and over-exploitation.

Dominican Republic
Large reef fish are scarce or absent due to over-fishing, and conch and lobster populations are also threatened.

Haítí
There is evidence of severe over-fishing in reports from the Dominican Republic that people from Haiti poach anything edible from the D.R. National Parks of Montecmsti and Jagua, leaving empty reefs.

Jamaica
Coral reef fish communities have been greatly altered and reduced by relentless fishing pressure, especially on the very narrow and accessible shelf of the north coast. All large species and most predators are absent or very scarce and the main target species left are small parrotfish and surgeonfish. Fish stocks are a little better on the south coast shelf and offshore banks.

Turks and Caicos Islands
An AGRRA survey at 6 separated sites in 1999 found that fish diversity was high and groupers were abundant and relatively large.

STATUS OF CORAL REEF FISHERIES

Bahamas
There is a well-developed commercial and export fishery plus a recreational/local consumption fishery. Government policy reserves the commercial fishing industry for Bahamian Nationals, and all boats fishing within the EEZ must be 100% Bahamian owned. Total recorded landings for 1999 were 4,954 metric tons valued at US$71.8 million, with half the weight and 85% of the value being spiny lobster tails. In addition, there were 866mt of snapper and 485mt of grouper mostly for export when the prices are sufficiently high. Recreational anglers come for pelagic gamefish, reef fish, and sportfish such as tarpon or bonefish that are caught and released. Widespread over-fishing is reported and awareness is growing about improved enforcement of existing regulations on fishing gear, closed seasons and size limits.

Bermuda
Commercial fish harvest is about 375mt per year, about 35% being reef species taken by hook and line. This is about 30% of the annual reef fish catch 40 years ago, prior to the banning of fish-traps. Recreational fisheries have much less impact, and there is no subsistence fishing. Dive tourism, for which reef fish are important, and other watersports, contribute significantly to the economy.

Cayman Islands
Coral reef fishing is limited to low volume recreational and subsistence fishing, which may have a relatively high impact, because of the limited habitat area. The increasing use of large, small-meshed fish-traps is causing local depletion and 4 spawning aggregations of Nassau Grouper (*Epinephelus striatus*) are intensively fished with hand lines, such that the average size and catch-per-unit-effort are declining. Fishing while using scuba gear is banned but 500 licensed local residents are permitted to use spearguns. Conch and lobster are also subject to intensive recreational and subsistence fisheries and are over-exploited, despite conservation regulations. The Department of the Environment and the Marine Conservation Board have recommended new measures to address all these issues but none has been implemented.

Cuba
Coral reef fishery resources are declining, due to habitat deterioration and over-exploitation in commercial and subsistence fisheries. They are still in better condition than those of Jamaica and other islands nearby. Annual exports of lobster are worth about $100 million. There is a very small, supervised trade in aquarium fish. Some recovery of conch stocks has been documented.

Dominican Republic
There is an important artisanal fishery (13,000mt in 1998), mostly for finfish, but also conch, lobster, other molluscs and crustaceans. Fishing pressure on the reefs has apparently decreased, partly because men have taken better paid work in tourism and through the adoption of FADs (Fish Aggregating Devices) which help fishers to exploit pelagic fish. However, there has been an increase on the use of hookah diving equipment to take conch and large fish at greater depths. Harvesting of aquarium and souvenir specimens also goes on.

Haiti
Most fishing is artisanal but stocks are depleted because of inappropriate methods. Aquarium fish exporters have recently decreased from 5 to 2.

Jamaica
An artisanal fishery operates from over 200 fishing beaches around the island and the Kingston Fisheries Terminal. The larger catch is made from the south coast, which has access to a wide island shelf and to Pedro and other banks, whereas the north coast shelf is narrow (<1km). The principal fishing gear is the Antillean z-trap, generally made with 3cm wire-mesh, but nets, lines and spears are also used. Nation-wide fishery data, taken at irregular intervals, show a marked decline in catch-per-unit-effort since the early 1960s. The stocks have been over-fished and catch rates are very low, partly due to a policy of subsidising the fishery in the 1960s and 1970s. Nonetheless, the latest survey (1996) estimates total catch at 14,000 tonnes: twice that of 1981, a result which is under review. There is an important commercial fishery by teams of hookah divers for conch, on the Pedro Bank.

Turks and Caicos Islands
Lobster and conch fisheries generate the only national exports. There are 5 processing plants which handled the 646mt of conch (US$1 million) and 314mt of lobster (US$2 million) in 1998. Finfish landings are on a much smaller scale, primarily for local consumption.

ANTHROPOGENIC THREATS TO CORAL REEF BIODIVERSITY

Bahamas
The population is less than 350,000 people, mostly concentrated in Nassau, Marsh Harbor and Freeport. The proximity to the Miami-Fort Lauderdale area has supported a billion-dollar tourist industry that is the envy of the wider Caribbean. But there is rapid degradation of coastal water quality and local destruction of habitats near the growing towns. Mangroves are systematically cleared for waterfront access and to control mosquito populations. There is:
- too much over-exploitation of fishes, lobsters, conch and other marine life; all are markedly declining in abundance as fishing efficiency and effort increases;
- too much pollution of nearshore waters, even in relatively sparsely populated areas, due poor wastewater treatment, solid waste disposal or runoff control; and
- too much of the coastline is being altered for development without consideration for marine habitat loss, especially mangroves and nearshore hard bottom habitats.

Bermuda
The resident population of 65,000 plays host to 500,000 tourists each year. Potential anthropogenic threats include ship groundings, oil spills, persistent anti-fouling poisons, shore-side development, contaminated runoff, sewage and industrial outfalls. The most serious impacts have resulted from ship groundings and the construction of the airport (loss of approximately 4% of reef area), and siltation from dredging and cruise ships. Accidental, but relatively minor, recreational damage comes from anchors, propellers and divers.

Cayman Islands
There are about 40,000 locals, along with 1.4 million tourists per year, 40% of whom go diving. Current anthropogenic threats derive mostly from large recent increases in population and economic development in the absence of any growth management plans or coastal area management policies. They include: dredging and filling of wetlands; coastal engineering projects; anchoring of cruise ships; and over-use of dive sites, in many cases exceeding 15,000 dives per year.

Cuba
The population is about 11 million, which has lead to extensive deforestation of watersheds since colonial times and land-based sedimentation which affects about 20% of the fringing reefs. Other continuing land-based sources of reef pollution include sugar mills, human settlements, cattle farms, yeast plants, and food and beverage processing factories. Commercial, subsistence and illegal fishing pressures continue. There is a small but growing tourist diving industry.

Dominican Republic
Increasing human population, now 8.7 million, and economic development underlie much of the stress on coastal ecosystems, through sedimentation, sewage and other terrestrial pollution from agriculture, mining, industry, shipping and tourism. Coastal habitats have been destroyed for tourism, not only in construction but misguided reconditioning of beaches causing more sediment damage. Over-fishing of reef resources is still a problem. New proposals for transhipment ports would require the destruction of reefs by dredging.

Haiti
There is little economic development for 7.5 million people but much artisanal activity. There are, however, increasing threats from near-shore road construction, sedimentation, pollution and over-fishing. There are no encouraging signs for the foreseeable future.

Jamaica
The population of 2.8 million is increasing, unemployment is high and the cost of living rising, therefore more people seek food and income by fishing, especially on the very accessible northern shelf. Rivers carry considerable runoff of suspended sediment, resulting from hillside clearance, and sewage pollution. Excess nitrogen is also channelled to the sea underground in the widespread limestone formations. The rapid growth of facilities for coastal tourism (about 1.5 million visitors per year) has also contributed to terrestrial runoff. There are proposals to site new industrial developments, including a transhipment port, in coastal areas.

Turks and Caicos Islands
There are 23,000 people with 73% on Providenciales (90km^2). The major human threats to coral reefs include: nutrient discharge from marinas and coastal development, fish processing plants, conch aquaculture and hotel sewage; heavy metal contamination from anti-fouling paints; damage to corals caused by snorkelers and divers; anchoring on coral reefs and seagrass beds; boat groundings; construction of tourism infrastructure and private jetties in the nearshore environment; uncontrolled fishing in the marine parks; and increasing visitor use of selected marine areas.

POTENTIAL CLIMATE CHANGE IMPACTS AND RESPONSES

If sea temperatures continue to rise, we can expect more coral bleaching events and more coral mortality. If sea-level rises by as much as 0.5m (and it has been measured in Cuba at about 0.25cm per year) there could be several adverse consequences. If *Acropora palmata* populations remain low, important reef breakwaters will be unable to keep up and there will be additional erosion on some beaches and other sheltered habitats such as mangroves. Low-lying human settlements will be flooded in all countries, which will include the poorer sections of some cities. The GEF/OAS project Caribbean: Planning for Adaptation to Climate Change (CPACC), is addressing these and other concerns in the CARICOM countries.

MARINE PROTECTED AREAS (MPAS) AND MANAGEMENT CAPACITY

Bahamas

There are 12 National Parks totalling 1,300km^2 of both land and sea bottom; about 10% of the land area of the Bahamas but less than 1% of the coastal shelf area. The Bahamas National Trust (BNT) developed by-laws in 1986 to protect and conserve marine life in all the Land and Sea Parks in the Bahamas. These were originally designed for recreation but were re-designated as marine replenishment areas and nurseries and a moratorium was placed on all types of fishing within park boundaries:

- Inagua National Park, Great Inagua (744km^2) - site of the world's largest breeding colony (60,000) of West Indian flamingos;
- Union Creek Reserve, Great Inagua (18km^2) - a tidal creek, important for sea turtles, especially the green turtle;
- Exuma Cays Land & Sea Park, Exuma (456km^2) - created in 1958, it later became the first marine fishery reserve in the Caribbean. It includes 10 large islands and numerous smaller cays in a 35.4km long section of the Exuma island chain;
- Pelican Cays Land & Sea Park, Great Abaco (8.5km^2) - contains extensive coral reefs;
- Peterson Cay National Park, Grand Bahama (6ha) - one of Grand Bahama's most heavily used weekend getaway spots. Small patch reefs and hardbottom communities are protected here;
- Lucayan National Park, Grand Bahama (16 ha) - gives access to the longest known underwater cave system in the world, with 10km of caves and tunnels already charted, and every vegetative zone found in the Bahamas, including a large mangrove area, but there is no reef habitat;
- Conception Island National Park, Conception Island - an important sanctuary for migratory birds, sea birds, and green turtles. Large reefs around this island are protected but there are no staff on site; and
- Tilloo Cay, Abaco (8ha). A nesting site for tropic birds, among others. Although there are reef resources around the cay, they are not protected as part of the park. This park also has no staff.

> ## Exuma - An Early Marine Protected Area
>
> The Exuma Cays Land and Sea Park (ECLSP) in the south central Bahamas, 80km south-east of Nassau, was established by the Government of the Bahamas in 1958, with by-laws allowing for a daily catch quota per boat. In the 1970s, commercial fishing escalated within the Exuma region, and many fishermen started using chlorine bleach to catch spiny lobsters. By the 1980s, fishing pressure had increased so dramatically that in 1986 the BNT declared the entire area a 'no-take zone', probably the first marine fisheries reserve in the tropical western Atlantic. The ECLSP is the only land and sea park in the Bahamas with a full-time staff: one warden, who is assisted by volunteers. The warden patrols the area in a boat each day to enforce the fishing regulations, assisted by an ECLSP Support Fleet of primarily visiting yachtsmen. Like the other marine parks, there were no management plan or clear objectives when it was granted protected status, apart from later no-take regulations. However, the ECLSP staff succeeded in developing many programmes and activities and, as these programmes grew and as the complexity of operating the ECLSP grew, it became increasingly apparent that the staff needed more management and funding support. The Park now needs additional living quarters for staff and Defence Force personnel, facilities to house researchers and volunteers and, most importantly, the capacity to generate a steady stream of income. The current concerns and issues of the ECLSP, typical of many marine parks, include: the lack of a long-term financial plan; conflicting perceptions of the mission of the Park (among Exuma residents, tourists, and the BNT); limited communication and outreach to local residents and businesses; and the incompatibility of recreational activities in the Park overwhelming the area's estimated carrying capacity.

Bermuda
There are 4 kinds of Marine Protected Area:

- Coral Reef Preserves, where it is illegal to remove any organism attached to the sea-floor, were established in 1966;
- Fishing is banned during the summer in Seasonally Protected Fisheries Areas, established in 1972 and most recently expanded in 1990, which cover about 20% of the reef platform;
- Similar legislation, starting in the mid-1980s, now bans fishing permanently at 29 popular dive sites; and
- the Walsingham Marine Park includes a marine area, with valuable seagrass beds and adjacent land (Bermuda National Parks Act, 1985).

The Department of Agriculture and Fisheries, which includes the Bermuda Aquarium, Museum and Zoo, has surveyed and mapped reefs through a Bermuda Biodiversity project, and is well equipped for reef monitoring. The Fisheries Division of the Department carries out reef fish censuses and the Bermuda Biological Station has maintained long-term reef monitoring of coral abundance, algal biomass, disease incidence, bleaching surveys and reef fish censuses at fore-reef terrace, rim reef and patch reef sites since 1991. Annual CARICOMP surveys have been carried out at two reef and two seagrass sites since 1992.

Cayman Islands

A system of Marine Protected Areas was established in 1986 and covers 34% of the coastal waters of the 3 islands. There are 3 levels of protection:

- Marine Park Areas ($15km^2$) where the taking of marine life, dead or alive, is prohibited, except for line-fishing from the shore;
- Replenishment Zone Areas ($52km^2$) where all but line-fishing is prohibited; and
- one Environmental Zone in the North Sound of Grand Cayman ($17km^2$), where all fishing, anchoring and in-water activities are prohibited.

The Department of the Environment is responsible for marine management and employs 5 Marine Park Enforcement Officers on Grand Cayman, one each on the sister islands, and 9 scientific staff running an extensive monitoring programme. Most notable is annual monitoring of 24 reef sites around the three islands since 1997, using photographic transects which are analysed digitally. CARICOMP also has surveys of Nassau Grouper aggregations and fisheries, conch and lobster and their recruitment, turtle nesting, and diver impacts. Water quality is monitored looking for evidence of pollution in and around Georgetown Harbour.

Cuba

The only legally protected reefs until recently were incidental to terrestrial Protected Areas. Now marine areas are receiving more attention and 20 MPAs are planned. The Ministry of Fishery Industry has declared 9 'no-take' areas, mostly on coral reefs. Cuba has the necessary professional and institutional capacity for coral reef research, monitoring and management, but has inadequate funding for implementation and enforcement. It has operated CARICOMP surveys at Cayo Coco since 1993 and, recently, surveys for AGRRA and Reef Check.

Dominican Republic

There are 6 MPAs, which cover the largest reef tracts and the most important nursery areas. Currently, they have no protection and no management, and there is intense fishing within them. Clockwise from the north-west they are:

- Parque Nacional Montecristi - the largest, least impacted coastal park, with diverse ecosystems;
- Humpback Whale Sanctuary - $38,000km^2$, including the Silver and Navidad Banks;
- Parque Nacional Los Haitises - in Samaná Bay, dominated by mangroves and estuaries;
- Parque Nacional del Este - the most studied MPA, an important nursery for conch and lobster;
- Parque Nacional Submarino La Caleta - the oldest MPA, now an important dive site; and
- Parque Nacional Jaragua - another important nursery for lobster.

The reefs in most of these areas were recently assessed by the Centro de Investigaciones de Biología Marina, Universidad Autónoma de Santo Domingo (which has maintained a CARICOMP monitoring site in Parque Nacional del Este since 1996), Fundación MAMMA, Inc. (a local NGO) and the National Aquarium.

The Dirección Nacional de Parques lacks qualified personnel to manage the MPAs and the Fisheries Department lacks enforcement officers and an effective extension programme. In addition, there are no appropriate penalties for violation of existing laws and confusion as to which institution should apply them. The development of tourism within the MPAs may help to bring better management of the marine resources.

Haiti
There are no MPAs and government currently has no capacity for management.

Jamaica
The management of National Parks is delegated to local NGOs. Marine Parks, Environmental Protection Areas with marine components, and Fish Sanctuaries are listed below, clockwise from the west:

- Negril Marine Park (1998) - managed by the Negril Coral Reef Preservation Society, within the Negril Environmental Protection Area;
- Montego Bay Marine Park (1989) - managed by the Montego Bay Marine Park Trust, includes the Bogue Island Fish Sanctuary;
- Ocho Rios Underwater Park (1966) - legally protected for 24 years, but still a 'paper park', the site of the classic T.F. Goreau study on the ecology of Jamaican coral reefs in 1959;
- Bowden Fish Sanctuary - associated with an oyster culture project;
- Palisadoes/Port Royal Protected Area (1998) - includes the Port Royal Cays; and
- Portland Bight Protected Area (1999) - managed by the Caribbean Coastal Area Management Foundation, includes 5% of Jamaica's land area and nearly half of the island shelf (1,350km^2).

There are plans to establish a Marine Park at Ocho Rios and another at Port Antonio. A voluntary fish sanctuary has been managed by the Discovery Bay Fishermen's Association in association with the Fisheries Improvement Programme since 1996, based at the Discovery Bay Marine Laboratory. The Pedro and Morant Cays Act (1909) provides some protection for those offshore banks.

Park management organisations have to satisfy the Natural Resource Conservation Authority (NRCA) of their competence and sustainability before they are granted management authority. The organisations do their own coral reef monitoring, or accept help from outside agencies. The government (NRCA) has staff and equipment for monitoring, and is managing the country's participation in the Coral Reef Monitoring component of CPACC.

Turks and Caicos Islands
The National Parks Ordinance (1975) created 3 levels of protected area. Sanctuaries have the highest level of protection: human entry is regulated. Nature Reserves may be used for

various activities, subject to ecological regulation. National Parks are open to the public for recreational use and development which might 'facilitate enjoyment by the public of the natural setting of the area' (including marinas) is permitted. There are 4 Sanctuaries, 5 Nature Reserves and 10 National Parks, which are wholly or partly marine. From the West, they are:

West Caicos
West Caicos Marine National Park ($4km^2$)

Providenciales
Princess Alexandra Land and Sea National Park ($26km^2$)
Northwest Point Marine National Park ($10km^2$)
Pigeon Pond and Frenchman's Creek Nature Reserve ($24km^2$)
Chalk Sound National Park ($15km^2$)
Fort George Land and Sea National Park ($5km^2$)

North Caicos
East Bay Islands National Park ($35km^2$)
Three Mary Cays Sanctuary ($0.13km^2$)

Middle Caicos
North, Middle and East Caicos Nature Reserve (International Ramsar Site; $540 km^2$)
Vine Point (Man O' War Bush) and Ocean Hole Nature Reserve ($8km^2$)

South Caicos
Admiral Cockburn Land and Sea National Park ($5km^2$)
Admiral Cockburn Nature Reserve ($4km^2$)
Bell Sound Nature Reserve ($11km^2$)
French, Bush and Seal Cays Sanctuary ($0.2km^2$)

Grand Turk
Columbus Landfall Marine National Park ($5km^2$)
South Creek National Park ($0.75km^2$)
Grand Turk Cays Land and Sea National Park ($1.6km^2$)
Long Cay Sanctuary ($0.8km^2$)
Big Sand Cay Sanctuary ($1.5km^2$

Management plans have been prepared for the 2 of the marine parks in Providenciales and that in West Caicos, by the Coastal Resources Management Project. This is expected to evolve into the National Parks Service in late 2000, and to assume responsibility for all protected areas, except for those leased to the National Trust. There is no active management of most of the protected areas outside of Providenciales.

GOVERNMENT LEGISLATION AND POLICY ON REEF CONSERVATION

Bahamas
There are 3 main government and non-government bodies responsible for coral reef resource protection and conservation in the Bahamas:

- The Bahamas Environment Science and Technology (BEST) Commission is responsible for the development of legislation to protect the environment and issue permits for development;
- The Department of Fisheries, within the Ministry of Commerce, Agriculture and Fisheries is aggressively taking responsibility not only for fisheries management, but also the establishment of marine fisheries reserves and coral reef monitoring programmes. It has a staff of 43, employing the greatest number of marine and fisheries scientists in the country, and participates in regional programmes such as CPACC; and
- The Bahamas National Trust, an NGO, was mandated, under the National Trust Act of 1959, with the responsibility and legal authority to manage the national parks of the country. The Government supports the park system through specific enforcement responsibilities carried out by the Bahamian Defense Force.

Bermuda
The management of marine resources is highly conservative and is mostly administered by the Ministry of the Environment through the Department of Agriculture and Fisheries. A comprehensive Green Paper on the Fishing Industry and Marine Environment in Bermuda in 2000 involved widespread public discussion and specialised working groups. The Marine Resources Board (a body advising the Minister on marine environmental matters) will recommend to the Minister revamped schemes for fishing licenses, more protection for corals, better management of sustainable resources, and improved concern for shoreline and coastal development, all with some strength in law. It is likely that the ensuing White Paper for proposed legislative initiatives (due in early 2001) will be a model for sustainable marine resource management for the region.

Cayman Islands
The Marine Conservation Law, passed in 1978, created the framework for marine conservation in Cayman.

Cuba
Since the Ministry of Science, Technology and Environment was created in 1994, environmental legislation has improved, including a Decree-law on Protected Areas (1999) facilitating the protection of coral reefs. There is collaboration between this Ministry, the Ministry of Fishery Industry, and the Ministry of Tourism in the development of protected areas, being a major step towards integrated coastal management.

Dominican Republic
An Environmental Law has only recently (August, 2000) been approved by Congress. Meanwhile, environmental legislation consists of over 300 environmental decrees, regulations and orders, administered by a large number of organisations. The newly formed

Subsecretaría de Gestión Ambiental and the Subsecretaría de Recursos Costero Marinos (both in the also newly formed Secretaría de Medio Ambiente y Recursos Naturales) face the task of managing this system, but there is considerable overlap of authority among institutions dealing with coastal issues and a lack of any central long-term vision on sustainable coastal area management.

Haiti

There are very detailed, though outdated, laws concerning coastal and marine resources. None is currently respected nor enforced. Both the Ministry of Environment and an NGO, the Fondation pour la Protection de la Biodiversité Marine, are working to raise awareness.

Jamaica

Until recently, the major law available for coral reef management was the Beach Control Act (1960). This was one of more than 50 environmental laws, administered by a multiplicity of ministries and other agencies. The enactment of the Natural Resources Conservation Authority Act (1991) started a process of rationalisation and, since it binds the Crown, obliged public sector entities to comply. That process continued in 2000 with the formation of the National Environmental Planning Agency (NEPA) by merger of the NRCA with the Town Planning Department and the Land Development Commission. In 1998, recognising the need for inter-sectoral collaboration in management of the Exclusive Economic Zone and coastal areas, government agreed to creation of the Council for Ocean and Coastal Zone Management. Its members are the heads of all relevant agencies and it reports, through the Ministry of Foreign Affairs, directly to cabinet.

Turks and Caicos Islands

There is no clearly defined national policy for conservation of marine resources, although some policy documents allude to conservation or sustainable development. The development trend in Providenciales, development proposals for West Caicos, current management of the national parks and the National Parks Regulation indicate that the maintenance of protected areas seem to favour recreational benefits for visitors and development opportunities for expatriates. These priorities, rather than broader conservation of coastal resources, have lead to the current type and intensity of uses within and adjacent to the parks. However, the recent UK White Paper on Progress Through Partnership proposes the following policy objectives for Overseas Territories:

- to promote sustainable use of the Overseas Territories natural and physical environment, for the benefit of local people;
- to protect fragile ecosystems such as coral reefs from further degradation and to conserve biodiversity in the Overseas Territories;
- to promote sustainable alternatives to scarce resources or species which are used for economic purposes; and
- to enhance participation in and implementation of international agreements by Overseas Territories.

The National Parks Ordinance (1975) and Regulations (1992) defined the protected areas and there are a number of other relevant Ordinances. There is no harmonisation among key pieces of legislation. The National Parks Ordinance needs revision, and will need more when the National Parks Service is established, especially in terms of the administration of regulations and in zone designations.

GAPS IN CURRENT MONITORING AND CONSERVATION CAPACITY

Bahamas
The National Park System needs more funding, staff and equipment. There is only one 'no-take' area among the MPAs. More Protected Areas are needed and several have been planned.

Bermuda
A gap in the current MPA structure is the lack of protection of critical habitats for juvenile reef fishes. Other conservation priorities are environmental education at all levels; better enforcement of conservation laws; and the enactment of new legislation.

Cayman Islands
Current funding for marine monitoring is relatively good, but could be advanced with by additional technical staff and training for the Department of the Environment. There are more serious legislative deficiencies and an absence of conservation planning or political support. The Marine Conservation Law of 1978 is outdated; there is no enabling legislation for the Department of the Environment; there is aggressive development but no development plan; and conservation is perceived as incompatible with economic development.

Cuba
The greatest deficiency for monitoring and conservation is the lack of funds. Protected area staff are few and poorly trained but co-management with tourist enterprises may help with the funding of MPAs. More environmental education is needed at all levels to increase the awareness and involvement of stakeholders, communities and decision-makers in coral reef issues.

Dominican Republic
The immediate need is for effective management of the MPAs, which will require more funding to employ and train staff, combined with education programmes. There is also a need to resolve sanitary and solid waste pollution, at least at the major tourist centres.

Haiti
There are very few people with any training in marine sciences and no formal capacity for coastal management. The current political situation has renewed the 'brain drain'.

Jamaica
The government, burdened with debt repayments, lacks resources for adequate monitoring and enforcement of environmental laws. Environmental education has been included in formal education system but this should be expanded at all levels to encourage: responsible use of coastal resources; resolution of conflict between tourism and fishery interests; and more appreciation of the benefits of integrated coastal management.

Turks and Caicos Islands
All conservation agencies are under staffed, and few staff have the requisite training and experience. All agencies have insufficient equipment to carry out daily operations and in some cases out-dated or inappropriate equipment. Some conflicts among key agencies need to be resolved, and rights and responsibilities in management of the protected areas need to be clarified. There should be mechanisms for key stakeholders and users of reef resources to provide input to monitoring and conservation.

CONCLUSIONS

Bahamas
This region can be considered one of the least impacted in the tropical Americas but that situation is changing rapidly.

Bermuda
Coral reefs are in reasonably good condition and are well managed, although a higher level of enforcement to protect reef fish stocks is desirable.

Cayman Islands
Coral reefs and their resources are in relatively good condition but are increasingly threatened by rapid development, population growth and intensive tourism. Proper management is now crucial to their continued health but is hampered by the lack of political support.

Cuba
Offshore reefs are in relatively good condition. General coral reef decline in Cuba derives from regional and global stresses e.g. diseases and increased temperature, while the main anthropogenic stresses are sedimentation, organic pollution and over-fishing. Priority should be given to reforestation, to the reduction of nitrogen and phosphorus pollution by waste-water and to better controls on fishing. All options should be explored for sustainable funding of MPAs and environmental management. On the social side, there should be more environmental education, especially among local communities and decision-makers, and more cross-sectoral integration.

Dominican Republic
Most reefs are degraded from the overgrowth of algae due to the lack of grazing caused by over-fishing and, in some areas, nutrient pollution, even in the MPAs. Other reefs, especially near towns or tourism sites, are degraded from other causes. Government capacity to manage the coastal zone is seriously in need of improvement.

Haiti
Near-shore reefs are suffering from excessive algal growth and depleted fish stocks. Restoration of management capacity will require funds, training and a stable government.

Jamaica

Overgrowth by algae is occurring everywhere due to the lack of herbivores and also due to nutrient pollution in many places. There has been some regrowth of opportunistic corals, partly facilitated by the spread of sea-urchins. The reefs are now less able to provide important services to fisheries, tourism and coastal protection.

Turks and Caicos Islands

Most reefs are in good condition, but the pressures from tourism are increasing. Proper conservation and monitoring of coral reefs depends on the enhancement of human and material resources, probably by increased donor support.

GENERAL CONCLUSIONS

- Coral reefs are well developed throughout this region.
- The status of reef communities has declined in all 8 countries recently, most important has been the loss of reef building elkhorn and staghorn corals (*Acropora palmata* and *A. cervicornis*), apparently from white-band disease. Many reefs are overgrown by macroalgae released from herbivore control by over-fishing and the mass mortality of the grazing sea-urchin *Diadema*. Other diseases have become more common, where reefs are stressed by human activity, and some are causing serious mortality. Sediments, nutrients and other pollutants have impacted reefs near human population centres.
- Reef condition is relatively good in the small low islands and bad nearshore in the high islands. The worst may be in Jamaica (where there is a little recovery) and the Dominican Republic, but there is little information about Haiti. Some deep reefs are in better condition than shallow ones. The best reefs are far from human influence, on the outer shelves of Cuba, offshore banks, or the more remote islands of the small-island groups. Bermuda reefs, which had no *Acropora* to lose, have probably changed least.
- The diverse and formerly abundant coral reef fish resources have been depleted to some degree in all countries. The loss of predators and herbivores has impacted on benthic communities.
- The impact of reef fisheries (and other human factors) is modulated by geography, demography and economics. Over-fishing is particularly acute in Jamaica, Haiti and the Dominican Republic, where narrow fringing reefs are easily accessible to numerous low-income fishers. Reef fish stocks dispersed over broad shelves are less depleted, as in Cuba, the Bahamas and the Turks and Caicos Islands. Higher standards of living in the Cayman Islands and Bermuda result in reduced fishing pressures and reef fish stocks are relatively healthy. However, the more valuable resources such as queen conch and spiny lobster are also being exploited by larger commercial interests, and are at risk throughout the region.
- Increasing human population and careless economic development drive other major anthropogenic impacts; sediment pollution, nutrient pollution, and habitat destruction. These derive from agriculture, human habitation, industry and, increasing tourism sector, which is the largest earner of foreign exchange in the

region. There are examples from all countries of hasty development designed to facilitate enjoyment of the coastal environment that has degraded it.
- Bleaching in the ENSO year 1998-99 occurred throughout the region but was severe only in Cuba and Cayman. Consequent mortality was not measured but could have been as high as 10%.
- The best managed systems of Marine Protected Areas and coral reef monitoring are in Bermuda, Cayman and Jamaica. In the last, operation of the National Parks has been transferred to selected local NGOs. In the Bahamas, all national parks are managed by an NGO and some MPAs receive active protection. The Turks and Caicos Islands has an extensive, well-established system and a management programme is being developed. The Dominican Republic has 6 National Parks, which are not managed. Cuba is just beginning to establish MPAs. Haiti has none with no plans to establish any.
- Conservation for tourism is now the major force driving MPAs and some tourist dive operators are cooperating with management e.g. installing mooring buoys at popular dive sites to reduce anchor damage. The concept of carrying capacity at these popular sites is hard to apply and localised damage is occurring. Both tourists and fishermen want to see more fish and all kinds of MPA may enhance fisheries, but cooperation and joint planning by all stakeholders are required.
- All countries inherited uncoordinated sector-specific environmental laws and most have, to differing degrees, introduced better integrated conservation laws and administration. But further harmonisation is needed in most countries. Local government policy is supportive of marine conservation in the Bahamas, Bermuda, Cuba, Jamaica and the Turks and Caicos Islands.
- More trained marine conservation staff are required in the Bahamas, Cuba, Dominican Republic, Haiti, Jamaica and the Turks and Caicos Islands. More environmental education is needed at all levels in fisheries, tourism, administration and among politicians. There should be more horizontal and vertical integration of coastal management, to involve all stakeholders.

RECOMMENDATIONS TO IMPROVE CONSERVATION OF CORAL REEF RESOURCES

A. International actions:
- all governments, agencies and institutions should act on the Current Priorities of the 1998 ICRI Renewed Call to Action and implement the 1995 Call to Action and the Framework for Action;
- all governments should quickly implement measures to reduce 'greenhouse' gas emissions, to a minimum of the Kyoto Agreement; and
- developed country governments and international funding agencies could assist the environment by relieving debt and allocating these resources to conservation.

B. National actions of general application:
The following recommendations, which are also in the ICRI documents cited above, apply to most countries in the region. They will require local action for which external funding may be necessary for most countries.

Education
Promote environmental education on marine and coastal issues, widely spread at all levels of society.

Research and Monitoring
Complete the assessment of coral reef status and maintain monitoring at key sites; investigate the role of anthropogenic stress in promoting coral diseases that are having catastrophic impacts on Caribbean coral reefs.

Capacity-building
Employ more professional and technical staff in national conservation agencies and provide them with better training and equipment; and support the work of conservation staff by involving community members as environmental wardens, particularly for education and enforcement.

Management
- Apply the principles of Integrated Coastal Management, notably: cross-sectoral and vertical integration; and the involvement of all stakeholders in planning and management.
- Modernise and harmonise environmental legislation.
- Work to reduce direct human impacts on coral reefs by: promoting sustainable fishing; educating across all levels; limiting the effectiveness of fishing gear; creating replenishment reserves; and licensing fishers.
- Reduce sedimentation: by appropriate agricultural practices, e.g. terracing; reforestation near water-courses; protection of mangroves and coastal wetlands; and better planning and control of coastal construction.
- Reduce the discharge of polluting nitrogen and phosphorus to coastal waters by: better wastewater management, including: tertiary treatment; or the use of nutrients to enhance plant growth in agriculture or aquaculture.
- Reduce solid waste and industrial pollution.
- Reduce the environmental impacts of tourism by working with the Caribbean Tourist Organisation and other industry leaders.
- Enhance the effectiveness of Marine Protected Areas: by involving fishery, tourism and other local interests; by creating and implementing management plans; and by seeking sustainable funding.
- Build on MPAs as centres of inspiration for proper management of the entire landscape.
- The Cayman Islands Government should: accept Marine Conservation Board and Department of Environment recommendations to amend current legislation and enact environmental Legislation; establish a Conservation Trust Fund with monies currently being collected as Environmental Protection Fees; develop a comprehensive development plan incorporating Integrated Coastal Management; and also develop a tourism management plan that incorporates environmental issues (these issues were raised by the Cayman Islands but would apply to many other countries);

Supporting Documents

P.M. Alcolado, R. Claro-Madruga, R. Estrada. Status of coral reefs of Cuba; T. Austin, G. Ebanks-Petrie. Status of coral reefs of the Cayman Islands; J. Barnes, Ann Glasspool, Brian Luckhurst, Robbie Smith, Jack Ward. Status of the coral reefs of Bermuda;

F.X. Geraldes, M.B. Vega. Status of the coral reefs of the Dominican Republic; F. Homer, D. Shim. Status of coral reefs in the Turks and Caicos Islands; ICRI Finance Survey, Bermuda Biodiversity Project; ICRI Finance Survey, Coastal Resource Management Project, Turks and Caicos Islands; K. Sullivan Sealey, E. Phillips. Status of coral reefs in the Bahamian Archipelago;

J.D. Woodley. Status of the coral reefs of Jamaica; J. Wiener. GCRMN report for Haiti (all available from the authors below).

Acknowledgements

For their co-operation and help in preparing this report, we would like to thank colleagues in our own institutions and elsewhere, particularly Judith Campbell, Mark Chiappone, Leandro Cho, Wesley Clerveaux, Michelle Fulford, Elena de la Guardia, Gaspar Gonzalez, Kristi Klomp, Judy Lang, Enrique Pugibet, Yira Rodriguez, Ruben Torres and John Tschirky.

Addresses

Jeremy Woodley, Centre for Marine Sciences, University of the West Indies, Jamaica. (woodley@uwimona.edu.jm);
Pedro Alcolado, Instituto de Oceanologia, Cuba. (alcolado@oceano.inf.cu);
Timothy Austin, Dept. of Environment, Grand Cayman, Cayman Islands. (Timothy.Austin@gov.ky);
John Barnes, Dept. of Agriculture and Fisheries, Bermuda. (jbarnes@bdagov.bm);
Rodolfo Claro-Madruga, Instituto de Oceanologia, Cuba;
Gina Ebanks-Petrie, Dept. of Environment, Grand Cayman, Cayman Islands. (Gina.Ebanks-Petrie@gov.ky);
Reynaldo Estrada, Centro Nacional de Areas Protegidas, Cuba;
Francisco Geraldes, Centro de Investigaciones de Biologia Marina, Universidad Autonoma de Santo Domingo & Fundacion Dominicana pro Investigacion y Conservacion de los Recursos Marinos, Inc., Dominican Republic. (mamma@codetel.net.do);
Anne Glasspool, Bermuda Zoological Society, Bermuda. (bamzcure@ibl.bm);
Floyd Homer, Coastal Resources Management Project, Ministry of Natural Resources, Turks and Caicos Islands. (fmhome@sunbeach.net);
Brian Luckhurst, Dept. of Agriculture and Fisheries, Bermuda. (bluckhurst@bdagov.bm);
Eleanor Phillips, Dept. of Fisheries, Ministry of Commerce, Agriculture and Fisheries, Bahamas. (eleanor@batelnet.bs);
David Shim, Coastal Resources Management Project, Ministry of Natural Resources, Turks and Caicos Islands. (daveshim@hotmail.com)
Robbie Smith, Bermuda Biological Station for Research, Inc., Bermuda. (robbie@bbsr.edu);
Kathleen Sullivan Sealey, Marine Conservation Science Center, University of Miami, Florida,

U.S.A. (sullivan@benthos.cox.miami.edu);
Monica Vega, Acuario Nacional & Fundacion Dominicana pro Investigacion y Conservacion de los Recursos Marinos, Inc., Dominican Republic;
Jack Ward, Bermuda Aquarium, Museum and Zoo, Bermuda.
Jean Wiener, Fondation pour la Protection de la Biodiversite Marine, Haiti. (jwiener@compa.net).

16. STATUS OF CORAL REEFS OF NORTHERN CENTRAL AMERICA: MEXICO, BELIZE, GUATEMALA, HONDURAS, NICARAGUA AND EL SALVADOR

PHILIP KRAMER, PATRICIA RICHARDS KRAMER,
ERNESTO ARIAS-GONZALEZ AND MELANIE MCFIELD

ABSTRACT

Recent large scale climatic events have had a tremendous impact on coral reefs of this region. The isolated, less-developed reefs of the Mexican Pacific suffered 40-50% coral mortality during the La Niña related cold-water events, following the 1998 El Niño bleaching event. The extensive, well-developed reefs on the Atlantic coast experienced unprecedented mass coral bleaching and mortality in 1995 and again in 1998, followed by

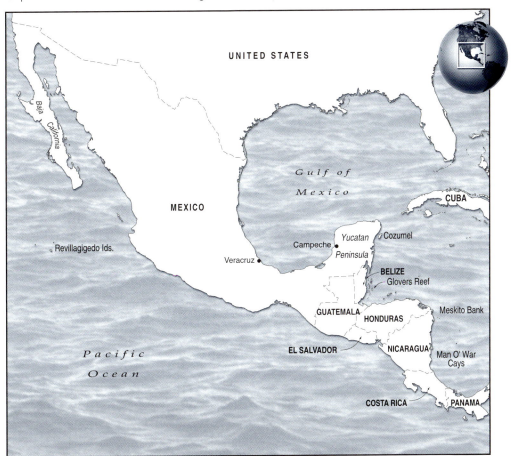

widespread damage from the intense Hurricane Mitch, also in 1998. These events heavily impacted reefs from the Mexican Yucatan to Honduras, causing losses in coral cover of 15-20% across the region with some losses as high as 75% in parts of Belize. Throughout large parts of the region there are intense fishing pressures and major threats to reefs from poor land-use practices and unregulated coastal development. Capacity to monitor and manage coral reefs varies enormously in the region, from advanced to virtually non-existent. Now, countries of the Mesoamerican region (Mexico, Belize, Guatemala, Honduras) are cooperating to conserve and manage their reefs and resolve cross-boundary management issues.

INTRODUCTION

Most striking about the coral reefs along the Pacific and Caribbean coasts of Northern Central America (NCA) is the dramatic difference in reef abundance, development and diversity. Fewer reefs exist on the Pacific coast, mostly in central Mexico, and these are usually small, less diverse, and isolated or patchy in distribution. These reefs are subject to frequent cold, nutrient-rich upwellings and El Niño events, and are sensitive to extreme disturbances and susceptible to local extinction due to their limited distribution and isolation. Numerous rivers flow onto the narrow steep shelf from southern Mexico south to Gulf of Fonseca, and Nicaragua further inhibiting coral reef development, while favouring the growth of mangroves. The next major reef area along the Pacific coast is to the south in Costa Rica.

In contrast, the well-developed reefs on the Caribbean side are extensive and contain the Mesoamerican Barrier Reef System (MBRS) which is the longest barrier reef system in the Western Hemisphere and the second longest in the world, extending over 1000km from the northernmost part in Yucatan, Mexico through Belize and east to the Bay Islands of Honduras. As well as the barrier reef, there are numerous patch, fringing, and atoll-like reefs, which are habitats for a high biodiversity of fishes, invertebrates, birds, plants, sea turtles, and mammals. Reefs along the Caribbean coast can be divided into several distinctive areas: SW Gulf of Mexico; extensive fringing and barrier reefs along the Yucatan and Belize coast; 4 unique atolls (Banco Chinchorro, Turneffe Island, Lighthouse, and Glover's Reef); small coral communities along mainland Guatemala and Honduras, and the Bay Islands; and extensive reef complexes in Nicaragua. Composed of luxuriant patch, fringing, and barrier reefs and four unique offshore atolls. These reefs are influenced by continental land masses and are also impacted to varying degrees by human pressures; coastal development, over-fishing, agricultural and industrial run-off, deforestation, land-use, and sewage pollution. In addition to these stresses, two major events in 1998 (El Niño mass bleaching and Hurricane Mitch) caused severe damage to reefs along the Mesoamerican Corridor (Yucatan, Belize, Guatemala, Honduras).

Although coastal communities depend heavily on the marine resources, the coastal zone is only a small part of the land area. Hence, national focus, priority and awareness of problems facing the coral reefs is much lower than other island nations where reefs dominate the resource base. This is also evident in the small amount of information available about the status of these reefs, except for a few well-studied reefs in Yucatan, Belize, and the Bay Islands. Large information gaps exist for the extensive reefs of Nicaragua, coastal coral communities along the mainland of Honduras and Guatemala, offshore islands and banks like Swan Islands and the Mysteriosa banks in Honduras, parts of the Belize reef

Country	Population x 1000 (growth)	Land area (km²)	Water (km²)	Coastline (km²)	Mangrove Coast (km²)*
Mexico	100,294 (1.73)	1,972,550	49,510	6760 P + 2900 C	5,246-14,202
Belize	235.8 (2.42)	22,690	160	386	730-783
Guatemala	12,335 (2.68)	108,890	460	400	160
Honduras	5,997 (2.24)	112,090	200	820	1,170-1,213
Nicaragua	4,717 (2.84)	129,494	9,240	307	600
El Salvador	5,839 (1.53)	21,040	320	307	352-450

Demographic and geographic summaries for the region. Population growth rate is given in parentheses after the total, with P the Pacific coast and C, Caribbean and low and high estimates for mangrove coast.

complex, and other reefs in the eastern Pacific coast. The following provides an overview of reef descriptions for each country, listed in order from north to south.

REEF DESCRIPTIONS

Mexico

There are 3 distinct coral reef areas in Mexico: 1) the Pacific mainland coast and Baja California; 2) Southwest Gulf of Mexico (Veracruz and Campeche Bank); and 3) the Yucatan Peninsula. Reef development and diversity on the Pacific side is much less than the Caribbean and often restricted by cool temperatures (e.g. on the western side of Baja California). There are 12 hard coral species in the Gulf of California, 15 species further south on the Mexican mainland, and 18 on the Revillagigedo Islands. Most of these species are rare or have limited distribution, and coral communities are usually small and patchy with low abundances of other invertebrates, like soft corals, sponges, crustaceans, and echinoderms.

There are about 20 reefs off Veracruz, in the Gulf of Mexico, which have adapted to high turbidity from coastal runoff. The coral diversity is low for the Caribbean with only 45 reef-building species. On the eastern side of the Yucatan peninsula, there is an extensive fringing reef along nearly 350km of coastline from Isla Contoy south to Xcalak, including the offshore islands and the Banco Chinchorro atoll. The southern part continues as the Belize barrier reef. At least 56 reef-building species are reported. Reef development along Quintana Roo varies considerably, is often discontinuous and can be divided into 3 zones, north, central, and southern. Those in the north area have overall low living cover (17%), are mostly denuded bare rock, with a few patches of high cover, and are dominated by stands of dead *Acropora palmata*. The central and southern areas contain more continuous shallow reefs and better developed platform reefs. Banco Chinchorro is a large (46km x 14km) atoll with well-developed reefs on both the broad windward shelf and narrow leeward shelf.

The largest populations and longest coastlines occur in Mexico. Coral reefs provide commercial, recreational and tourism value for coastal communities, mainly Mestizo (Amerindian-Spanish), Amerindian, and some white. The main towns are near La Paz, Huatulco and Puerto Vallarta (Pacific) and Veracruz, Cancun and Cozumel (Atlantic). Tourism, concentrated in Huatulco, Puerto Vallarta and the northern portion of Quintana Roo, is the major economic activity and is expanding rapidly both to the north and south of Quintana Roo.

Belize

The second longest barrier reef in the world extends for 250km and covers 22,800km^2, as a unique assemblage of lagoon patch reefs, fringing reefs, and offshore atolls. Like Mexico there are three distinctive areas, northern, central, and southern barrier reefs. The Northern reefs are well developed and continuous from the Mexican border to Caye Chapel, while discontinuous and less developed south to St. Georges Caye. The Central Reefs are continuous and are considered to be the best developed. Southern reefs are discontinuous and less well developed. Numerous patch reefs, dominated by *Montastraea annularis*, are found throughout Belize and the unique rhomboidal-shaped shoals and reefs occur towards the south of the central barrier. Located 7-45km off the barrier reef, the three atolls (Lighthouse, Turneffe, and Glovers) have very different reefs on the leeward versus windward sides. Lighthouse and Glovers atolls are structurally similar having deep lagoons with numerous patch reefs, whereas the protected Turneffe Island has extensive mangroves in its shallow lagoon.

This is the most sparsely populated nation in Central America (about 9 people km^2), and coastal resources are important for commercial and artisanal fishing, tourism (e.g. snorkeling, diving, fishing), aquaculture, cultural resources, and limited shipping. Tourism and the export of marine products have increased significantly in the last few years. Belize has a diverse mixture of ethnic cultures (Mestizo, Creole, Maya, Garifuna, as well as Asian, European and Mennonites). Less than 50% of the population occurs in the coastal zone and it has actually decreased in the last few years.

Guatemala

Most of the coastline borders the Pacific, with only a narrow coast in the Caribbean, which contains many mangroves, seagrass beds and coastal lagoons. Large flows from the Motagua, Sarstun and Dulce Rivers limit reef development to a few isolated corals and small patch reefs in the Gulf of Honduras. Few if any reefs occur on the Pacific. The local coastal communities of Mestizo (or Ladino) and Amerindian use the coastal resources for food and transport and on the Caribbean, they also rely on obtaining fish from adjacent Belize waters.

Honduras

Coral development is restricted on Caribbean coastal areas by elevated runoff from high rainfall running off the mountainous terrain, with only few scattered, poorly developed coral communities only around Puerto Cortes, La Ceiba and Trujillo. The only real reef development is around the offshore Bay Islands which consist of 60 small islands and several large islands that form 4 main groups: Roatan, Utila, Guanaja, and Cayos Cochinos. Most of these reefs fringe the drop-off and grow seaward down to 9-12m depth, and then drop sharply to 75m. Discontinuous well-developed reef buttresses are dominated by rich growth of *M. annularis* on most of the Bay Islands. There is also a discontinuous shallow fringing barrier reef of *A. palmata* and *Agaricia tenuifolia* on the northern sides of Roatan, Guanaja, and parts of Cayos Cochinos and southeastern Utila. These are sometimes exposed at low tides and much of the coral is dead and covered with dense turf algae. The Swan Islands, an isolated group of small islands to the far northeast, are surrounded by a fringing reef, and small fringing and patch reefs occur near the Mosquitia Cays and Banks. There are no reefs on the Pacific coast. Indigenous Mestizo, Garifuna and Miskito communities rely heavily on subsistence fishing, especially those living in coastal villages on the Caribbean. The Bay Islands are a well known tourist destination for scuba diving.

Nicaragua

Very little is known on the distribution and condition of the extensive reefs growing on the broad carbonate bank extending out from the Caribbean coast. Close to the coast (out to 10km), coral reef development is limited due to high sedimentation with less than 10% coral cover. The greatest reef development occurs in four main areas: Miskito Bank; Man O'War Cays; Crawl, Taira, Pearl, and Set Net Cays; and Little and Big Corn Islands, and is generally classified into nearshore (to 25km), mid or central shelf, and shelf edge. Miskito Cays, 50km offshore, are mangrove islands surrounded by fringing coral reefs, and extensive seagrass beds, with the most abundant coral on seaward edges. Complex reef development includes patch reefs, large pinnacles and fringing reefs, but little is known about these reefs, especially the Man O'War Cays. The Pearl Cays are shallow reefs with thickets of *A. palmata* on the windward coast, and the Corn Islands are the largest islands, with a series of three fringing reefs on the northeast side and numerous patch reefs. *Acropora palmata* and *M. annularis* are the major reef building corals, averaging 25% coral cover. Reef development is limited on the leeward side of the islands. On the Pacific, there are virtually no corals, except for a few isolated patches of individual Pocilloporid corals and scattered gorgonians.

The country is mostly agricultural with a small manufacturing industry, thus the coral reefs are largely ignored. Only 10% of the population lives along the Caribbean coasts, primarily indigenous Miskitos, Creole and some Garifuna. These people are closely tied to their marine resources and harvesting is restricted within their traditional land and sea territories. The largest coastal communities are on the mainland near the myriad of coral cays and fishing grounds around the Miskito Reefs. The Miskito Indians are the primary users of the Miskito Coast Marine Reserve (MCMR) surviving mainly through subsistence fishing.

El Salvador

There are few natural resources because it is small with only 307km of coast on the Pacific. Some coral communities have been reported at Los Cobanos, but little is known about the extent and current status. The smallest country in Central America has a population of almost 6 million (primarily Mestizo), and is one of the most densely populated countries in the whole Caribbean region.

STATUS OF THE CORAL REEFS

The coral reefs of this region have experienced an increasing frequency and intensity of disturbances in recent years, with a few areas receiving recurring or coinciding disturbances. Prior to 1998, the principal disturbances were hurricanes, coral diseases, and recent mass coral bleaching (1995 and 1997). A history of moderate to severe hurricanes has impacted localised areas in Mexico and Belize and recovery from these storms has been variable. White band disease probably devastated *Acropora* populations since the early 1980s including many areas in Belize where they used to be the primary shallow reef builder. The 1983 die-off of the grazing sea urchin *Diadema antillarum* also damaged the region. Mass bleaching was first reported in 1995, with variable, but limited effects, followed by the major bleaching event and hurricane of 1998, which had profound impacts. Prior to 1998, most of the reefs were routinely described as being in good condition.

1998 CORAL BLEACHING IN THE MESOAMERICAN BARRIER REEF SYSTEM (MBRS)

There were few large-scale bleaching events in the north Central American region compared to other areas in the Western Atlantic because high temperature stresses were infrequent and other environmental stresses were relatively minor. For example, coral bleaching was reported for much of the Caribbean during 1983 and 1987, and the first well-documented mass bleaching event in Belize occurred in 1995 where 52% of coral colonies bleached, although only 10% of the corals suffered subsequent partial mortality (10-13% loss of coral cover). These impacts in 1995 were also observed in Cayos Cochinos, Honduras, where 73% of scleractinian corals and 92% of hydrocorals bleached and slightly higher mortality was reported. A less severe bleaching event was reported in 1997, although the extent of damage is not known.

Then there were major disturbances during 1998. First high sea-surface temperatures appeared during August and intensified during September. Reports of intense bleaching (>50% of colonies) started in the Yucatan in August/September, followed by reports in Belize (September) and Honduras (September/October). Coral mortality was first reported in the Yucatan in early October, particularly on Agaricia tenuifolia colonies. Reports of massive bleaching and mortality of A. tenuifolia and Millepora spp. in the central and southern Belize barrier soon followed. Then severe Hurricane Mitch passed over in October; sea surface temperatures decreased and recovery of some branching corals was reported while massive corals continued to remain bleached into 1999. Extensive surveys conducted indicated that the 1998 bleaching event affected the entire MBRS region, and was possibly more severe than mass bleaching in 1995. Shallow reef corals tended to either die immediately or recover more rapidly from bleaching compared to deeper depths, where coral recovery from bleaching proceeded more slowly. The magnitude of the bleaching was evident in deep fore reef sites where significant remnant bleaching was observed up to 10 months after the initial bleaching. Specific findings from this study showed:

- Regional average of 18% recent coral mortality on shallow reefs, 14% on fore reefs;
- Up to 75% recent coral mortality on localised patch and barrier reefs in southern Belize;
- Highest mortality observed for A. tenuifolia (>35%), M. complanata (28%), and Montastraea annularis complex (25-50%);
- Regionally high recent mortality and disease on Montastraea annularis complex;
- Low to moderate levels of recent mortality in Acropora palmata from bleaching;
- Remnant bleaching still evident on fore reefs 10 months later (up to 44% of corals still bleached); and
- High incidence of coral disease following the bleaching event on Belize shallow reefs (black band) and Honduras and Belize fore reefs (white plague).

In 1998, a sequence of catastrophic disturbance events impacted the region: unprecedented coral bleaching was documented throughout the region as well as elevated incidences of disease and in late October 1998, Hurricane Mitch impacted much of the coast from Nicaragua northwards to Yucatan. The synergistic effects of these events are expected to have long-term ecological consequences for the coral reefs.

Mexico

These reefs are among the best studied in the region. The main disturbances to Pacific reefs are strong El Niño events and extremely low winter water temperatures. Since most Pacific corals are found as small isolated communities, they are susceptible to disturbance and local extinctions can result, i.e. *Leptoseris papyracea* is probably extinct along the Mexican mainland and the distribution of *Porites sverdrupi* has been reduced to only 500km. Atlantic reefs of the SW Gulf, particularly the nearshore reefs near Veracruz receive intense anthropogenic impacts from agricultural, industrial, and urban wastes, over-fishing, oil spills, and ship groundings. Extensive mortality of *Acropora* has been documented at Isla de Sacrificios, whereas the offshore reefs of Campeche Bank are remote from direct human impacts, although occasional oil leaks and damage from the oil facility have some impacts. Coral cover ranged from 2 to 28% in 1981, but was 1 to 6% in 1993 at the CARICOMP monitoring site of Puerto Morelos. Much of the loss was due to Hurricane Gilbert in 1988 and the mass coal bleaching in 1995, although this was not measured. The largest decline in coral cover was on the back reef, from 28 to 5%, and the reef crest 27 to 6%. Despite these losses prior to 1998, these reefs showed evidence of recovery with little extreme bleaching or algal overgrowth, and some signs of re-establishing *Diadema* populations. AGRRA surveys (1999) in Veracruz found 17% average coral cover, low recent mortality, low coral disease and bleaching, evidence of coral recruits, few *Diadema*, and low macroalgae.

The ENSO event impacted reefs in the Gulf of California (18% coral mortality) and Bahia de Banderas (more than 70%). Oaxaca reefs were affected by the La Niña of late 1998 (more than 70% mortality). On the Atlantic coast, the 1998 bleaching and Hurricane Mitch impacts were lower in Mexico than in Belize or Honduras; only 4 of 27 reefs surveyed after the event had higher than normal recent mortality, averaging 11% on fore reefs and 7% on shallow reefs. The southern area was impacted by bleaching, disease, and minor hurricane impacts while the north and central region had low disturbance. In 1999 AGRRA surveys in the central and southern regions of the Mexican Caribbean, there was 12% average coral cover (10m depth), 37% total coral tissue mortality, greater average algal turf cover than macroalgae and crustose coralline algal cover, and evidence of coral recruitment, although there was an almost total absence of *Diadema*.

Belize

While there have been extensive studies on several select reefs in Belize, much of the barrier reef has not been studied. Some of the earliest disturbances reported were severe storms and hurricanes, such as Hurricane Hattie (1961) which reduced living coral cover by 80% and destroyed spur and groove structure along sections of the barrier reef. Prior to the mid-1980s, *A. cervicornis* was the dominant reef-builder on the fore-reef slope until populations were devastated, presumably by white band disease i.e. coral cover near Carrie Bow Cay shallow fore reef dropped from 30-35% in the late 1970s to 12-20% now. These losses were followed by increases (up to 60%) in fleshy macroalgal cover, primarily

Lobophora. *Agaricia tenuifolia* colonized dead *A. cervicornis* rubble and replaced it as the dominant coral species. Similar transitions between coral and algal communities have been seen in other areas of the Belize barrier reef. Patch reefs on remote Glovers Reef also changed during the last 25 years suffering a 75% loss of coral cover, 99% loss of *A. palmata* and *A. cervicornis*, and over 300% increase in macroalgae. These changes are probably linked to the reduction in herbivores (especially the urchin *Diadema*), and death of corals by white band disease. Although the acroporids were almost wiped out on the patch reefs, massive corals showed little mortality. A study of 12 deep fore reef sites throughout the barrier reef and atolls found the mean live coral cover to be 28% in 1997. The first mass coral bleaching event documented in Belize was in 1995 and resulted in about 10% of corals suffering mortality from bleaching and disease. Another bleaching event in 1997 affected the Snake Keys area, but details are lacking and it may have been salinity, not temperature-induced.

Belize experienced major impacts from the 1998 bleaching and hurricane events, with the majority of reefs (72 of 80 surveyed) showing significant damage: 46 with moderate disturbance; 26 had severe damage; and only 8 had low damage. Some of the highest mortality in the region was found on shallow reefs in the south with coral tissue losses twice as high (24%) on shallow reefs than fore-reefs (12%). Damage was less in the north, but some shallow reefs were affected, while deep fore-reef sites had mainly bleaching damage. The central area had moderate to severe disturbance at most exposed shallow sites, especially from the hurricane, while protected sites had more disease. Pre- and post 1998 disturbance studies on Glovers Reef found a reduction in coral recruitment on windward fore reefs and shallow patch reefs. Signs of recovery on reefs were seen in AGRRA surveys of 35 sites in July 2000, with increased recruitment on shallow reefs, including *Acropora*, more *Diadema* urchins, and little disease, bleaching or recent mortality. Two back reefs in the north, however, had high incidences of black band disease and mortality of *M. annularis*, particularly at Cay Chapel, a privately owned island under intense tourism development including a jet airstrip and 18-hole golf course. The 12 deep fore reef sites surveyed in 1997 were resurveyed in 1999 finding an average of 49% decrease in live coral cover (from 28% to 14%). The loss in live coral was greatest in the south (62%), followed by the North (55%), atolls (45%) and Central (36%), which were probably somewhat protected from the hurricane by the presence of the atolls.

Guatemala
There have been no surveys of the distribution and condition of coral communities. The reefs were probably heavily impacted by Hurricane Mitch, especially from storm run-off and the 1998-bleaching event.

Honduras
Disturbance to reefs in the Bay Islands has been caused by hurricanes, bleaching, coral diseases, over-fishing, and extensive coastal development. In Roatan, coral cover varies with reef zone: 8% on the reef crest; 18% on the shallow fore reef; 28% on the deep fore-reef; and 17% on spur and grooves (average of 21%). Coral cover in Sandy Bay West End Marine Reserve ranges from 24-53%, although macroalgal cover varies from 19-57%, and the sites with high algal cover (>50%) were near outfalls of sewage and sediment. Utila and Guanaja have low coral cover (5-20%). Cayos Cochinos has many small patch reefs

Coral Disease in the MBRS

Past information on the extent of disease incidence in the region is incomplete. A few well-documented studies suggest that the extensive mass mortality in the early 1980s of Acropora cervicornis was attributed to white band disease in Belize. In Mexico, Acropora palmata ramparts along the coastline apparently died in the early 1980s, possibly from white band disease. White band disease was also reported in Roatan and Cayos Cochinos although the extent was not determined. Black band disease increased in Cayos Cochinos after the 1995 bleaching event, with 34% of M. faveolata colonies infected; many bleached corals (particularly *M. annularis* complex) in Belize also became infected with black band after the 1995 bleaching event. Results from a large-scale survey of the MBRS found a high incidence of disease following the 1998 bleaching event; higher than reported for other areas of the Caribbean. The survey also documented:

- Black band was the most common disease on shallow reefs, white plague most common on fore reefs, low to rare occurrences of yellow band or other diseases;
- Fore reefs in Honduras (10% of colonies) had more disease than Belize (5%) and Mexico (3%);
- Belize shallow reefs (6%) had more disease than Mexico (3%) or Honduras (2%).
- 12 shallow and 11 fore reef corals were infected with disease - M. annularis complex most affected (10-22%)
- Very few acroporids were found with active diseases

The large proportion of *M. annularis* complex colonies in the MBRS infected and dying from disease is of particular concern, especially in Belize shallow reefs and deep fore reefs in the Bay Islands. AGRRA surveys conducted at 35 sites in 2000 suggest the overall incidence of disease has decreased compared to 1999, yet localised shallow sites had high levels of black band disease (especially Cay Chapel). Although disease incidence was lower, the high mortality this year suggests infections of white plague and black band in 1999 were responsible for much of the observed mortality.

with low coral cover, small colonies and signs of significant past disturbance, although deep reefs have higher cover (14 to 40%) and abundant macroalgae (*Lobophora variegata* and *Dictyota cervicornis*). Surveys in 1985 reported that all reefs in Roatan but one, were healthy, but white band disease was present. White band disease was reported on *A. cervicornis* in Cayos Cochinos probably triggered by the 1995 bleaching event where 34% of *M. annularis* and *M. faveolata* colonies were infected with black band disease.

Most reefs surveyed in Honduras (38 of 44) were damaged from the 1998 disturbance events; 25 had moderate disturbance, 13 had severe disturbance, and 6 had low disturbance. Recent mortality was similar on shallow (16% average) and forereefs (17%).

Localised shallow reefs, primarily in Guanaja, were damaged by the hurricane, while bleaching and disease affected deep reefs in the Bay Islands. Coral cover at 9 sites in Utila, Roatan, and Guanaja ranged from 13 to 33%, with 50 to 80% of coral colonies showing some partial mortality between 34 to 73% of the surface. Algae covered ranged between 52 and 85%.

Nicaragua

The coast is fairly remote with limited access, therefore it has not received intensive human use or scientific attention. Increased deforestation has resulted in higher sediment loads which are

---

IMPACTS OF HURRICANE MITCH ON THE MBRS

This region has a long history of hurricanes damaging the coral reefs; some major, some minor. Hurricane Gilbert (1988) caused severe damage to shallow reefs along the Yucatan peninsula; Hurricane Haiti (1961) and Greta (1978) were two of the most significant storms to hit the central coast of Belize; and Hurricane Fifi (1974) devastated the coast of Honduras. But then came Hurricane Mitch; a Category 5 hurricane with sustained wind speeds over 250km per hour, which battered the Caribbean coast and parts of Honduras, Nicaragua, El Salvador, and Guatemala between Oct. 27 - Nov. 1, 1998. It killed over 11,000 people and destroyed more than 50% of the infrastructure in Honduras. Hurricane Mitch is considered one of the largest and deadliest tropical cyclones in the Caribbean this century. Moreover, it stalled for over 48 hours over the Bay Island of Guanaja, and hurricane force winds destroyed houses and denuded pine and mangrove forests. Large storm swells smashed on the Belize barrier and southern Yucatan. The extent and type of damage to coral reefs varied depending on location, species present, and architectural complexity. Results from a large-scale survey to assess Hurricane Mitch's impact on barrier and fore reefs at 151 sites found:

- Greatest damage on Belize barrier (29% of shallow corals and 5% of fore reef corals damaged);
- Guanaja (22%) had more damage than Roatan (13%), Utila (8%), and Cayos Cochinos (5%);
- Southern Yucatan (11% shallow, <1% fore reef) had minimal damage;
- Almost 80% of corals damaged at NE Glovers Reef, the highest in the region;
- Localised shallow reefs in Belize and Honduras had 50-70% of corals damaged;
- Acropora tenuifolia and M. complanata were affected the most; and
- There was major reduction of reef structure on many shallow reefs.

Secondary impacts were likely from extensive runoff of low salinity, high nutrient, sediment-laden water into the Gulf of Honduras as far north as Glovers Reef, and possible contamination from high quantities of pesticides and fertilisers in runoff. Decreased water temperatures in Honduras and southern Belize were reported after the hurricane. Loss of revenue in fishery and tourism industries was also reported.

seriously degrading nearshore reefs, and over-fishing and damaging fishing practices also contribute to the degradation. Most shallow reefs around the populated Corn Islands are degraded due to discharge of untreated sewage i.e. on 5 CARICOMP transects at 12-15m in 1994-96, bottom cover was dominated by algae (44% - mostly *Dictyota* and *Padina*); live coral (25% - mostly *M. annularis, Agaricia, Porites*, and *Millepora alcicornis*); and minimal sponge and soft corals (5%). There was no evidence of coral bleaching or gorgonian diseases and *Diadema* urchins were seen. In the Miskito Reserve, coral recruitment was high, coral disease and bleaching were low, algal diversity was high but macroalgal overgrowth was low, probably due to urchin grazing. Nothing is known of the status of the well-developed *A. palmata* reefs on the windward eastern edge of the Pearl Cays, which are subject to coastal runoff. Little is known of damage from Hurricane Mitch, but high damage is anticipated because it went over the Bay Islands and resulted in massive runoff of sediments. Anchor damage is minimal because the fishing boats are small. Overall, the reefs of Nicaragua have probably lost about 10% of coral cover over the last 10 years, but coral cover of 25% is comparable to other areas of the Caribbean.

El Salvador
Little is known about the coral communities at Los Cobanos.

POTENTIAL CLIMATE CHANGE EFFECTS ON CORAL REEFS

Perhaps the greatest future threats to the reefs in this region are from bleaching events. The 1990s decade was the warmest on record with the most extreme El Niño events. Models for the next 100 years predict that warming will continue and coral bleaching will become more frequent and severe, probably becoming an annual occurrence over much of the region by 2015 and seriously damaging these coral reefs. Continued Hurricane impacts are also expected. Thus, increased management efforts will be needed to protect the coral reefs from additional anthropogenic stress which would compound these global effects, prevent recovery from acute disturbances, and further increase the probability that these reefs will experience significant shifts in community structure, including continued losses of live coral.

STATUS OF CORAL REEF FISHERIES AND FISHERIES

Northern Central American reefs have traditionally been an important source for food, but the intensity and frequency of fishing is increasing at an alarming rate. Marine fish captures are largest in Mexico, are steadily increasing in Nicaragua (259mt in 1992) and Guatemala (92mt in 1993), while there is an overall decline in landings in Belize and Honduras between 1996-1998. These declines are attributed to lowered populations, over-fishing, changing economic circumstances, illegal fishing, destructive fishing methods like gill nets, and lack of enforcement. While fisheries management is non-existent in some areas, there are several examples of effective management to use as models for other areas. Two of the most exploited reef species are lobsters and queen conch, which both rely on healthy reefs and are vulnerable to over-fishing.

Mexico
There are at least 245 reef fish in Atlantic Mexico: 68% of these in the Gulf of Mexico; and 92% along the Yucatan. Herbivores dominate on Gulf of Mexico reefs, carnivores are more

STATUS OF CONCHS AND LOBSTERS

Throughout this region, queen conch *(Strombus gigas)* and spiny lobster *(Panulirus argus)* have important economic, social and cultural values. But over-exploitation, illegal fishing, poor enforcement, and lack of trans-boundary management over the last 30 years have resulted in declining populations and decreases in catches. The Mexican Atlantic has the greatest conch catches, while the largest lobster catches are from Nicaragua. Management of these fisheries varies throughout the region, although illegal fishing is common and trans-boundary issues are largely ignored.

Mexico: Extensive over-exploitation in the late 1970s caused conch stocks to collapse, leading to fishery closures in Yucatan (1988) and seasonal closures in Quintana Roo (1991). Shallow water conch and lobster populations have declined and deeper water populations are at risk since collection using scuba and hookah is permitted. Larger lobster populations are found in Bahia de la Ascencíon, Espirtu Santo Bay and Banco Chinchorro. Lobster regulations include minimum size (145mm) restrictions, 4 month closure, and a ban on collecting berried females.

Belize: Lobster is the biggest, most important fishery and years of over-fishing have altered both lobster and conch populations. The populations are skewed towards smaller lobsters, while many conch are legal size but not sexually mature; yet the overall status of lobster and conch is not known. Regulations include closed seasons for lobster (Feb 15-June 15) and conch (July-Sept), gear restrictions (no scuba), prohibited take of egg-bearing lobsters, and minimum size restrictions for lobster and conch (18cm). There is extensive illegal fishing and the use of baited gill nets to harvest lobsters damages reefs.

Honduras: Lobster and conch are important commercial and artisanal fisheries. Conch are collected by free or scuba diving and harvesting restrictions include a closed season (March-August) and minimum size (22cm); there are no regulations for the lobster fishery. Historically, Honduras had the largest lobster catches, but these have drastically declined. Few large lobsters exist and populations of conch and lobsters are only found in deeper waters. Lobster and conch collecting is prohibited in Cayos Cochinos Reserve, and they are overfished in the Bay Islands, whereas the status on the offshore banks is unknown.

Nicaragua: Lobster is the most important fishery by Miskito artisanal, industrial, and pirate industrial fishermen. Fishing effort has doubled in the last 10 years, yet lobster populations are not considered to be heavily over-fished by the locals. Little is known about conch. There are no regulations to conserve lobster or conch and illegal fishing by foreign vessels (Honduras) is the greatest threat.

Management recommendations include closed seasons, size and weight restrictions consistent with size or sexual maturity, gear restrictions, protection of nursery and spawning grounds, establishment of no-take zones, limited entry or quotas, gathering more information on status and distribution, alternatives to fishing, and measures to improve trans-boundary management of larger, regional fisheries, including greater enforcement and harmonisation of regional regulations.

Country	Conch	Lobster	Fish
Belize	252	502	111
Guatemala	--	--	213
Honduras	490 (1996)	306	160
Mexico	3,293	613	93,291
Nicaragua	162	3,729	4,088

The harvest for the major marine resources - strombid conch, lobster, and fish 1998 in metric tons (mt)

abundant in the Caribbean, and important families include Scaridae, Pomacentridae, Labridae, Acanthuridae, Lutjanidae, Haemulidae and Serranidae. There is higher species diversity, more abundant herbivores, and 2-3 times higher predator biomass in the Sian Ka'an Biosphere Reserve, than on adjacent unprotected reefs. Mexico has an advanced fishing industry, concentrated mainly on the Pacific side. Coral reef fisheries concentrate on snappers, groupers and parrotfishes, but fisherman often avoid reef areas to prevent net damage and instead fish for larger fishes found near rocky points. Harvesting for reef invertebrates (e.g. sea cucumbers and molluscs) and aquarium fish (angelfish, butterflyfish) is high and the fishery is poorly regulated. Most reef fishing targets groupers, grunts, barracudas, snappers, snook, and Atlantic Spanish mackeral on reefs in the Gulf of Mexico and the Yucatan. Fishing regulations near Veracruz in the SW Gulf of Mexico have been fairly successful at limiting fishing to subsistence and recreational fishing, but fishing on the Campeche Bank is intense and the resources are heavily exploited, particularly lobsters and groupers. Six spawning aggregations on the Yucatan have been commercially fished for grouper (*Epinephelus* spp.) for over 50 years and catches continue to increase, yet these aggregations are considered to be over-fished. Regulations exist for the amount, size and depth of collection of black coral, but it is over-exploited and commercially threatened. A lack of enforced fishing regulations, increases in tourism, and lack of alternatives to fishing threaten the status of coral reef fishes in unprotected areas. Yet, awareness on the need to develop sustainable fisheries is increasing with actions underway to designate new protected areas (Xcalak), and implement tighter fishing regulations at existing ones (Banco Chinchorro, Sian Ka'an, Cozumel, Contoy Island).

Belize
There are more than 317 reef fish species with higher fish density on shallow reefs (primarily surgeonfish) versus deeper fore reefs (primarily parrotfish). Densities on shallow reefs are generally higher at Glovers, Lighthouse, and Central Barrier than Turneffe and the northern barrier. Abundance at deep sites is very similar, although Lighthouse has the lowest. Belize has a small but expanding commercial and subsistence fishing industry. Lobster is the most important fishery, followed by marine fishes, conchs, and *Penaeus* spp. shrimps. Now shrimp mariculture accounts for more than the value of all fisheries combined. At least 10 spawning aggregations are currently over-exploited, the largest one is at Gladden Spit with

at least 20 species. Little is known about black coral collection, sport fisheries (e.g. snook, bonefish), small scale harvesting of sponges and the alga *Euchema*, and collecting for the small aquarium trade. Fish are exploited throughout Belize, with Gladden Spit and Sapodilla Cayes being most vulnerable to over-fishing and substantial illegal foreign fishing. The Hol Chan Marine Reserve has effectively protected fish populations for over 13 years, with more species, higher abundance, and larger sizes of commercial species compared to non-protected areas. Belize has the infrastructure and legislation to implement effective marine fisheries regulations, but the challenge will be to enforce such regulations and reduce illegal fishing.

Guatemala

Commercial fishing is not well developed, although apparently, abundant resources exist. On the Pacific side there are small-scale local fisheries for fishes, sharks, rays, and skates, and the yellowleg and *Penaeus* shrimps. On the Caribbean side the main fish include manjua, shrimp, red snapper, mutton snapper, billfish, jack, tarpon, and snook. Very little is known about the fish resources and level of exploitation on the Caribbean coast. Some areas are believed to be over-exploited i.e. Amatique point and the coastline east to the Honduran border. Many local people rely heavily on fishing (legal and illegal) and buying from fisherman in southern Belize and Honduras.

Country	1980		1990		1997	
Belize	131	$141	183	$241	31	$191
El Salvador	246	$164	188	$252	119	$303
Guatemala	6	$7	347	$66	67	$391
Honduras	59	$65	281	$609	700	$1,405
Mexico	960	$1,995	27,272	$35,718	27,943	$52,646
Nicaragua	544	$735	234	$291	1,675	$6,625

Total marine fish exports in metric tons and the value in US$1000

Honduras

There are over 294 reef fish species with at least 226 in Cayos Cochinos Biological Reserve. Abundant fish include parrotfish, damselfish, surgeonfish, and wrasses, bar jack, and yellowtail snapper, but at least 34 species have been over-fished. The fishing industry consists of small-scale local fisherman and industrial fishing fleets. On the Pacific coast, marine fishes, crabs and *Penaeus* shrimps are important, while on the Caribbean, *Penaeus* shrimps, lobster, marine fish, and conch are the major fisheries. There is intense fishing along the mainland coast and gill nets are often used, yet the status of the coastal fisheries is unknown. There is a long history of large industrial fishing trawlers exploiting all reef fish around the Bay Islands resulting in a dramatic collapse of many fish populations. Fishing regulations at the Cayos Cochinos Reserve prohibit industrial fishing vessels and commercial vessels and limit fishing to the traditional Garifuna communities. Total fishing effort and landings decreased in 1998-1999, probably due to Hurricane Mitch. Fish are exploited along the north coast and Bay Islands (except Cayos Cochinos), but the status of fisheries on the offshore banks is not known. Despite hosting the largest commercial fleet of shrimp trawlers along the Caribbean coast of Central America, Honduras' capacity for fisheries management remains weak, and data collection on stocks and factors limiting productivity is unreliable or non-existent.

Nicaragua

There is little information on reef fish in the Caribbean, although the species composition is probably similar to other areas. Fish have always been an important local and domestic food source, but extensive fishing for shrimp and lobster in the 1980s threatens fish stocks. Lobsters are the most important fishery, as well as fish and shrimp. USAID reports from 1996 suggest there are few commercially important fish on shallow reefs, with the most abundant being small yellowtail snappers. Illegal fishing by foreign fisherman is a serious threat to the small artisanal fishery, which appears to be sustainable, but industrial fishery is not sustainable. There is a growing aquarium fish trade at Corn Island that may expand into other areas. Endangered green turtles are heavily fished for food (14,000 turtles/year are harvested) and immediate action is needed to conserve them. There is great opportunity to develop and implement fisheries regulations before extensive over-exploitation destroys stocks of lobster, shrimp, reef fish, and turtles. A first step would be to reduce the amount of illegal fishing by neighbouring countries.

El Salvador

There is no reef fishery, but there is a small offshore shrimp (*Penaeus* spp.) and marine finfish fishery.

ANTHROPOGENIC THREATS TO CORAL REEF BIODIVERSITY

These are the greatest threats to coral reef biodiversity, with the potential to reduce species diversity and richness, abundance, habitat quality and quantity, productivity, and critical habitats such as spawning, breeding or foraging sites. After the dramatic impacts from the 1998 mass bleaching event and Hurricane Mitch, the long-term recovery of reefs will not only depend on natural recovery processes, but also on the ability of governments and managers to reduce the level of cumulative anthropogenic disturbances. Continued human damage to reefs will have severe ecological and socioeconomic consequences.

As part of the Mesoamerican Barrier Reef System (MBRS) conservation initiative with Mexico, Belize, Guatemala, and Honduras, and preparation of a regional GEF/World Bank project to conserve and sustainably manage the MBRS, a 'Threat and Root Cause Analysis' was conducted to identify the main threats to the ecology of coral reefs as: 1) inappropriate coastal/island development and unsustainable tourism; 2) inappropriate inland resource and land use and industrial development; 3) over-fishing and inappropriate aquaculture development; 4) inappropriate port management, shipping and navigation practices; and 5) natural oceanographic and climate meteorological phenomena and how these interact with the other threats. These countries share many of the same resources, therefore many issues, problems and concerns about the use and management of these resources extend beyond country boundaries. These trans-boundary issues have demonstrated the need for mechanisms to address existing and emerging threats. The Threat and Root Cause Analysis identified several trans-boundary threats to coral reef resources including:

- Agricultural/Industrial Runoff (e.g. Aguan, Motagua, Dulce, New Rivers etc.);
- Country boundary (Honduras & Nicaragua, Belize & Guatemala);
- Land based pollution, contamination (Chetumal Bay etc.);

- Maritime Transport/Port, pollution (e.g. Gulf of Honduras);
- Migratory/Endangered species (e.g. sea turtles, manatees);
- Sedimentation, contamination (e.g. Gulf of Honduras);
- Tourism (regional);
- Unsustainable fishing (regional); and
- Illegal fishing (e.g. widespread illegal foreign fishing in Belize and Nicaragua)

In the Mexican Pacific, the main threats are excessive sedimentation from deforestation, anchor and other diver-related damages, and illegal fishing. The remoteness of the Caribbean coast of Nicaragua and history of civil unrest has limited development except by the indigenous communities. Specific threats to coral reefs include over-fishing, coral extraction, oil pollution, deforestation, soil erosion and sedimentation, and water pollution. Natural hazards such as earthquakes, volcanoes, landslides, and severe hurricanes also pose threats to reefs, although these are periodic rather than chronic sources of disturbance. Water quality around coastal communities and inhabited cays has declined due to untreated sewage, industrial activities, fish processing plants and maritime transport. Sedimentation and eutrophication from coastal deforestation along the Honduran coasts affect the Miskito Coast Marine Reserve downstream, in addition to local and intense foreign fishing of lobster, fish and turtles. Extensive poaching of Nicaragua's marine resources by Hondurans has been reported.

In El Salvador current threats to the coastal zone include deforestation and soil erosion, water pollution, soil and water contamination from disposal of toxic wastes, and frequent and often destructive earthquakes and volcanic activity.

Coasts / Threats	AGM	CMX	CBL	CGU	CHO	CNI	PMX	PES	PNI
Agricultural runoff	+		+	+	+	+		+	+
Aquaculture development			+	+	+	+		+	
Coral extraction (curio trade)	+	+	+		+	+	+		
Deforestation	+	+	+	+	+	+	+	+	+
Destructive fishing	+	+	+	+	+				?
Diving activities	+	+	+		+		+		
Dredging	+	+	+						
Fish extraction	+	+	+	+	+	+	+	+	+
Garbage pollution	+	+					+		
Heavy metal pollution	+			+				+	?
Industrial activities	+	+	+	+	+	+			?
Maritime activities	+	+	+	+	+	+		+	?
Oil pollution	+		+	+	+			?	+
Over-fishing	+	+	+	+	+	+	+	?	?
Sedimentation/siltation	+	+	+	+	+	+	+	+	+
Sewage pollution	+	+	+		+	+		+	?
Tourism activities	+	+	+		+		+		?
Urban development	+	+	+	+	+			+	+

A listing of the anthropogenic threats to coral reef biodiversity affecting the coasts of this region: AGM: Atlantic Gulf of Mexico; CMX: Caribbean Mexico; CBL: Caribbean Belize; CGU: Caribbean Guatemala; CHO: Caribbean Honduras; CNI: Caribbean Nicaragua; PMX: Pacific Mexico; PSA: Pacific El Salvador; PNI: Pacific Nicaragua.

CURRENT MPAS, MONITORING PROGRAMMES AND CONSERVATION MANAGEMENT CAPACITY

The conservation and sustainable use of marine resources is becoming a higher priority in countries of the region and programmes or regulations are being developed and implemented to address land use and development, fisheries exploitation, pollution control, and tourism. Mexico and Belize have several well-developed management programmes to conserve coral reef resources, while continuing progress is being made in Honduras, Guatemala, and Nicaragua. One of the most significant steps towards protecting coral reefs was the signing of the Tulum Declaration (1997) by the leaders of Mexico, Belize, Guatemala, and Honduras which was an agreement to work towards regional conservation to ensure the integrity and future management of the 'Mesoamerican Barrier Reef System' (MBRS). The objectives of the MBRS Project with Mexico, Belize, Guatemala, and Honduras, are to enhance protection of vulnerable and unique marine ecosystems of the second longest barrier reef in the world. The project, financed by the GEF with support from the World Bank, is to assist the countries involved to strengthen and coordinate national policies, regulations, and institutional arrangements for marine ecosystem conservation and sustainable use. Specific conservation targets identified by the governments include strengthening a regional system of protected areas, regional fisheries management and conservation of key species and habitats.

Current Marine Protected Areas

There are over 100 marine and coastal protected areas in North Central America providing protection for over 45,000km^2 of marine habitats and resources and most have been designed specifically to address issues from over-exploitation of fishery resources, protection of critical habitat, and pollution and degradation of resources related to excessive use. Identifying and limiting access to within limits of human carrying capacity is an important MPA conservation management tool. Mexico and Belize have both governmental and NGO managed MPAs, while Guatemala and Honduras have delegated more fully to NGO management.

The effectiveness and conservation approach varies with each MPA. There have been calls for assessments of the effectiveness of MPAs, and Belize has undertaken such a study. Overall, the effectiveness of their management was found to be 'moderately satisfactory' with the main problem being with administration. Although some MPAs have been relatively successful at reducing human impacts on coral reefs, many others remain as 'paper parks', mainly due to the lack of financing for management activities. Developing reliable funding mechanisms and building the human capacity to manage the resources remain the two greatest challenges for the region's MPAs. A frequently overlooked consideration in the design of marine protected areas is the 'regional perspective', including trans-boundary issues and threats, physical linkages between MPAs, and the lack of cooperation and coordination between countries. Strengthening mechanisms to promote trans-border coordination in the protection of system wide resources and ecological processes will be essential for protecting coral reefs in this region.

CONSERVATION OF THE MBRS

The Mesoamerican Barrier Reef System (MBRS) has been recognised as one of WWF's Global 200 ecoregions with outstanding biodiversity, whose protection is vital for the conservation of global biodiversity. The MBRS also offers numerous benefits for coastal inhabitants and visitors including subsistence, recreational, and commercial fishing, tourism, snorkeling and diving, as well as providing structural protection against storms and erosion. Yet years of indiscriminate and unsustainable use, such as unregulated coastal development and over-exploitation of fishery resources, now threatens the balance of this unique ecosystem.

Recognising the ecological, aesthetic, cultural, and economic value of the MBRS, the Presidents of Mexico, Guatemala and Honduras and the Prime Minister of Belize, signed the Tulum Declaration in June 1997, launching the Mesoamerican Barrier Reef Initiative (MBRI). The primary objective is to promote the conservation and sustainable use of the MBRS and to ensure its continued contribution to the ecological health of the region and the livelihood of present and future generations. An Action Plan for the MBRS was developed at the formation of the Mesoamerican Barrier Reef Congress in November 1997, which was jointly organised by the Central American Commission for Environment and Development (CCAD), the World Bank, the World Wide Fund for Nature (WWF), and the International Union for the Conservation of Nature (IUCN). With support from Global Environment Facility and World Bank, a regional strategy to protect the MBRS is currently being prepared based on actions to:

- Integrate policies and legislation to strengthen coordination at the regional level;
- Promote conservation;
- Promote sustainable use;
- Broaden environmental education and awareness;
- Develop a regionally compatible ecosystem/biodiversity monitoring program and information

Since the signing of the Declaration in 1997, significant steps have been made towards developing a regional conservation strategy to ensure the integrity and future management of the MBRS including:

- 1st and 2nd MBRS Regional Project Preparation Workshops (July & September 1999 in Honduras and Belize)
- Status of the MBRS: Impacts of Hurricane Mitch and 1998 Bleaching on Coral Reefs Report
- Threat and Root Cause Analysis Report
- Guidelines for Developing a Regional Monitoring and Environmental Information System Report
- Sustainable Fisheries Management Report
- Establishment and Management of Marine Protected Areas Report

- Environmental Education and Public Awareness Report
- Harmonisation of Policies and Legislation and its Implications for the MBRS Report

AGRRA Capacity Building for Science and Management in the MBRS, Workshop 1999
WWF Report on Preliminary Meeting of Experts (June 1999, Belize City)
WWF/RSMAS- Development of a regional GIS database for MBRS
WWF/CCAD Workshop: Ecoregion Planning: Ecological vision & conservation priorities (April 2000)

Several other efforts are progressing at all levels (grassroots, national, regional, and international) to address the threats endangering the MBRS. Other organisations involved in the MBRS initiative and conservation in the entire NCA region include Food and Agricultural Health Organisation, World Wildlife Fund for Nature (Conservation of the MBRS Ecoregion), PROARCA-COSATAS/USAID (Regional Environmental Project for Central America), The Nature Conservancy, University of Rhode Island/Coastal Resource Center (URI/CRC), AGRRA, CARICOMP, UNEP, GCRMN, UNDP, GTZ, USAID, DANIDA, and Inter-American Development Bank. Contact: Marea Hatziolos The World Bank, Mhatziolos@worldbank.org.

Country	#of Marine Protected Areas	Approximate size of protected areas (km^2)	Comments
Mexico	44 (marine & coastal)	>40,000	Most recent MPA is Xcalak, Yucatan- a community based marine reserve (2000) and Puerto Morelos (2000)
Belize	12 (marine)	>2117.5	Most recent MPA is Gladden Spit, one of the largest spawning aggregation areas for snappers, groupers (2000).
Guatemala	4	170	Established the first Manatee Protected Area.
Honduras	25	4300	Cayos Cochinos Biological Reserve (government and NGO support) and Sandy Bay Reserve (NGO based).
Nicaragua	4	1200	Miskito Cay Marine Reserve- community based MPA.
El Salvador	2	52	
Total	91	>47,739	

A summary of Marine and Coastal Protected areas in North Central America

Coral Reef Monitoring

Monitoring capacity and activities vary considerably across the region, with some regional monitoring programmes including GCRMN, AGRRA, REEF, Reef Keeper and Reef Check, and numerous local and national programmes. Most monitoring and research has been conducted recently at select locations in Belize and Mexico, and more recently in Cayos Cochinos and Roatan (limited), Honduras. Monitoring is almost non-existent in Guatemala and Nicaragua.

A recent review in the MBRS showed that a lack of a regional monitoring perspective, and poor coordination and data sharing among monitoring programmes were hindering conservation efforts. Monitoring capacity is often based in a NGO or academic sector with limited ties to decision making. Many government agencies lack the funding and capacity to implement extensive monitoring programmes to assist in policy formulation. Therefore MBRS Initiative focuses on improving monitoring capacity, developing an integrated regional perspective, and improving access to environmental information throughout the region. Only Mexico and Belize had national geo-referenced databases and an integrated environmental information system. But a new regional Geographical Information System (GIS) database has been created to develop a more comprehensive and regional approach to monitoring (RSMAS/WWF), and to include ecological, physical, and socioeconomic data as part of a distributed data based monitoring and information system.

Mexico

Regional monitoring programmes include CARICOMP, AGRRA, REEF, Reef Keeper and ReefCheck. Institutions involved with local monitoring include SEMARNAP, UNAM, ECOSUR, Amigos de Sian Ka'an, CINVESTAV, Universidad Quintana Roo, Centro Ukana I Akumal, EcoSur, University of Miami, INVESTAV, INVEMAR, University of Miami/RSMAS, University of Maine, University of Rhode Island/CRC, and Dauphin Island Sea Lab. On the Pacific coast, ongoing monitoring is done by the Universities: Baja California (La Paz), Guadalara (Puerto Vallarta) and del Mar (Puerto Angel). Several scientific studies are ongoing by Mexican and international academic institutions.

Belize

Regional monitoring programmes include CARICOMP, AGRRA, CPACC and REEF. Institutions involved with local monitoring include Coastal Zone Management Authority, Belize Dept. of Fisheries, Belize Audubon Society, University College of Belize, The Nature Conservancy/TIDE, Greenreef, University of South Florida (M. McField), CARICOM Fisheries Unit, CFRAMP, Wildlife Conservation Society, University of Miami/RSMAS. There are numerous scientific studies by foreign academic institutions.

Guatemala

With few coral reef resources, most monitoring revolves around forestry, ecotourism, environmental education, capacity building, and sustainable land development. Groups involved include Fundary (Fundacion mario Dary), Association for Research and Social Studies (ASIES), PROARCA/COSTAS, WWF, USAID, FUNDAECO Izabel, CONAP, CECON (Centro para Estudios Conservacionistas), Sierra de Sta. Cruz, and Sierra de las Minas.

Honduras
Groups involved with monitoring include Cayos Cochinos Marine Reserve, Bay Island Conservation Association, Biodiversity of Protected Areas Project (PROBAP), Wildlife Conservation Society, WWF, AGRRA, CARICOMP, Inter Development Bank (BID), REEF, USGS, and NOAA.

Nicaragua
Groups involved with monitoring/research include Mikupia, Miskito Reef Mapping Project, USAID, CARICOMP, and MARENA.

GOVERNMENT POLICIES, LAWS AND LEGISLATION

Mexico has signed all major international agreements dealing with coastal and marine biodiversity, while the other countries are party to a number of international agreements. The most significant regional plan for increasing government capacity to protect coral reefs is the Declaration of Tulum. The Cartagena Convention is also important to protect and conserve fragile ecosystems, as countries agreed to prevent, reduce and control pollution from ships, dumping, sea-bed activities, airborne, and land-based sources and activities within the Wider Caribbean. National government capacity varies with each country, yet there is an overall lack of infrastructure and weak institutional framework that has lead to poor implementation of international agreements and a lack of enforcement of national environmental regulations.

Mexico
There is extensive legislation and a solid institutional capacity to manage natural resources. The Secretariat of the Environment, Natural Resources and Fisheries (Secretaria del Medio Ambiente, Recursos Naturales y Pesca - SEMARNAP) is the Federal Government agency responsible for managing natural resources and fisheries. SEMARNAP oversees the following administrative units that have jurisdiction over the management of coastal and marine ecosystems:

- Federal Attorney for Protection of Environment (PROFEPA) - Enforcement of environmental law;
- National Institute of Ecology (INE)- Protected areas, wildlife management, pollution control, environmental zoning and environmental impact assessment;
- National Institute of Fisheries (INP)- Fisheries research;
- National Water Commission (CNA)- Water management;
- Undersecretariat of Fisheries- Fisheries management, aquaculture and fisheries infrastructure; and
- Undersecretariat of Natural Resources- Federal marine-coastal zone, soil conservation, forestry.

Other Federal agencies with jurisdiction over coastal and marine ecosystem management include: the Navy Secretariat; the Transportation and Communications Secretariat; the Tourism Secretariat; the Governance Secretariat and the Health Secretariat. The Governance Secretariat administers federal islands and cays, including the management of island ecosystems. The Health Secretariat has jurisdiction over issues affecting health and subsequently natural resources. There is extensive legislation including key regulations such as: Declaration of Tulum 1997; Ecology Law 1988; Fisheries Law 1992; National Waters Law 1992; Federal Ocean Law 1986; Port and Navigation Law; and National Properties Law. Mexico has an extensive and effective protected areas program, with numerous ecological zoning programmes responsible for regulating coastal activities.

Belize

There is the legal and institutional policy framework to manage coral reefs, but there is a lack of enforcement and monitoring capacity. The system of MPAs is not financially sustainable, and relies on international support. Administrative units involved with the conservation and sustainable use of coastal marine resources include:

- Fisheries Department - management of fisheries, responsibility for marine reserves; plans to delegate more to NGOs
- Forestry Department – responsibility for National Parks, most delegated to NGOs
- Coastal Zone Management Authority/Institute – monitoring and research, conservation and MPA planning, policy development, financing of many MPAs
- Marine Protected Areas Committee - promotes communication among MPAs;
- National Coral Reef Monitoring Group- proposed coordination for national reef monitoring efforts;
- Barrier Reef Committee - oversees MBRS project and World Heritage sites;
- Coastal Zone Advisory Committee - shares information and advice between governmental departments and NGOs
- Land Utilisation Authority - regulates land use;
- Department of Environment - enforces environmental regulations;
- NEAC (National Environmental Appraisal Committee) - approves/disapproves EIAs
- Belize Tourist Board - oversees tourism industry.

The pertinent laws include: the Fisheries Act 1948; Environmental Impact Assessment 1995; Environmental Protection Act 1992; Water and Sewerage Act 1971; Declaration of Tulum 1997; and Coastal Zone Management Act 1998.

Guatemala

There are few laws or regulations that pertain directly to coral reefs, although several may benefit reefs indirectly: Declaration of Tulum 1997; Decrees Issued by the Congress of the Republic on protected areas and reefs; Elimination and Disposal of Sewage and Residual Water; Law for the Rational Exploitation of the Fisheries Resources of the Country; Law of Protection and Improvement of the Environment; Regulatory Law of the Areas of Territorial Reserves of the State of Guatemala; and Rules Concerning an Environmental Impact Assessment.

Honduras

There are several laws and regulations affecting coral reef resources, although the enforcement of these is sometimes lacking. Important regulations includes Declaration of Tulum; Biodiversity Convention; Central American Component for the Protection of the

Environment Convention, Environmental Protection; Climatic Change Convention; Conservation of Biodiversity in Central America Convention; General Law of the Environment; Fisheries Law, Fisheries Resolutions; Law for Planning and Development of Tourism Zones; and Merchant Marine Law.

Nicaragua
There is no national legislation or institutional framework dedicated to conserving coral reefs. The Instituto Nicaraguense de Recurosos Naturales y del Ambiente (IRENA) is in charge of conservation and wildlife. In 1991, 23 coastal Miskito communities made an agreement with the Nicaraguan Ministry of Natural Resources and international conservation groups to establish the largest coastal marine protected area - the Miskito Coast Protected Area (MCPA). The agreement supports true community-based marine management where the local Miskito communities have the right and responsibility to manage their own marine resources. In addition, guidance from an advisory team, assistance to provide enforcement of the MPA, and national and international financing assistance is included. In 1993, a national integrated coastal zone management plan (PAA-NAC) was approved to promote the sustainable use of coastal resources, but implementation has been slow. Other existing regulations need updating and modification, while several proposed laws are waiting for approval (e.g. Fishery and Aquaculture Law). There are a few small NGOs, but few collaborative efforts between them or public/private partnerships to manage coastal resources.

El Salvador
There is no legislation for coral reefs probably due to the lack of reef resources.

INFORMATION GAPS, MONITORING AND RESEARCH NEEDS

Significant information is lacking on key issues required to improve our understanding and ability to manage coral reefs and minimise human impacts. Based on existing information of the status and threats to coral reefs in the region, these are four target areas that require further focus:

1) **Sustaining fishery resources**
    - Fishing activity (effort, catch)
    - Status of fisheries resources (population, life history data)
    - Location, size and exploitation of nursery and spawning areas
    - Economic alternatives to fishing
    - Larval dispersal, juvenile settlement, adult migration patterns
2) **Conserving coral reefs**
    - Status and distribution of coral reefs
    - Physical oceanography data (currents, wind, gyres)
    - Water temperature trends and patterns
    - Information on other coral benthos, particularly keystone species
3) **Sustainable development**
    - Current and projected land use
    - Land carrying capacity
    - Tourism levels and potential sustainable expansion
    - Agricultural development and impacts

4) Improving Water Quality
- Status of water quality
- Sources of contamination (point and non point)
- Water discharges and flows
- Measures to minimise water pollution and improve water quality

CONCLUSIONS

- Reefs are poorly developed on the Pacific compared to the well-developed reefs on the Atlantic coast which contain some of the most outstanding reefs in the Caribbean, including the second longest barrier reef in the world, luxuriant patch, fringing, and barrier reefs, 5 offshore atolls, and important habitat linkages to extensive adjacent mangroves, seagrasses, bays, and lagoons. Significant information gaps remain on the development and condition of reefs in Nicaragua, mainland Honduras and Guatemala, offshore areas like Swan Islands and Mysteriosa banks and other reefs in the eastern Pacific coast and more long-term monitoring is needed throughout the region.
- Major anthropogenic threats to regional reefs include: inappropriate and unsustainable land use; tourism, and industrial development; over-fishing and impacts from aquaculture farms; pollution from inappropriate sewage treatment, waste disposal, agricultural runoff, and other land-based sources; and absence of sound port management, shipping and navigation practices, in addition to global threat of climate change.
- Development along the coastlines varies from highly developed tourist hot spots in Huatulco and Puerto Vallarta (Pacific Mx) and Quintana Roo (Atlantic Mx) to small local indigenous fishing villages in the Miskito Cays, Nicaragua.
- The intensity and frequency of fishing pressure is increasing for important fishery species like conch, lobster and grouper which are over-exploited in much of the region. Failure to implement new (or enforce existing) fishery regulations, fragmented management of trans-boundary resources, and the absence of economic alternatives to fishing continues to impede the ability to properly conserve fishery resources.
- Coral reefs in the region have declined in the last 25 years due to coral diseases and mortality of the sea urchin *Diadema antillarum*; mass coral bleaching is a relatively recent event (1995, 1997, 1998), while hurricane impacts have a long history in the region. The distribution and status of many other coral reef resources is not well known at the regional scale.
- The 1998 coral bleaching event and Hurricane Mitch severely damaged many reefs in the MBRS, particularly shallow reefs in Belize. Extensive losses of individual coral species (e.g. *Agaricia tenuifolia*, *Millepora complanata*, *Montastraea annularis* complex, *Acropora* spp.) are of particular concern as they are major reef builders.
- The long-term ecological consequences of the 1998 bleaching and Hurricane Mitch will depend on recovery processes (ability of corals to recruit, adapt, persist, etc), future disturbance events (intensity, duration, and frequency), and the degree to which human impacts are minimised.

- Pacific reefs in Mexico suffered 18-70% coral mortality related to the 1998 El Niño event and subsequent La Niña event.
- A variety of capable localised monitoring and research programmes exist especially in Mexico and Belize, but the lack of a regional monitoring capacity and poor data sharing among programmes has hindered a better understanding of these systems and conservation efforts. A regional monitoring program for the MBRS is currently being developed to address these deficiencies and preliminary efforts are currently underway.
- Overall management and conservation awareness is advanced in Belize and Mexico, while capacity continues to grow in the other countries. Yet, the ability to address and minimise anthropogenic threats is highly variable and completely absent in some areas. Few mechanisms are in place to promote trans-border coordination and cooperation to reduce the escalating scale of human impacts. Poor or inconsistent enforcement of existing regulations is a continuing problem.
- Over 90 marine and coastal protected areas encompass more than 45,000km^2 of marine resources, but many of these lack financial sustainability and trained personnel to be effective. Mexico and Belize have the legal and institutional framework to implement MPAs, while the other countries often lack legislation, and all are in need of financial sustainability.
- A significant step towards protecting coral reefs was the signing of the Tulum Declaration (1997) by 4 neighbouring countries; a conservation strategy for the Mesoamerican Barrier Reef. International efforts are underway to strengthen and coordinate national policies, management programmes, monitoring efforts, and other actions aimed at conservation and sustainable use.

RECOMMENDATIONS

- A lack of understanding of ecological processes contributing to reef health limits our ability to conserve them, thus it is important to improve knowledge of these processes. A high priority is to identify and characterise the extent and condition of coral reef habitat and associated organisms, especially in less known areas as Nicaragua, mainland Guatemala/Honduras, offshore islands and banks like Swan Islands and Mysteriosa banks, and reefs in the eastern Pacific. Other data gaps include physical oceanographic and climate meteorological phenomena and how these interact with anthropogenic threats, and the ecological connections between reefs and adjacent coastal ecosystems and watersheds.
- Coral reefs that warrant immediate protection need to be identified, particularly those with high biological production; biodiversity hot spots; endangered, imperilled, or rare species; nursery and breeding areas; sources of larval corals, fish and other important reef organisms; and those at risk of pollution or other human impacts.
- Unsustainable and unregulated development of coastal resources must be eliminated throughout the region by developing effective land use strategies, implementing/enforcing appropriate land use regulations, eliminating inappropriate uses of resource, and minimising unsustainable tourism and industrial development. Sources of land based pollution need to be identified

and actions to reduce the impacts need to be developed. Mechanisms to reduce inappropriate port management, shipping and navigation practices must be created. The importance of addressing trans-boundary threats is emphasised.
- Unsustainable fishing practices throughout the region must be reduced by: enforcing existing fishery regulations and improving compliance; designing new regulations that reflect an ecosystem approach to management and are harmonised regionally; designating marine fishery reserves; developing financial mechanisms to sustain marine protected areas; eliminating destructive fishing practices; developing economic alternatives to unsustainable fishing; and promoting economic incentives for conservation.
- The institutional infrastructure and financial sustainability for implementing and enforcing national and international government policies needs improvement in Mexico and Belize and full development in the other countries. Policies and legislation to strengthen cooperation and coordination on conservation and sustainable use of coral reefs at the regional level need to be developed.
- There is an immediate and attainable requirement to develop the infrastructure and capacity to carry out an interdisciplinary regional program for collecting, exchanging, and using important ecological and socioeconomic information as well as monitoring the regional status of coral reefs and their resources.
- Raising environmental consciousness on the importance of coral reefs and the need to conserve their resources is the highest priority that needs to be incorporated into all of the above recommendations. Local, national, and regional education/awareness programmes need to be developed along with incentives to promote conservation and to take action to use coral reefs in a sustainable manner.

ACKNOWLEDGEMENTS AND SUPPORTING DOCUMENTATION

Authors
Philip A. Kramer, University of Miami/RSMAS.USA. pkramer@rsmas.miami.edu
Patricia Richards Kramer, University of Miami/RSMAS. USA. tkramer@rsmas.miami.edu
Ernesto Arias-Gonzalez, CINVESTAV-Univ. Merida, MX earias@mda.cinvestav.mx
Melanie D. McField, University of South Florida. USA/BZ. mcfield@btl.net

We thank R. Ginsburg, M. Hatziolis, H. Reyes Bonilla, M. Vierros, P. Dulin, P. Sale, and C. Wilkinson for their comments.

Invaluable information was gained through communications with many experts in the region including M. Alamilla, A. Arrivillaga, B. Aspra, J. Azueta J. Barborak, J. Bezaury, T. Bright, A. Fonseca, C. Garcia Saez, R. Garza-Perez, J. Gibson, D. Gomez, J. Gonzalez Cano, A. Izurieta, N. Jacobs, E. Jordan-Dahlgren, M. Jorge, M. Lara, R. Loreto, S. Marin, C. Padilla, J. Robinson, R. Rodriguez, A. Salaverria, G. Smith, G. Vincent, B. Wade, and N. Windevoxel. The Mexican National Report by J.P. Carricart-Gavinet, J.Carriquiry, D. Gutiérrez Carbonell, M. Lara Pérez, G. Horta Puga, G. Leyte, P. Medina, H. Reyes, and J.M. Vargas, provides further information. Additional information came from unpublished reports of S. Andrefouet et al., C. Bouchon et al., Huitric and McField, G. Horta Puga; Nunez-Lara et al.; A.M. Garduno; and C. Gonzalez-Gandara.

Many exceptional resources were used to compile this report and full citations can be found on the AGRRA website www.coral.aoml.noaa.gov/agra/. Other general resources include:

> www.fao.org;
> www.worldbank.org;
> www.reef.org
> www.odci.gov/cia/publications/factbook/;
> www.ims.wcmc.org.uk; and
> www.fao.WAICENT/FAOINFO/FISHERY/STATIST/FISOFT/fishplus.asp.

KEY DOCUMENTS

FAO. 2000. Conservation and sustainable use of MBRS. Threat & Root Cause Analysis. Report # 00/008 CP-CAM.

Gibson, J., M. McField and S. Wells. 1998. Coral Reef Management in Belize: an Approach through Integrated Coastal Zone Management: Ocean & Coastal Mgmt. 39:229-244.

Guzman, H.M.(ed.). 1998. Cayos Cochinos Archipelago, Honduras. Revista de Biologia Tropical 46.

ICRI. 1998. Status of coral reefs in Mexico and United States Gulf of Mexico. NOAA No. FAO-A-40AANF703477 (CD-ROM).

Jameson S.C., L.J. Trott, M.J. Marshall, M.J. Childress. 2000. Nicaragua: Caribbean and Pacific coasts. In: Sheppard C (ed) Seas of the Millennium: an environmental evaluation, Elsevier Science, Amsterdam, Chapter 32 pp 505-518, Chapter 33, pp519-531.

Kramer, P.A. and P.R. Kramer. 2000. Ecological status of the Mesoamerican Barrier Reef System: impacts of Hurricane Mitch and 1998 coral bleaching. Final report to the World Bank.

UNESCO. 1998. CARICOMP –Caribbean coral reef, seagrass, and mangrove sites. Coastal region and small island papers 3, UNESCO, Paris.

USAID. 1996. Recommendations and Reports for the Management of Fisheries in the Miskito Coast Marine Reserve of Nicaragua. Environmental Initiative of the Americas.

17. Status of Coral Reefs in the Eastern Caribbean: The OECS, Trinidad and Tobago, Barbados, The Netherlands Antilles and the French Caribbean

Allan H. Smith, Mark Archibald, Trish Bailey, Claude Bouchon, Angelique Brathwaite, Ruleta Comacho, Sarah George, Harold Guiste, Mark Hastings, Philmore James, Cheryl Jeffrey-Appleton, Kalli De Meyer, Andre Miller, Leonard Nurse, Clive Petrovic and Paul Phillip

Abstract

The islands of the eastern Caribbean vary widely in their legislative and administrative frameworks, and in their approaches to reef management and protection. However, many of them are faced with a common set of threats and impacts: sedimentation of inshore water from development of coastal areas; overfishing due to increasing demand from local populations; and tourism. Some islands have established effective management institutions and monitoring programmes that are either well supported or are capable of generating their own revenues, while in others management is hampered by inadequate legislation or lack of enforcement of existing legislation, and insufficient human and financial resources. In recent years, the promotion of initiatives such as GCRMN, and the availability of training and technical assistance has allowed a number of countries with limited resources to begin implementing monitoring activities, based on the selection of methods suited to their current capacity, such as the Reef Check protocol. In a region where many islands are faced with the task of balancing the needs of coastal communities, the expansion of water-based tourism, and the maintenance of a limited resource base, participatory approaches to planning and management appear to offer an effective alternative to traditional centralised management.

Introduction

A Node of the Global Coral Reef Monitoring Network was established in early 2000 for the islands of the Organisation of Eastern Caribbean States (OECS), Trinidad and Tobago, and Barbados. It is coordinated by CANARI with support from UNEP-CAR/RCU (Caribbean Regional Coordinating Unit). A node for the French Caribbean Islands has also been established in Guadeloupe, coordinated by the Université Antilles-Guyane and the Direction Regionale de l'Environnement (DIREN). Despite the economic importance of reefs for all these islands, especially for tourism and fisheries, there is little quantitative information on trends in the status of reef resources on which to base management decisions in many of the states.

Antigua and Barbuda
Reefs cover an estimated 25km^2, with large bank reefs predominating, particularly off the north and east coasts of Antigua, as well as Middle and Cades Reefs off the southwest coast. Patch reefs build up from the lagoon floors in Barbuda from depths of 20m. Fringing reefs occur close to the shore of both islands, most facing channels with high surf and strong currents, such as those off Judge Bay, Halfmoon Bay and Rendezvous Bay.

Dominica
The island is characterised by steep topography, which limits the area available for reef development. The most significant areas of coral community growth on the volcanic rock are primarily in the north of the island, including Marigot, Woodford Hill, Calibishie, Pointe Round, and lesser reefs in the south at Soufriere and Scott's Head.

St. Lucia
The most extensive coral reefs are found on the south and east coasts, whereas the best known reefs for tourism, research and monitoring are on the west coast and consist primarily of veneers on volcanic rock.

Grenada
There are mostly patch or fringing reefs along the east and south coasts of Grenada, and at Grande Anse and Moliniere on the west coast, and around the islands of the Grenadines.

Montserrat
There is a high-energy, erosion-prone coastline with volcanic beaches scattered along the coast. There are series of small patch reefs interspersed with sand or sediment, most relatively close to shore, and a few patch reefs in deeper water.

British Virgin Islands
The 60 small islands and rocks of the British Virgin Islands occupy just over 150km^2 of land on a shelf of over 3,000km^2. This archipelago sits on the eastern part of the Puerto Rican Bank with average depths from 10m to 30m, with maximums just over 50m. While mangrove and seagrass communities are found in sheltered bays of some islands, they account for a small percentage of the marine habitats. Most of the Bank consists of sand and numerous rock outcrops covered by coral reefs. These vary from small isolated patches of a few square metres to the extensive Horseshoe Reef of Anegada covering approximately 77km^2.

Barbados
The island is surrounded by a narrow 2-3km wide shelf, with 5 distinct sections: west; south-west; south-east; east; and north, each with particular characteristics. The total area covered by bank reefs is 4.9km^2, while fringing reefs cover 1.4km^2. The west coast is relatively calm with little or no surf and characterised by a sloping shelf on which fringing reefs grow out to 300m from the beach and to a depth of 10m. Extending from these reefs are the patch reefs, which terminate at 30m depth, and seaward of these patch reefs are two bank-barrier reefs approximately parallel to the shore. The inner reef is 700-900m offshore and the outer reef between 1300m and 1km. The south west coast is periodically exposed to Atlantic swells, with a moderate surf which pounds the relict fringing reefs,

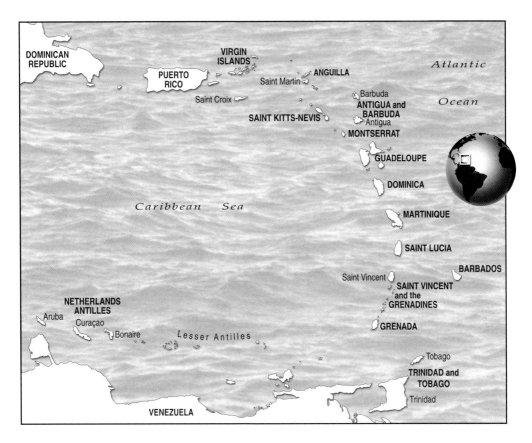

then there is a gently sloping sandy shelf and extensive patch reefs in depths of 6-15m. A bank reef runs parallel to, and approximately 1000m from the shore, and is continuous with the west coast bank reef. The south-east coast is fully exposed to the Atlantic swells with heavy surf. There are no actively growing fringing reefs, however there is a flat, shallow bank reef, 400-800m from shore running roughly parallel to it. The east coast is fully exposed to Atlantic swells and consists of limestone pavements in the southern areas as a gently sloping featureless ramp from the beach off the Scotland District out to 40-60m. Further north (off Pico Tenerife), the bottom type changes abruptly as there is a submarine canyon. North of the canyon, the shelf is terraced and carbonate rich, while to the south it is more terrestrial and the slope is more gentle. The north coast has a rock terrace which extends seaward from the base of 100m cliffs out to 40-60m, ending in a ridge parallel to shore with another terrace and isolated ridges and mounds further seaward. The north west coast is particularly well sheltered and supports the most extensive and diverse hard coral reefs on the island.

THE FRENCH CARIBBEAN

Guadeloupe
The Caribbean coast of Guadeloupe has a narrow island shelf and high levels of volcanic sedimentation, therefore there are no true coral reefs. However, there are most diverse coral communities on the rocky bottoms, particularly around the Pigeon Islets. The Grand Cul-de-Sac Marin in the north is enclosed by a barrier reef, 20km long, which encloses a 10km wide lagoon covering 15,000 ha. This barrier protects extensive seagrass beds in the lagoon and mangroves on the coast from oceanic swells. On the Atlantic coast of the island there are fringing reefs, which are more developed on the eastern part of the Basse-Terre (Petit Cul-de-Sac Marin). The surrounding islands are either devoid of coral reefs (Les Saintes) or possess narrow fringing reefs (Marie-Galante, La Désirade). A shallow, 3 0m deep island shelf (100 x 70km), has the islands of Saint-Barthélémy, Saint-Martin and Anguilla and several small islets with the bottom covered by sediments, which are periodically resuspended by hurricanes. This resuspension has probably limited the development of coral reefs around Saint-Barthélémy and Saint-Martin, which have only narrow fringing reefs.

Martinique
Reefs are absent on the leeward side of Martinique from the northwest and west coasts because of the narrow shelf and high levels of volcanic sediments washing off Montagne Pelée. Nevertheless, the rocky bottoms have flourishing coral communities. Fringing reefs are the only types on the west and south coasts near Sainte-Luce. The northern part of the Atlantic coast has no coral reefs from the north to the Presqu'île de la Caravelle, due to the steep slope and volcanic sediments. Around the Presqu'île de la Caravelle, there are some fringing reefs, and further south, a barrier reef system extends for about 25km. The barrier is separated from the shore by a lagoon 3 to 8km wide, with variable depth from a few metres to 30m with extensive seagrass beds. Fringing reefs have developed along the rocky shores behind this barrier.

THE NETHERLANDS ANTILLES

Bonaire and Curaçao
These are small oceanic islands approximately 70km north of Venezuela. They lie on a northwest – southeast axis with the east coasts exposed to persistent trade winds. The shelf around each island is narrow extending up to 300m seawards, then the bottom shelves off steeply onto a sandy terrace at 50–60m, and then drops off again. The islands are surrounded by continuous fringing reefs which are much better developed on the leeward than on the windward shore, particularly in shallow water.

STATUS OF CORAL REEF BENTHOS

Antigua and Barbuda
There has been a recent trend of deterioration as reports from 1986 described the reefs as exceptional in their variety, beauty and health. In 1998, overall reef condition was considered generally poor, with live coral cover averaging 20% or less at all sites examined except the north of Barbuda. For both islands, the offshore reefs have higher coral species richness and abundance than the inshore sites. Turbidity of inshore water and elevated algal cover on reefs are linked to the impacts of coastal development, with sedimentation being a major influence on reef condition. Hurricane Luis in September, 1995 followed by Marilyn two weeks later caused additional stress and damage. Coral bleaching during the 1998-bleaching event was relatively mild.

Dominica, St. Kitts and Nevis, St. Vincent and the Grenadines
No recent information was available.

St. Lucia
Live coral cover in 1999 ranged from 31% to 50% at 4 Reef Check sites established in the Soufriere Marine Management Area (SMMA). Bleaching was common in 1998 but did not result in high levels of mortality, however, in November 1999, the unusual track of Hurricane Lenny resulted in severe wave action on the leeward coast of the island, severely damaging coastal infrastructure. A Reef Check survey in Malgretoute in June and December 1999 showed live coral cover had declined from 50% to 25% at 3m depth, and from 35% to 17% at 10m. The site was previously characterised by a high density of the urchin *Diadema antillarum* and low algal abundance at 3m, but the Hurricane resulted in a decline from 1.4 to 0.1 urchins per m^2 and a proliferation of filamentous algae. Recently there has been an unusually high incidence of white band disease on reefs in the SMMA, resulting in a living coral loss of over 3% between 1997 and 1998. Data from west coast reefs indicate live coral cover frequently greater than 50% prior to the series of storms, beginning with Tropical Storm Debbie in 1994. Sedimentation from that storm reduced coral cover by 50% at some sites, particularly near large river mouths.

Grenada
During the 1980s many shallow reefs around Grenada and the Grenadines were degraded and became overgrown with algae, presumably resulting from a combination of sewage and agrochemical pollution and sedimentation from coastal development. No recent quantitative data are available on reef status.

Montserrat
A survey of species richness conducted in 1995 and 1996 identified 37 hard coral species, 17 octocorals, 87 other invertebrates, 3 seagrasses, 67 fish and 37 algae. Live coral cover ranged from 20 to 45%.

BVI
Overall reef condition remains relatively good, with localised deterioration associated with areas of high population, shoreline development, and rapid growth of marine activities such as yachting, snorkeling and diving. The passage of numerous storms since 1995, and

most recently Hurricanes Jose and Lenny in late 1999, caused severe damage to dive sites at Norman Island, Peter Island, Salt Island, Cooper Island, Ginger Island and Virgin Gorda.

Barbados
Quantitative surveys have been conducted since 1982, initially on the west coast, with the south and east coasts included in 1987. West coast fringing reefs are considered to be in poor condition and patch reefs on the southwest coast deteriorated significantly during the study period. From initial surveys of the Atlantic coast it appears that reefs are generally in good condition with high faunal diversity.

The French Caribbean
Mapping using remote sensing revealed that only 15 to 20% of the marine communities of Martinique and Guadeloupe still have flourishing coral communities. Coral cover on Guadeloupe varied between 14% on reef flats, to 45% on outer slopes. On one site on Pigeon Island, coral cover dropped from 46% in 1995 to 26% in 1999, and equally worrying, are the large numbers of partially diseased colonies among survivors varying from 11 to 56%, with an average 19 to 53% of surfaces being dead. There is similar degradation of the coral communities on Martinique Island. These losses are probably due to a combination of both natural and anthropogenic factors:

- Hurricanes: These develop in the north tropical Atlantic from July to October, and Martinique has been damaged by hurricane swells of David (in 1979) and Allen (in 1980), but there have been no recent impacts. In 1989, Guadeloupe suffered from hurricane Hugo. In 1995, Hurricanes Luis and Marilyn crossed Saint-Barthélémy and Saint-Martin and also caused damage in Guadeloupe and waves from Hurricane Lenny in 1999 smashed up on the Caribbean coast of Guadeloupe;
- *Diadema antillarum* Disease: The loss of the sea urchins in 1983 is still evident as over growth of algae, however urchin populations are now recovering;
- Coral bleaching: There was major bleaching in 1998, when seawater temperature exceeded 29°C during September and October. Most hard corals, anemones, zoanthids and symbiotic octocorals were affected. On Martinique, 59% of the coral colonies were affected with an average of 69% of the surface bleached. In Guadeloupe, the impact was similar (56 % of the colonies affected with 80% of bleached surface). However subsequent mortality of corals was relatively low with 20 to 30% of the bleached colonies dead in 1999, except for the coral *Diploria labyrinthiformis*, which had 80% mortality. There was new bleaching on the reefs of Guadeloupe in September 1999, affecting 50% of the corals, but the effects were magnified when Hurricane Lenny passed close by in November;
- Heavy siltation: The reefs receive large quantities of sediment from deforestation, mangrove clearing and poorly planned development, particularly affecting reefs in enclosed bays;
- Pollution: Measurement of pollution concentrations is minimal with the only assessments in the Bay of Fort-de-France. Heavy metals and pesticides have been found in the sediments and animals, and the importation of 1,500 to 2,000 metric tons of pesticides per year into Martinique and Guadeloupe potentially threatens their coastal ecosystems. Wastewater treatment is largely ineffectual and organic pollution from the growing human populations is a major threat to the coral communities.

- Algal proliferation: The invasions of algae (mostly *Sargassum*) in 1984 after the die off of *Diadema* have continued with the probable causes being eutrophication of the coastal waters from the city of Fort-de-France and a lack of algal grazers. Most reefs on Martinique suffer from algal proliferation of *Turbinaria* on the reef fronts, *Sargassum* on the fore reef zones and *Dictyota* in the lagoons. A similar situation exists in Guadeloupe, Saint-Martin and Saint-Barthélémy;
- Tourism: The Pigeon Islets, on the leeward side of Guadeloupe, are one of the most famous scuba diving spots of the French West Indies, and between 60,000 and 80,000 divers visit the islets each year with less experienced divers causing physical damage to the corals.

THE NETHERLANDS ANTILLES

Bonaire and Curaçao

The general structure and zonation pattern of the reefs of the leeward (western) shore is as follows:

- a shallow terrace extends 30-150m from the shore to the drop off which starts at 10-15m. The terrace has a characteristic zonation (shallow to deep) of *Acropora palmata*, *Acropora cervicornis* and a mixed zone with *Montastraea annularis* being the dominant stony coral and a variety of gorgonians.
- from the drop off the fore reef slopes at 30-60 degrees to a sediment covered platform at a depth of around 50m. From 10-25m the dominant corals are usually *Montastrea annularis* and *Agaricia agaricites*. At greater depth the dominant species are the plate forming *Montastrea cavernosa* and *Stephanocoenia michelinii*.

Approximately 55 other species of coral can be found on the reefs, and these vary by habitat. The density and distribution of gorgonians, and sponges differ between areas. Over 250 species of fish have been recorded on Bonaire's reefs.

Atypical reef structures include:
- shallow water spur and grooves;
- to the north of the island the western facing shore is often covered by large coral heads, several metres in diameter, in shallow water;
- to the north of the island the southern facing shore the reef slope shows buttress formation and is steeper than average with the deep water sediment platform occurring below 100m;
- on Klein Bonaire the north-eastern coast has virtually no shallow water terrace and the drop off begins at 2-5m;
- at some sites there is an intermediate sediment platform at 25-30m and a secondary reef formation to the seaward (double reef).

On the windward (eastern) shore the terrace extends generally 100-200m off shore to a depth of 12m. It is covered primarily with crustose coralline algae and *Sargassum*, and also some gorgonians. The reef slope is generally far less steep than on the leeward shore with less coral cover and abundant brown algae.

STATUS OF CORAL REEF FISHES

In most of the English-speaking islands, reef fisheries are artisanal and are mainly based on the use of fish traps (fishpots), but there are few data on reef fisheries as greater attention has been paid to pelagic fisheries. In St. Lucia, reef fishery landings were estimated in 1993 to be 116 metric tons, but this is probably an underestimate. Catches include as many as 40-55 species, the majority of which are considered to be over-fished, based on the absence of large, mature fish, and a predominance of small, quick maturing herbivores. Detailed studies of trends in reef fish stocks have been conducted in the Soufriere Marine Management Area since 1995, and show an increase in both abundance and size of some key species. One small reserve showed a doubling in total biomass of commercially important species in 2 years, with abundance being significantly higher for parrotfishes and snappers. Fishermen believe that catches have increased in adjacent Fishing Priority Areas, where the total weight of food fish had doubled by 1999. Species richness increased by 20% over the same period.

Coral reefs represent the primary fisheries resource in Antigua and Barbuda, where demersal fish make up the major proportion of total fish landings. The spiny lobster is the most important species economically, especially in Barbuda from where most of the catch is exported. Reef fish populations have declined significantly in recent years because of natural disasters and over-fishing, indicated by a decrease in size of landed fish and a proliferation of algae on many reefs.

In September 1999, a fish kill affected reef species on Barbados, Martinique, Grenada, St. Vincent and the Grenadines, and Tobago, and in some islands it was attributed to factors related to the fresh water plumes from the Orinoco and Amazon Rivers. In most islands the specific cause was not determined, but in Barbados a high incidence of a *Streptococccus* pathogen was believed to be responsible.

In the BVI, populations of all commercially important species appear to be over-fished, particularly conch, spiny lobster, and most species of groupers and snappers, due to growing local populations and a booming tourism industry. Commercial catches of all species in 1998 were estimated at 819,330kg. The decline in demersal fish and lobsters due to trap fishing over the last 10-20 years has also been reported by local dive guides and recreational divers and there are anecdotal reports of popular dive sites teeming with lobster and large reef predators only a decade ago. Recreational hook and line fishing is minimal and collection of ornamental fish for the aquarium trade is prohibited.

The French Caribbean

There are 228 fish species in 59 families, with the average fish density on Guadeloupe varying between 119 and 550 fish per 100m^2, with the highest numbers being in the protected zone of Pigeon Island. The estimated fish biomass in Guadeloupe and other islands of the French West Indies varies between 368 and 1893kg per hectare. Fish stocks in all the islands are over-exploited and large fish are relatively rare (groupers, snappers, parrotfish). Fishing on Guadeloupe in essentially traditional with 2034 professional fishermen in 1998, working from 1062 fishing boats. There are also about a thousand people who fish regularly, but are not licensed. The annual demand is 15,800mt of fish, whereas the total landings for Guadeloupe, Saint-Barthélémy and Saint-Martin is about

8,500mt. There are 40,000 Caribbean traps around these islands and about 20,000 are lost each year in Guadeloupe during the hurricane season. As these are built with wire netting, they continue to catch fish for months. The traditional fishery on Martinique includes 900 professional fisherman plus many non-official fishermen, with 848 fishing boats in 1997. The total landing of 6,000mt is not enough to supply the local demand estimated at 16,000mt. There are even more traps (50,000) around the island. There was major fish mortality in September 1998 on the Atlantic coast of Martinique and dead reef fish from all trophic levels washed ashore, but there were no obvious changes to the structure of the reef fish community.

The Netherlands Antilles

There are only approximately 20 commercial fishermen on Bonaire, although practically everyone 'goes fishing'. Since commercial fishermen target pelagic fish species (tuna, dorado, wahoo) and fish predominantly with hook and line their impact on reef fish populations is negligible. Fish caught are generally consumed locally, only big eye scad may be exported to Curaçao. As a consequence reef fish are abundant and diverse and biomass is high.

Artisanal fishing practices include hook and line fishing, some rod fishing, and the use of kanasters (fish traps), trai (throwing nets) and reda (encircling nets). Spearguns and spears, although illegal, are still in sporadic use. Few fish traps are used on Bonaire's reefs due to conflict with recreational divers. Throwing nets are used almost exclusively to catch bait fish and encircling nets are used to catch big eye scad. Target fish include grouper and snapper; parrotfish are not considered edible. Illegal poaching of turtle continues and may amount to as many as 20 turtles/month. Reef fish populations, however, do not appear to be adversely effected by fishing pressure at the current level, but an exception is conch which have been fished to the point of collapse. Collection for the aquarium trade is banned.

ANTHROPOGENIC THREATS TO CORAL REEF BIODIVERSITY

Most of the anthropogenic threats and impacts reported were common to most of the islands represented here:

- Increased sedimentation from shoreline development, often for tourism infrastructure, road construction, destruction of mangroves, and dredging. In the BVI, even on the more remote islands with small populations and limited construction, grazing by feral goats has reduced vegetation cover resulting in erosion and sedimentation;
- Eutrophication from the input of agricultural chemicals and sewage, resulting in proliferation of algae and increased turbidity;
- Physical damage from yacht anchors and divers, although the rapid expansion in permanent mooring installation will greatly reduce this;
- Over-fishing.

In many cases, the common theme was a lack of enforcement of existing legislation to reduce these impacts.

In contrast, the major impacts on reefs in Montserrat were from massive volcanic eruptions, particularly in 1995 and 1996, which resulted in the deposition of large quantities of ash on reefs on the south and southwest coasts. A 1995-1996 survey indicated that reef areas not directly affected by volcanic ash were relatively pristine with no indication of the human impacts that affect reefs elsewhere in the region.

CURRENT AND POTENTIAL CLIMATE CHANGE IMPACTS

A regional programme, Caribbean: Planning for Adaptation to Global Climate Change (CPACC), is being implemented in 12 countries, including Antigua and Barbuda, St. Kitts and Nevis, Dominica, St. Lucia, Barbados, St. Vincent and the Grenadines, Grenada, and Trinidad and Tobago. The objective is to support the countries in preparing to cope with the adverse effects of climate change, particularly sea level rise. Specific activities include establishing a sea level and climate monitoring network and databases, resource inventories, vulnerability and risk assessments, and policy formulation.

CURRENT MPAS AND MONITORING/CONSERVATION MANAGEMENT CAPACITY

Antigua and Barbuda
The established MPAs are the Diamond Reef, Salt Fish Tail, Palaster Reef, and Cades Bay Marine Parks. Cades Bay is the most recent, established in 1999, and includes beaches, reefs, seagrass beds and mangroves.

Dominica
Soufriere/Scotts Head Marine Reserve is managed by the Local Area Management Authority. The Cabrits National Park includes a marine component. Permanent photoquadrats were established but are not monitored systematically.

St. Lucia
The Maria Islands Nature Reserve, established and managed since 1986, includes reefs, and will become part of a proposed Pointe Sable National Park, which will include some of the largest reefs on the southeast coast of the island. The Soufriere Marine Management Area (SMMA) was established in 1995, and includes an 11km stretch of coastline, with zoning that includes Marine Reserves, Fishing Priority Areas, and Multiple-use areas, and is managed by the Soufriere Marine Management Association. This includes representatives of all agencies and user groups with direct management responsibility for activities in the area. User fees from yachting and diving directly support management. The establishment and operation of the SMMA is a significant development in the management of multiple uses of reef resources in the region, and demonstrates the potential for institutions based on a process of participation, negotiation, conflict management, and the devolution of authority. The Canaries - Anse la Raye Marine Management Area, north of the SMMA on the west coast, was declared in 1999 and is becoming operational in 2000 with management by the SMMA.

Barbados
Folkstone Marine Park has not functioned effectively since its establishment in 1981 and a participatory study is underway to improve its management. Two offshore breakwaters, at Rockley and Reed's Bay, have attracted a variety of marine life and have been designated as Marine Protected Areas. A third MPA in Carlisle Bay consisting primarily of artificial reefs will be upgraded to a Marine Park.

BVI
The Wreck of the Rhone National Park and the Baths are managed by the National Parks Trust but little protection is provided. Commercial fishing is legal but limited due to the high level of use by tourists. Horseshoe Reef was designated a protected area but limited fishing permits have been issued for the area. Neither the National Parks Trust nor the Conservation and Fisheries Department have the resources or legal mandate to control activities in these areas. A monitoring programme for reefs, mangroves and seagrass habitats is being implemented as part of the CPACC programme, and Reef Check surveys have been conducted since 1998.

The French Caribbean
The marine reserve of the Grand Cul-de-Sac Marin, Guadeloupe was created in 1987 and covers about 3,700ha, including coral reefs, seagrass beds and mangroves. This MPA is managed by the Parc National de la Guadeloupe, and is also a Man in the Biosphere (MAB) Reserve and a RAMSAR site. On the Atlantic side of Guadeloupe, the nature reserve of Petite-Terre was created in 1998 for 990ha by the Office National des Forêts (ONF). A 1,200ha marine reserve was established in 1996 on Saint-Barthelemy, which is managed by an NGO. A nature marine and terrestrial reserve was created on the northern and the eastern sides of Saint-Martin, in 1998, covering 3,060ha, and also managed by an NGO. There is a project to establish a marine and terrestrial reserve around the islets of Pigeon, which are already protected from fishing. There are no marine reserves in Martinique but 4 marine zones have been decreed for protection: the Baie du Trésor, the Baie du Robert, the Ilet à Ramier, where all fishing activities are prohibited for 3 to 5 years, and Sainte-Luce where ship anchoring is forbidden.

Bonaire
The Bonaire Marine Park, which includes all the waters around Bonaire and Klein Bonaire from the shore line to a depth of 60m, was established in 1979 and has been under continuous active management since 1991. It manages 2700ha of coral reef, seagrass and mangrove ecosystems and management includes maintenance of Park infrastructure and public moorings, provision of outreach and education, research, monitoring and law enforcement. The Bonaire Marine Park also acts formally and informally as an Advisory Body and the Marine Park is protected under the Marine Environment Ordinance (A.B 1991 Nr.8). Additionally in November 1999 it was declared a National Park and is managed by STINAPA Bonaire, a local non-governmental, not for profit organisation which also manages the island of Klein Bonaire and the Washington Slagbaai Land Park. The Marine Park has a manager and 4 rangers and is financially self supporting through the levying of diver admission fees. Monitoring continues to be conducted by visiting scientists, Park staff as well as volunteers. Bonaire is a CARICOMP site, implements Reef Check annually and has also set up AGRRA monitoring stations.

Curaçao
The Curaçao Marine Park encompasses 20km of coastline from Oostpunt to Willemstad. Despite being established 1983 it still has not been legally designated. Only the collection of coral and spearfishing are banned but these bans are rigorously enforced. Curaçao has had a CARICOMP site since 1994.

GOVERNMENT POLICIES, LAWS AND LEGISLATION

The Fisheries Acts and Regulations for Dominica, Antigua and Barbuda, St. Kitts and Nevis, St. Lucia, St. Vincent and the Grenadines, and Grenada are similar, through a harmonising of fisheries legislation. As an example, in St. Lucia, the legislation prohibits the use of dynamite and poisons, collection and sale of, and damage to corals and sponges, and allows for the creation of Marine Reserves, with regulation of activities in Marine Reserves without a permit, and allows the establishment of Local Fisheries Management Authorities. Additional national legislation has been established specifically for MPAs and marine environmental management, such as the Marine Areas Act (1982) and National Parks Act (1984) in Antigua and Barbuda, the National Conservation and Environment Protection Act no. 5 (1987) in St. Kitts and Nevis. Although members of the OECS, the UK Territories do not adopt the harmonised acts but they can establish ordinances, such as the BVI's National Marine Park and Protected Area Ordinance no. 8 of 1979.

In Barbados, two sets of legislation were passed in 1998 and proclaimed in 2000. The Coastal Zone Management Act provides a comprehensive statutory basis for coastal management and planning in Barbados. It seeks to coordinate and update the existing fragmented statutes relevant to coastal management and makes provision for critical areas of concern not covered by current legislation. The Act provides the legal basis for the preparation of a Coastal Management Plan, which establishes and clearly sets out the Government's coastal management policy and technical guidelines for the use and allocation of coastal resources. The Act specifically deals with protection of marine resources, for example destruction of corals and fouling of the foreshore. It also encompasses the designation of Marine Protected Areas and Marine Parks. The Marine Pollution and Control Act focuses on the quality of the marine waters on the south and west coasts of the island. Overall, the legislation seeks to prevent, reduce and control pollution from various sources, and recognises that much of the marine pollution originates from land based sources and activities. Both these Acts seek to ensure that the provisions of the Coastal Zone Management Plan are consistently applied, and there is conformity in the Acts in the establishment of coastal water quality standards which are to be enforced.

The French Caribbean
There is legislation regulating fishing activities (decree #094-77 bis; 1998) for Guadeloupe, Saint-Barthélémy and (French) Saint Martin. Fishing and collecting of any non-commercial marine plants or benthic invertebrates, and fishing or collecting using scuba are all prohibited. Similar laws exist for Martinique. A minimum mesh size for the Caribbean trap is 38mm, as are mesh sizes of nets, and non-professional fishermen are prohibited from selling their catch. There are also regulations controlling the harvest of sea urchin, queen conch, and spiny lobsters, with either stipulations on size limits, harvest seasons or quotas to license holders. Specific laws have been formulated to control the sale of potentially ciguatoxic fish species, and to protect sea turtles and their eggs. These regulations are relatively well accepted in Guadeloupe and Saint-Barthélémy, but not on Saint-Martin, where a lack of coordination with the Dutch part of the island (Sint-Maarten) makes enforcement difficult.

The Netherlands Antilles

The Bonaire Marine Park is protected under the Marine Environment Ordinance (A.B 1991 Nr.8). Plans are still underway to have legislation passed protecting the Curaçao Marine Park.

GAPS IN CURRENT MONITORING AND CONSERVATION CAPACITY

Any analysis of the monitoring and management capacity within government and non-governmental agencies in the 8 islands in the eastern Caribbean shows that capacity to maintain monitoring programmes to guide management varies greatly, but that the need for improved management of reef resources is considered to be important. Islands with marine research institutions, such as Barbados, and Trinidad and Tobago, are able to maintain viable programmes, either for national management or for the regional CARICOMP programme. Where monitoring is the responsibility of a government Fisheries Department, activities are constrained by a shortage of manpower or a lack of the expertise required to participate in regional initiatives, and a need to concentrate on fisheries management priorities. The availability of the Reef Check protocols have been effective in allowing both government agencies and NGOs in some islands to begin monitoring where it has not previously been possible. There is little monitoring of reef fish populations in most islands, except for monitoring of catches at some landing sites, however there is often under-sampling due to the variety of species involved and the grouping of species. Length frequencies for reef species are rarely included in the fisheries data collection systems.

Antigua and Barbuda
There are no baseline data and monitoring is currently limited to qualitative observations. Monitoring and control are difficult as MPAs are not yet marked. There is no government budget allocated for monitoring and conservation, and long-term funding is needed to implement the programme that is planned by the Fisheries Division.

Dominica
Effective monitoring has been hampered by lack of resources.

Barbados
Gaps include the monitoring of algal blooms, studies of coral recruitment and growth, and research on the status of reef fish stocks. Conservation capacity is hindered by the inadequate number of people qualified to carry out the necessary research and monitoring activities, and the lack of a strong tradition in community based management approaches. Traditionally, coral reef management has been conducted primarily by central government and communities have not been invited to participate.

Montserrat
No data have been collected on the status of reefs and associated fisheries since a survey in 1995-1996 that was conducted during a peak period of volcanic activity, mostly due to limited human and financial resources. A number of sites have been identified as suitable MPAs but none has been designated.

BVI
Monitoring has been most consistent and detailed at sites around Guana Island, where it started in 1991, and coral cover has not declined since then. The sites did not experience the effects of storms seen at many other site. A monitoring programme began at the same time at 4 sites selected for the nature of the impacts on them, including overfishing, anchor damage from yachts, damage from divers and snorkelers, sewage from yachts and onshore sources, and sedimentation. Data collection was discontinued in 1992 but the planned resumption should provide valuable information on trends over the past decade. Isolated studies and surveys by people from outside the Territory have recently come to light and results should be available for future status reports.

The French Caribbean
Reef monitoring has mainly focused on Guadeloupe, and a monitoring programme will be extended to Martinique and Saint-Barthélémy in 2001.

The Netherlands Antilles
Bonaire has an actively managed Marine Park which has been in continuous operation since 1991. Much of the Park's success can be attributed to the emphasis placed on outreach and education and an ongoing policy of stakeholder participation. Conservation efforts have been largely successful as can be seen by the conferment of National Park status. The most difficult task of the Marine Park is balancing conservation needs against the continued development of Bonaire and exploitation of the island's marine resources for tourism. Much more research and monitoring could be achieved with more funding. It will be difficult for the Curaçao Marine Park to move forwards until it is legally designated. Lack of political support for the Park have hampered progress since its establishment. Nevertheless there is a good knowledge base and the Park is well staffed.

CONCLUSIONS AND RECOMMENDATIONS FOR CORAL REEF CONSERVATION

The following recommendations were made:

- enhancement of the information base on trends in coastal systems and increased sensitisation of the public is needed, and particularly for policy-makers, regarding the impacts of human activities on coastal ecosystems;
- strengthening and broadening Integrated Coastal Management frameworks is required beyond the responsibilities of government Fisheries Departments;
- sharing of responsibilities for monitoring is essential by involving user groups and coastal communities, and this will enhance capacity for monitoring to guide management;

- determination of the carrying capacity of reefs for recreational uses is essential for the management of MPAs;
- improvement of legislation, and increased enforcement of existing legislation for reef conservation
- improvement of the mapping and inventory of coastal and marine habitats and resources is required for sustainable planning and management of protected areas;
- improvement of the linkages between management agencies and the private sector is essential. The involvement of the diving sector in monitoring and management in a number of areas is an encouraging development;
- improvement of regional linkages among management agencies is essential for sharing solutions to common problems and for controlling trans-boundary issues;
- there is a need to adopt participatory approaches and encourage stakeholder involvement in the planning and establishment of institutions for managing the use of reef resources, particularly where the diversity of uses and limited resource base lead to conflicts among user groups.

There is apparently little quantitative information on the status of reefs for a number of islands, despite the interest in implementing monitoring programmes since the 1980s. In some cases, this was because the methods proposed required more resources than were available, or could be allocated by the fisheries departments with the responsibility for monitoring. It is certainly apparent that more information is needed on the status of reef fisheries in many islands.

Monitoring and research on reef resources commonly focus on reef areas that have been selected for the establishment of MPAs or other forms of management, and usually represent the most spectacular examples in the country with the greatest potential for generation of tourism revenue. There is a need for studies of human interactions with other less important reefs, which are unlikely to be designated MPAs, but which are of major importance to coastal communities whose cultural and economic well-being depends on them.

AUTHOR CONTACTS AND KEY DOCUMENTS

Allan H. Smith, Caribbean Natural Resources Institute, St. Lucia; Mark Archibald, Claude Bouchon, Yolande Bouchon-Navaro, Universite des Antilles et de la Guyane, Guadeloupe; Ruleta Comacho, Kalli De Meyer, The Coral Reef Alliance, Bonaire; Philmore James, Cheryl Jeffrey-Appleton, Fisheries Division, Antigua and Barbuda; Trish Bailey, Association of Reef Keepers, BVI; Angelique Brathwaite, Andre Miller, Leonard Nurse, Coastal Zone Management Unit, Barbados; Sarah George, Department of Fisheries, St. Lucia; Harold Guiste, Fisheries Division, Dominica; M. Hastings, Clive Petrovic, H. Lavity Stoutt Community College, BVI; Paul Phillip, Fisheries Division, Grenada.

Brown, N. 1997. Devolution of authority over the management of natural resources: the Soufriere Marine Management Area, St. Lucia. CARICAD and CANARI. CANARI Tech. Rep. 243. 21pp.

Fenner, D. 1998. Reef topography and coral diversity of Anse Galet Reef, St. Lucia. Caribb. Mar. Stud. 6:19-26.

Laborel J, (1982) Formations coralliennes des Antilles françaises. Oceanus 8 (4):339-353

THE GCRMN IN THE EASTERN CARIBBEAN

Two GCRMN nodes have been established in the Eastern Caribbean. One is coordinated by the Caribbean Natural Resources Institute with support from the Regional Coordinating Unit of the Caribbean Environment Programme of UNEP, and covers the OECS states, Trinidad and Tobago, and Barbados. During the first year, activities have included an assessment of the capacity for reef monitoring in 8 countries, a regional workshop in St. Lucia on recent developments in monitoring methods, a series of national training activities and technical assistance focusing on the Reef Check monitoring protocol. For further information contact Alessandra Vanzella-Khouri, UNEP-CAR/RCU, Jamaica, (avk.uneprcuja@cwjamaica.com) or Allan Smith, CANARI, P.O. Box VF383, Vieux Fort, St. Lucia, tel: + 758 454 6060, <canari@candw.lc> or <smitha@candw.lc>. The second is a Node for the French Caribbean Islands, established in Guadeloupe in 1999 and is coordinated by the Université Antilles-Guyane and the Direction Regionale de l'Environnement (DIREN). For further information contact Claude Bouchon, Universite Antilles-Guyane, Laboratoire de Biologie Animale, B.P. 592 / 97159 Pointe-a-Pitre, Guadeloupe (France), tel. + 590 93 87 15, <claude.bouchon@univ-ag.fr>.

18. STATUS OF CORAL REEFS IN SOUTHERN TROPICAL AMERICA: BRAZIL, COLOMBIA, COSTA RICA, PANAMA AND VENEZUELA

JAIME GARZÓN-FERREIRA, JORGE CORTÉS, ALDO CROQUER, HÉCTOR GUZMÁN, ZELINDA LEAO AND ALBERTO RODRÍGUEZ-RAMÍREZ

ABSTRACT

Five countries have joined a recently created Node of the GCRMN for Southern Tropical America (STA), covering coral reef areas of the Eastern Pacific and the Western Atlantic. Coral reefs in this region are not extensive because of a strong continental influence, but they support important biodiversity reservoirs and an expanding tourism industry. Most coral reefs in STA have undergone major changes in the last 30 years, in particular during the 1980s, due to natural and anthropogenic agents. There have been important losses of live coral cover in many reef areas and transitions to algal dominated reefs. Nevertheless, considerable levels of coral cover can still be found at numerous locations on the Caribbean (means between 20-40%) and Pacific (means above 40%) coasts. Bleaching events appear to have increased in frequency, but decreased in severity, throughout the 1990s, which could be related to the global warming phenomenon. The 1997-98 strong ENSO event in the Pacific generated only moderate bleaching and low coral mortality

throughout the STA. It is predicted that coral bleaching will become a more frequent event during the next decade, although the level of bleaching and related coral mortality will probably continue to be low. Reef monitoring in the STA needs to be expanded and maintained in the long term, because few of the coral reef areas are being covered by current programmes. In order to properly address coral reef decline, specific and more rigorous government policies and laws for coral reef sustainable management must be developed during the next few years, as well as effective protection of marine natural parks and reserves. Finally, funding for coral reef research, monitoring and management throughout the STA needs to be considerably increased by both national governments and international agencies.

INTRODUCTION AND BIOGEOGRAPHY

A regional Node for Southern Tropical America was formed as part of the Global Coral Reef Monitoring Network (GCRMN) in 1999 via the 'Instituto de Investigaciones Marinas y Costeras' (INVEMAR) - a Colombian marine research institution on the Caribbean coast. This Node includes 5 countries: Costa Rica, Panama, Colombia, Venezuela and Brazil, with 3 of these having reefs in the Caribbean and the Pacific. All are cooperating to evaluate reef status and monitoring capacity, and develop coordinated monitoring activities throughout the region, with the support of UNEP-CAR/RCU.

The coastal marine environment in the tropical areas of South America is characterised by strong continental influences, which introduce large amounts of sediments that inhibit the development of extensive coral reefs along most shores. Rainfall in this region is amongst the highest in the world and there are many large rivers, including the Amazon, Orinoco and Magdalena rivers. Therefore, both water turbidity and sedimentation are high and reduce coral growth in most coastal areas. In addition, there are several important cold-water upwelling areas (e.g. Perúvian shore, the Gulf of Panama, eastern Colombian Caribbean, and eastern Venezuela) that also reduce reef development. The best coral reefs occur along the Caribbean coasts of Panama as well as islands off Colombia and Venezuela. In general, coral reef development is less along the Pacific side than the Caribbean, with the best Pacific reefs of this region on the Costa Rica-Panama coast. The following presents a synopsis of the biogeographic features of each.

Brazil
There are only sparsely distributed, discontinuous coral reefs along approximately 2,500km of the Western Atlantic coastline. Reef development is poor and with low species diversity (e.g. only 18 hard coral species), but 10 of these species are endemic to Brazil. There are five major coral reef areas: (1) Touros-Natal, in the northeast, which is an extensive line of simple coastal knoll and patch reefs; (2) Pirangi-Maceió, also in the northeast, with similar linear coastal reefs but more developed and with more species diversity; (3) Bahia de Todos los Santos-Camamu; (4) Porto Seguro-Cabrália; and (5) Abrolhos Region. The last three are further east and have extensive fringing and bank reefs with mushroom-shaped pinnacles ('chapeiroes') of simple coral species composition.

Colombia
This is the only 'South American' country with both Caribbean (1700km) and Pacific (1300km) coasts and coral reefs. There are about 2,700km^2 of coral reefs in the Caribbean

waters which are sparsely distributed among 26 discrete areas, but comprise three main groups: (1) the mainland coast with fringing reefs on rocky shores, such as the Santa Marta and Urabá areas; (2) reefs on the continental shelf around offshore islands, such as the Rosario and San Bernardo archipelagos; and (3) the oceanic reef complexes of the San Andrés Archipelago in the Western Caribbean. The complexes illustrate the best-developed coral formations, including atolls, banks, barrier reefs, fringing reefs and patch reefs. They collectively comprise more than 75% of Colombian coral reefs. Reefs on the Pacific coast are very poor, with Gorgona Island being the only site with large coral formations. There are a few reef patches in Ensenada de Utría and the oceanic Isla de Malpelo, the latter is 350km off the coast and has coral growing down to 35m depth.

Costa Rica

There are about 100km^2 of coral reefs along the Caribbean and Pacific coasts of Costa Rica. The 212km long Caribbean coast consists mainly of high energy sandy beaches, with corals growing only in the southeast where a few poorly developed fringing reefs on fossil carbonate outcrops in three areas: (1) Moin-Limón, which is adversely affected by a large port; (2) Cahuita Natural Park, which includes the largest and best studied fringing reef on the Caribbean coast; and (3) Puerto Viejo-Punta Mona, which has a few minor coral formations. In contrast, Pacific reefs are more abundant and distributed along the 1160km coast, although they have low coral diversity and are not well developed. The principal Pacific reefs are near Santa Elena, Bahía Culebra, Isla del Caño and Golfo Dulce, but also in the oceanic Isla del Coco that is 500km offshore. Around 60 hard coral species are known from the Caribbean, while only 18 are recorded from the Pacific side.

Panama

There are 2,490km of coastline along both the Caribbean and Pacific shores, and includes approximately 290km^2 of reefs. Most reefs (99%) are in the Caribbean, where there are more hard coral species (64 species) than the Pacific (23 species). Caribbean coral reefs occur along most of the coast, with three major areas: (1) the western coast (Bocas del Toro-Rio Chagres) which has the highest coral cover of all the Caribbean reefs in Panama; (2) the central coast (Colón-Isla Grande) which is near the major industrial area and the most degraded of these Caribbean reefs (less than 4% coral cover); and (3) the eastern coast (San Blas or Kuna-Yala territory), with the most extensive and diverse reefs in Panama. Most of Panama's Pacific reefs occur on islands near the coast, comprising two major reef areas: (1) the Gulf of Chiriqui, with the best fringing reefs and the most representative for the eastern Pacific region; and (2) the Gulf of Panamá, including reefs on Las Perlas archipelago, Taboga and Isla Iguana.

Venezuela

Reefs occur mainly in three Caribbean areas out of total 2,875km of Caribbean and Atlantic coastline in Venezuela: (1) the Morrocoy National Park and adjacent reefs (San Esteban, Turiamo and Ocumare de la Costa), which have the best developed coral formations on the continental coast (more than 30 coral species and reef growth to 20m); (2) the Mochima National Park and adjacent reefs (Coche and Cubagua islands), with more than 20 coral species growing down to 14m depth; and (3) the oceanic islands found more than 100km offshore, with the best reef formations of Venezuela occurring at Los Roques archipelago (e.g. 57 coral species of and reef growth to more than 50m depth), as well as the Aves, Orchila and Blanquilla islands.

STATUS OF THE CORAL REEFS

Most coral reefs in this region have undergone major changes in the last 30 years, particularly during the 1980s. Some of these changes were caused by natural agents (i.e. coral disease outbreaks, bleaching, ENSO events), but others are clearly related to human impacts. There have been considerable losses of live coral cover in many reef areas and recovery has occurred in only a few places. In contrast, marine algae are now very abundant, and dominate many reef surfaces. Diseases of corals and other benthic animals (gorgonians, sea urchins) appear to have increased dramatically and are now common on Caribbean reefs. Mass mortalities of dominant species, such us the *Acropora* spp. in Caribbean reefs, and other branching and foliose corals, have resulted in significant changes to coral community composition in many shallow water reefs.

	1970s		1980s		1990s	
	1st half	2nd half	1st half	2nd half	1st half	2nd half
A. Caribbean reef sites						
• Cahuita (Costa Rica)	?	?	40%(1980)	?	10%(1993)	3%(1999)
• Kuna-Yala, San Blas (Panamá)	?	?	40%(1983)	22%(1988)	15%(1993)	13%(1997)
• SW Isla de San Andrés (Colombia)	75%* - ?	?	?	?	39%(1992)	26%(1999)
• NE Bahía de Chengue (Colombia)	?	?	53%(1985)	?	34%(1993)	35%(1999)
• Playa Caimán, Morrocoy (Venezuela)	?	?	?	?	43%(1995)	5%(1996)
B. Pacific reef sites						
• Isla del Caño (Costa Rica)	?	?	33%(1984)	50%(1989)	?	10%(1999)
• Isla Gorgona (Colombia)	?	70%(1979)	15%(1983)	50%(1989)	78%(1995)	56%(1999)
• Isla Malpelo (Colombia)	65%(1972)	?	?	?	?	45%(1999)
• Ensenada de Utría (Colombia)	?	?	?	33%(1989)	?	33%(1996)

Comparisons of percent live coral cover over the last three decades at several reef localities of southern tropical America clearly indicate the decline in many areas, particularly the Caribbean (* calculated using a 36% of recent coral mortality estimated in 1992; ?, no data).

Brazil
There have been no coral reef monitoring programmes until now, thus there is little information about the status and recent changes of the reefs. Most nearshore reefs show clear signs of degradation and only those reefs in areas with low urban development are still pristine. These include those in the northeastern Tamandaré area, around Camamu Bay in the State of Bahia, and offshore reefs on the continental shelf. The only recent natural disturbance was coral bleaching reported in 1993 in São Paulo, and the 1998 coral bleaching in north Bahia and Abrolhos with low levels of coral mortality. Bleaching levels of 80% or more were reported in important species such us *Agaricia agaricites, Mussismilia hispida*, and *Porites astreoides*.

Costa Rica
There is evidence of a dramatic reef decline at Cahuita Natural Park along the Caribbean coast, which is the only site with long-term data. Live coral cover decreased from 40% in the early 1980s to 10% in the mid 1990s. CARICOMP reported only 3% coral cover 1999,

with algae the dominant bottom cover (60%). Coral cover at other Caribbean sites was also low (5-10%) and has been that way since the 1980s. Reefs near Punta Mona are still considered healthy. There was extensive bleaching and subsequent mortality on some Pacific reefs during the 1982-1983 El Niño Southern Oscillation (ENSO) event, such that reefs at Isla del Caño lost up to 50% of their living coral and coral cover at Isla del Coco was reduced to less than 5%. There was extensive bleaching again in the 1992 and the 1997-98 ENSO events, but coral mortality was low. Severe phytoplankton blooms also affected reefs at Isla del Caño in 1985, which caused a mass coral mortality down to 3m depth. There has also been coral mortality from exposure during extreme low tides at Isla del Caño. However, most coral reefs on the Costa Rican Pacific coast are in good condition with live coral cover averaging above 40%, based on measurements at Santa Elena, Bahía Culebra, Península de Osa and Golfo Dulce during the 1990s.

Colombia

There are distinct signs of damage to coral communities at all Columbian reef areas during the last 3 decades, but especially during the 1980s. Major reef degradation occurred on both Pacific and Caribbean coasts, including remote continental and oceanic reef areas, as well as reefs near urban centres. The mean coral cover is generally around 20-30% for most Caribbean reefs, whereas algae are dominant (40-70% cover). It is thought that both natural and anthropogenic stresses have caused the overall degradation of Columbia's reefs.

Coral mortality levels around 50% were observed in the 1990s on the oceanic Caribbean island of San Andrés (which is a densely populated area) but similar levels were noted for

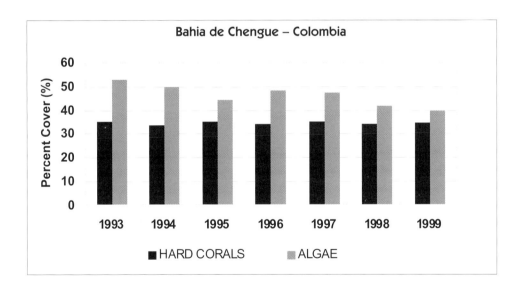

Variations in coral and algae cover between 1993 and 1999 in reef communities at 9-12m depth of the eastern side of Bahía de Chengue (Parque Natural Tayrona), Caribbean Colombia.

several nearby unpopulated atolls (Courtown, Albuquerque, Serrana and Roncador). The San Andrés archipelago is the only Colombian reef area directly impacted by hurricanes. At Islas del Rosario, live coral cover loss was about 20% between 1983 and 1990. CARICOMP data from Chengue Bay (Tayrona Natural Park) since 1993 show that coral cover has been stable following major degradation of the 1980s. This is the only long term monitoring site in Colombia, and coral cover has remained around 35% with algae cover remaining around 50%.

The 1997-98 ENSO event had little effect on Colombian Caribbean reefs, with less than 5% coral bleaching and insignificant coral mortality. The mean incidence of coral disease in several Caribbean and Pacific sites in 1998 and 1999 (through the SIMAC monitoring programme) was less than 5% for all areas except San Andrés island (8.6%). The most frequent diseases in the Caribbean side were 'dark spots, white plague and yellow band'. Pacific coral reefs had high bleaching during the 1982-83 ENSO event but have largely recovered, particularly the dominant *Pocillopora* communities. Live coral cover in Gorgona Island declined from 70% to 15% during the 1982-83 event, but recovered to near 60% levels in 1998. Seawater temperatures were high along the Colombian Pacific coast during the 1997-98 ENSO although bleaching levels were moderate and mortality was low. The maximum bleaching was on Gorgona Island affecting about 20% of the corals, and a decline in live coral cover from 62% to 56% in 1999. The oceanic island of Malpelo suffered similar coral decline with coral cover declining from 65% in 1972 to 45% in 1999 at El Arrecife, and this reduction was probably related as well to the 1982-83 El Niño event. A high incidence of white band disease was observed on *Pocillopora* corals of Malpelo in June 1999. At La Chola reef in Utria, the mean coral cover (33%) did not change between 1989 and 1996, but there was a major increase in macro-algal cover.

Panama

Reefs on the Caribbean coast have lost considerable live coral cover in the central and eastern areas, whereas there is still high coral cover on western reefs, e.g. coral cover can be up to 75%, with a mean above 25% on Bocas del Toro archipelago. Reefs along the central coast have been affected by heavy industrial and urban development, for example

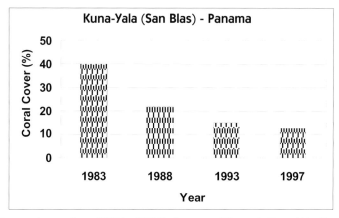

Changes in live coral cover from 1983 to 1997 in the area of Porvenir, Kuna-Yala (San Blas), Caribbean coast of Panama

there is now less than 6% of coral cover at Bahía Las Minas and about 13% at the Parque Nacional Portobelo. In the San Blas (eastern coast) area, coral cover dramatically declined from 40% down to below 15% due to natural and anthropogenic causes. Many reefs in the Pacific were destroyed by the 1982-83 ENSO event and very little recovery has been observed. Mortality ranged from 50 to 100% in some areas, particularly in the Gulf of Panamá. Several reefs in Las Perlas archipelago now have less than 2% coral cover, while at Isla Iguana (west from Las Perlas) coral cover is above 30%, probably the highest along the Pacific coast of Panamá.

Venezuela
There is little recent information about the status and changes in the coral reefs. Coral reef health and other animal communities collapsed in Morrocoy Natural Park during the 5 years to 1996, with coral cover reducing from 43% to less than 5%. This mass mortality was related to a climatic and oceanographic anomaly that resulted in a severe phytoplankton bloom followed by sudden oxygen depletion, but also to chemical pollution. Many diseased coral colonies (2-11%) were observed in the summer of 1999 in Morrocoy, with yellow band, white plague and dark spots being common diseases. The oceanic islands of Venezuela have the most pristine reefs e.g. coral cover at the Archipelago de Aves ranged from 75-80% in the mid 1970s, but there are no recent assessments of this area. AGRRA surveys of the Los Roques Archipelago during 1999-2000 showed coral cover near 60% for several sites, with mean coral cover of 27%. Coral diseases (yellow band, black band, white plague, dark spots and white band) were observed in the archipelago in several coral species during 1999.

STATUS OF CORAL REEF FISHES

Data on reef fisheries and the status of reef fish populations are extremely scarce in this region, however the general consensus is that that coral reefs fish communities have been changed markedly and populations of important commercial species are severely depleted, particularly in the Caribbean. These changes are probably related to the general degradation of the reefs and over-exploitation of fish stocks. In Costa Rica, the aquarium trade has resulted in reductions in populations of target species of *Thalassoma lucassanum, Pomacanthus zonipectus* and *Holacanthus passer*. Recent fish counts during SIMAC monitoring programme in Colombia on Caribbean reefs showed few commercial species and snappers and groupers were absent or had densities less than 1 fish per 60m^2. However, populations of the three-spot damselfish (*Stegastes planifrons*) apparently increased and are now resulting in significant coral mortality with high fish densities at all sites (5.8-34.1 ind/60m^2). Fish counts on Gorgona Island (in the Pacific) showed that the snapper *Lutjanus viridis* and the parrotfish *Scarus ghobban* were the most abundant target species. On the Caribbean coast of Venezuela, 1999 there were abundant *Stegastes planifrons* (mean density 19.2 ind/60m^2) in the Morrocoy Natural Park, whereas the parrotfish *Scarus iserti* was the most common at Los Roques archipelago (9 ind/60m^2). Surveys have showed that important commercial species are now rare in continental reefs, whereas they are still abundant on most oceanic islands of Venezuela.

ANTHROPOGENIC THREATS TO CORAL REEF BIODIVERSITY

A broad spectrum of anthropogenic stresses is causing degradation of the coral reefs throughout the regions of southern tropical America. It is anticipated that these threats will probably continue; however, there are insufficient data and direct evidence to demonstrate clear cause and effect relationships in this region.

Sedimentation
Increased sedimentation appears to be the most damaging stress recorded for all coastal areas in the region. Negative impacts including losses in coral cover, reproduction and growth were reported at: Cahuita reef, Caribbean coast of Costa Rica; Golfo Dulce, Pacific side of Costa Rica; Santa Marta area in the Colombian Caribbean; and Morrocoy Natural Park, Venezuela. However, the only historical record of sedimentation rates are at the Cahuita site, where sediment concentrations continue to increase and live coral cover continues to drop. In most areas, deforestation is the major cause of sedimentation and nutrient pollution increases as logging and land clearing for agriculture continue in inland forest and mountain areas. Until more diversified and equitable economic alternatives are found for most all of the countries in this region, it is anticipated that agriculture and logging threats to reefs will remain unabated.

Sewage Pollution
This is another important source of nutrients which has increased considerably in recent years due to high population growth rates in the coastal and inland cities, all of which discharge untreated domestic waste. Sewage pollution poses a major threat to most coastal ecosystems in the region. While there is anecdotal evidence of negative impacts on coral reefs, no conclusive studies have been performed.

Resource Extraction and Tourism
Coral and fish exploitation is also a threat to coral reefs in coastal areas from subsistence and commercial fisheries, as well as tourism-related activities. Continued fishing pressure, often with the use of damaging extractive methods e.g. dynamite, has resulted in the over-exploitation of the main commercial species throughout the region. Reef tourism has increased enormously and is becoming a major economic alternative for all countries. For example, tourists visiting the Abrolhos National Marine Park in Brazil increased over 400% between 1980s and 1990s, and tourism numbers in Cahuita Natural Park, Costa Rica grew from 31,000 to 100,000 over the same period. However, tourism has also generated a wide array of threats, including coral extraction, over-exploitation of fish resources, damage to coral colonies by swimmers and divers, anchors and boat propeller damage, increased sewage and garbage, increased sediment loads from construction etc.

Oil Pollution, Mining and Coastal Development
The major impact from oil pollution and mining was observed at Bahía Las Minas along the Caribbean coast of Panamá, where reefs have been affected since the 1960s and coral cover has declined to less than 4%. Entire reefs have also been destroyed in the San Blas region (Caribbean Panamá) during mining for construction and landfill. Coral mining is an important activity in many Brazilian coastal cities where corals were used to construct fortresses centuries ago and continue to be mined for modern beach resorts now. Coastal and riverine developments that have changed river flows and discharges have resulted in

A summary of natural and anthropogenic threats to coral reefs of southern tropical American countries within the last three decades. CCR, Caribbean Costa Rica; PCR, Pacific Costa Rica; CPA, Caribbean Panamá; PPA, Pacific Panamá; CCO, Caribbean Colombia; PCO, Pacific Colombia; CVE, Caribbean Venezuela; ABR, Atlantic Brazil. The threats and causes of coral reef degradation were rated as 0=no threat; 1=low threat; 2=localised low threat; 3=average threat; 4=localised major threat; 5=general major threat.

Threats	(and score)	CCR	PCR	CPA	PPA	CCO	PCO	CVE	ABR
A. Natural									
1. Coral bleaching	(28)	3	5	5	5	3	3	2	2
2. Algal proliferation	(27)	5	2	5	3	3	0	4	5
3. ENSO events	(21)	3	5	3	5	0	3	0	2
4. Earthquakes	(18)	5	3	4	4	0	2	0	0
5. Phytoplankton blooms/red tides	(16)	0	4	2	4	0	2	4	0
5. Global warming	(16)	1	1	3	3	3	3	1	1
7. Diadema mortality	(13)	5	0	5	0	3	0	0	0
8. Low tide exposure	(12)	0	2	1	1	0	3	0	5
9. Upwellings	(11)	0	2	2	3	1	1	2	0
10. Coral disease outbreaks	(10)	1	0	2	2	3	1	1	0
11. Hurricanes	(4)	0	0	0	0	4	0	0	0
12. Acanthaster proliferation	(2)	0	1	0	1	0	0	0	0
NATURAL IMPACT RATINGS		23	25	32	31	20	18	14	15
B. Anthropogenic									
1. Deforestation	(37)	5	5	5	4	5	4	4	5
2. Overfishing	(33)	5	5	5	5	5	1	4	3
3. Increased sedimentation/siltation	(30)	5	4	4	2	4	2	4	5
4. Tourism activities	(29)	5	5	4	3	4	1	4	3
5. Urban development	(28)	4	4	4	4	4	0	4	4
6. Fish extraction	(25)	3	4	4	4	5	2	1	3
7. Sewage pollution	(24)	4	4	4	4	4	0	0	4
8. Garbage pollution	(23)	3	3	5	5	3	0	0	4
9. Changes to river beds	(20)	3	3	4	2	4	2	0	2
10. Oil pollution	(18)	3	2	4	2	2	2	1	2
11. Coral extraction for curio trade	(17)	0	2	4	4	2	2	0	3
11. Diving activities	(17)	3	3	4	3	2	0	0	2
13. Dredging	(16)	0	0	4	4	2	0	4	2
14. Industrial development	(14)	0	0	4	3	0	0	4	3
15. Nautical activities	(13)	1	2	2	2	2	2	0	2
16. Heavy metal pollution	(12)	1	1	4	4	0	0	0	2
16. Dynamite fishing	(12)	0	0	0	0	4	4	0	4
18. Coral mining for construction	(7)	0	0	2	1	2	0	0	2
19. Military activities	(5)	0	0	3	2	0	0	0	0
ANTHROPOGENIC IMPACT RATINGS		40	47	70	58	54	22	30	55
TOTAL IMPACT RATINGS		63	72	102	58	74	40	44	70

considerable degradation of coral reefs in Colombia for the last few centuries. During the 17th century, Spanish colonists dug the Canal del Dique, changing the Magdalena River to discharge into the Bahía de Cartagena, and turning the bay to an estuary, thereby eliminating extensive coral communities. Additional branches of the Canal del Dique have been opened recently to the Bahía de Barbacoas, resulting in turbid waters and eutrophication that have damaged the coral reefs of the nearby Islas del Rosario. Accelerated damage to coral reefs due to urban and industrial development was also reported on the central coast of Venezuela (Puerto Francés-Carenero), where more than 80% of the associated invertebrate species have disappeared since the mid 1980s.

CURRENT AND POTENTIAL CLIMATE CHANGE IMPACTS

The impacts of climate change have not been properly assessed in the region, although there has been limited data collection on the occurrence of coral bleaching events. Bleaching events were first recorded in the region during the first half of the 1980s. Bleaching appears to have increased in frequency but not severity throughout the 1990s. All bleaching events coincided with elevated sea surface temperatures, illustrating a possible correlation with global climate change impacts. The most damaging bleaching episode was along the Pacific coast during a severe ENSO (El Niño Southern Oscillation) event in 1982/83 with coral mortality of 50-100% in most reefs. There was a second strong ENSO event on the Pacific side in 1997/98, but the level of coral bleaching was moderate and resultant mortality was low. On the Caribbean side, the most severe bleaching event was also in 1982/83, with high levels of coral mortality at several localities, but unfortunately this event was poorly documented. Mild bleaching occurred in 1997/98 at several Caribbean and Atlantic localities, with less than 10% of the corals bleached in most cases and insignificant mortality. As global warming is predicted to increase, and based on the recurring patterns of recent bleaching in the region during the last two decades, it is predicted that coral bleaching will become a more frequent event during the next decade; although the level of bleaching and related mortality during each episode will probably be low. However, the long-term negatives impact of these recurrent bleaching events may be high if the increased frequency counteracts natural recovery capacity of corals.

Bleaching in Region	1982/3	1987	1990	1992	1993/4	1995	1997/8	1998	1999
Costa Rica - Caribbean	Severe			Mild				Mild	
Costa Rica - Pacific	Severe	Interm		Mild			Mild		
Panamá - Caribbean	Unkn					Mild	Mild		
Panamá - Pacific	Severe			Mild		Mild	Mild		
Colombia - Caribbean	Severe	Interm	Mild			Mild		Mild	Mild
Colombia - Pacific	Severe						Mild		
Venezuela - Caribbean	Unkn					Mild	Mild		
Brazil - Atlantic					Mild			Mild	

The incidence of coral bleaching events appears to be increasing over the last two decades in the coasts of southern tropical America. The relative severity of those events was: Mild: <20% mortality; Intermediate (Interm): 20-50% mortality; Strong: >50% mortality, and Unknown (Unkn)

CURRENT MPAs AND MANAGEMENT CAPACITY

There is a broad variety in types of Marine Protected Areas (MPAs) in this region, covering a great diversity of marine habitats, including many with coral e.g. 23 in the Caribbean-Atlantic and 15 in the Pacific. Unfortunately most of these MPAs are 'paper parks' due to lack of management planning, operational capacity, especially with regard to regulatory development and enforcement. Most MPAs are managed by government agencies and lack reef monitoring programmes. In many cases the effectiveness of MPAs is undermined due to the poor economic status of the government and community i.e. limited funds for management and poverty in local communities. In some areas, education, awareness and communication are paramount issues, especially in places where there are conflicts over rights between traditional users and government over fisheries and tourism activities.

Brazil
All protected areas and marine parks with coral reefs have management plans and most have conservation programmes in action. Even the coastal reefs along the state of Bahia are protected by law, although not included in the parks. However, legislation enforcement is not very effective and few reefs are fully protected. Existing MPAs include the following, but none have extensive coral reef monitoring in place.
Manuel Luiz Bank - State Marine Park since 1991: but no monitoring in place;
Coral Coast Environmental Protected Area - Marine Protected Area since 1998: no monitoring;
Atoll das Rocas - Biological Reserve since 1979: no monitoring;
Fernando de Noronha Archipelago - National Marine Park since 1988: no monitoring;
Paripueira District - Marine Park since 1993: no monitoring;
Abrolhos - National Marine Park since 1983: no monitoring;
Pinaunas Reef - Municipal Marine Protected Area since 1997; and
Recife de Fora - Municipal Marine Protected Area since 1998.

Colombia
There are 7 protected areas in Colombia with coral reefs: 4 in the Caribbean are comparatively larger and more complex; 3 in the Pacific are smaller. The Pacific reserves have fewer management problems and are better conserved. All extractive or damaging activities, as well as construction for development, are prohibited within the National Parks. However, infrastructure and resources are still very scarce for effective control and enforcement of regulations. Consequently, some illegal activities, such as dynamite fishing, still occur sporadically within the Caribbean protected areas. Management plans have been produced for many of the reserves, but unfortunately those plans need to be more effectively implemented.

Caribbean:
- Parque Nacional Natural Tayrona: National Park since 1969: CARICOMP monitoring since 1992, SIMAC and Reef Check monitoring since 1998;
- Parque Nacional Natural Corales del Rosario y San Bernardo National Park since 1977: SIMAC and Reef Check monitoring since 1998;
- Parque Nacional Natural MacBean-Isla de Providencia - National Park since 1995: no monitoring

- Los Corales del Archipelago de San Andres, Providencia, Santa Catalina y Cayos – Special Management Area since 1996: SIMAC, CARICOMP and Reef Check monitoring since 1998.

Pacific:
- Parque Nacional Natural Isla Gorgona - National Park since 1983: SIMAC and Reef Check monitoring since 1998.
- Parque Nacional Ensenada de Utría - National Park since 1986: no monitoring;
- Santuario de Fauna y Flora Isla de Malpelo - Sanctuary of Fauna and Flora since 1955: no monitoring.

Costa Rica

There is no extraction of corals or live rock within the Costa Rican marine protected areas, but artisanal fishing continues, in some cases due to social problems and in others due to lack of control.

Caribbean:
- Parque Nacional Cahuita - Natural Park: CARICOMP monitoring since 1999, tourism starting to be regulated;
- Refugio Nacional de Vida Silvestre Gandoca-Manzanillo - National Wildlife Refuge: monitoring will start in 2000.

Pacific:
- Parque Marino Ballena - Marine Park: no monitoring;
- Parque Nacional Manuel Antonio - National Park: no monitoring;
- Area de Conservación Marina Isla del Coco - Marine Conservation Area: tourism being regulated, monitoring plans being designed;
- Area de Conservación de Osa - Conservation Area: no monitoring;
- Area de Conservación Guanacaste - Conservation Area: no monitoring;
- Reserva Absoluta Cabo Blanco - Absolute Reserve: no monitoring;
- Reserva Biológica Isla del Caño - Biological Reserve: reef monitoring since 1984; tourism being regulated.

Panama

Protected areas in Panama include the best coral reefs on both of the Caribbean and Pacific sides, but the total area of reefs inside parks and reserves is not known. There are no adequate inventories of marine fauna and flora in the protected areas and information about the marine biota is very scarce. There are 3 levels of protection for the reserves (National Parks, Wildlife Refuges and Indian Reserves), however, the use of resources is generally the same for all as there are no management objectives and plans.

Caribbean:
- Reserva Indígena Kuna-Yala - Indian Reserve since 1953: reef monitoring between 1992-1997;
- Reserva Indígena Ngöbe-Buglé - Indian Reserve since 1997: no monitoring;
- Parque Nacional Portobelo - National Park since 1976: Reef Check and Reef Keeper monitoring since 1997;

- Parque Nacional Isla Galeta - National Park since 1999: STRI monitoring since 1970;
- Parque Nacional Marina Isla Bastimentos (Bocas del Toro) - Marine National Park since 1988: CARICOMP monitoring since 1999. *Pacific:*
- Refugio de Vida Silvestre Isla Iguana - Wildlife Refuge since 1981: no monitoring;
- Refugio de Vida Silvestre Isla Taboga - Wildlife Refuge since 1984: no monitoring;
- Parque Nacional Coiba - National Park since 1991: no monitoring;
- Parque Nacional Cerro Hoya - National Park since 1984: no monitoring;
- Parque Nacional Golfo de Chiriquí - National Park since 1994: no monitoring;

Venezuela

The Instituto Nacional de Parques (INPARQUES) manages the national parks in Venezuela, of which Los Roques is the only island and marine area. There are other oceanic islands such as La Blanquilla, La Orchila and Refugio de Fauna Isla de Aves, which are military bases with restricted access, but not managed by INPARQUES.

- Parque Nacional Mochima - National Park: no monitoring;
- Parque Nacional Morrocoy - National Park: CARICOMP monitoring since 1992;
- Parque Nacional San Esteban - National Park: no monitoring;
- Parque Nacional Archipiélago de Los Roques - National Park: no monitoring.

GOVERNMENT POLICIES AND LEGISLATION

Despite the great importance of coral reefs in the region and the clear evidence of accelerated degradation, there are few specific government policies and laws to promote the study (including monitoring), sustainable use and conservation of coral reef systems in this region. Most protection takes place through the inclusion of reef areas within natural parks and other reserves; however, most of these are 'paper parks' and effective protection is limited to a few example areas. Only Colombia and Panamá have developed specific national regulations that prohibit the extraction, exploitation and trade of hard corals (Resolution 1002 of 1969 from INDERENA in Colombia and Res. J.D.-033 of 1993 from IRENARE in Panamá) and reef fishes for the aquarium trade (Executive Decree 29 of 1994 in Panamá). In Brazil, there is also a specific local law (State Constitution, Article 215, Chapter VIII) which declares coral reef areas for permanent protection along the coast of the state of Bahia.

The Convention on International Trade in Endangered Species of Wild Fauna and Flora (CITES) has been ratified by Colombia (Law 17, 1981), Costa Rica, Panamá (Law 14, 1977), and Venezuela. Other international treaties pertinent to the conservation of coastal resources have been ratified in the region, such us the Convention for the Protection of the Marine Environment in the Southeastern Pacific Coastal Zone (Colombia: Law 45, 1985), the Convention for Protection and Management of the Marine Environment in the Wider Caribbean Region (Colombia: Law 56, 1987; Costa Rica), the Convention on Global Climate Change (Colombia: Law 64, 1994; Costa Rica), and the Convention on Biological Diversity (Colombia: Law 165, 1994; Costa Rica and Venezuela).

Recent efforts in several countries have been made to organise national systems and legislation for the management and conservation of the environment and biodiversity.

Colombia created the National Environmental System (SINA) and the Ministry of the Environment through Law 99 of 1993. Panamá created a general law for the environment and the National Environmental Authority through Law 41 of 1998. Venezuela is working on a Law of Biodiversity which is about to be approved which will enable the creation of a Biological Diversity National Agency to manage and preserve the biological diversity and the genetic resources.

GAPS IN CURRENT MONITORING AND CONSERVATION CAPACITY

There has been reef monitoring in the region for 20 years and all countries have recognised the need to implement coordinated national monitoring programmes. However substantive monitoring programmes are limited to a few sites with a need for more comprehensive monitoring and expansion throughout the region. A lack of funding is the principal restriction on attempts to improve national capacities for conservation of the marine resources. Although there are numerous national parks and reserves with coral reefs, and government agencies for the environment within these countries, infrastructure and resources are insufficient for effective control and enforcement of regulations

The CARICOMP programme has been very important in stimulating and increasing monitoring capacity in the Wider Caribbean, but there are only 1 or 2 monitoring sites per country. 'Reef Check' activities have contributed to an expansion of monitoring coverage in

COUNTRIES ATTRIBUTES	COSTA RICA	PANAMÁ	COLOMBIA	VENEZUELA	BRAZIL
Marine research institutions	2	2	7	6	4
Active reef researchers	4	6	21	6	21
Reef monitoring programmes	CARICOMP CIMAR	CARICOMP STRI Reef Check Reef Keeper	CARICOMP SIMAC Reef Check	CARICOMP	None
Monitored reef localities	4	9	6	1	0
Reef monitoring stations	13	21	22	3	0
Years of reef monitoring	15	20	9	9	0
Monitored parameters	16	15	15	13	0
1999 monitoring investment (US$)	23,000	182,000	34,000	10,000	0
Funding capacity for monitoring	Low	Low	Low	Low	Low
Professional capacity for monitoring	High	High	High	High	High
Logistic capacity for monitoring	Low	High	High	Medium	High

Current reef monitoring capacity in countries of the Southern Tropical America region shows a wide disparity in capacity and expenditure on monitoring.

some countries (e.g. Colombia and Panamá), although the depth and quality of data is less rigorous due to the use of volunteer, non-professional personnel and a rapid protocol. Brazil is the only country that has not yet implemented any substantial reef monitoring. All countries in the region have high professional capacity and generally good logistics to implement the required monitoring, but funding is limited to develop and maintain national monitoring programmes. Colombia is the only country that has a national integrated coral reef monitoring programmes (SIMAC). Two workshops since 1998 were used to design and implement SIMAC, while simultaneously gathering data from Caribbean and Pacific sites. However there is a risk that current economic problems in Colombia will interrupt the SIMAC programme, although several Colombian institutions have made commitments to maintaining the monitoring system over the long-term. Furthermore, the system is still in an early stage, covering only 5 of nearly 30 Colombian reef localities.

Conclusions

The coastal environment in Southern Tropical America is characterized by strong continental influences (river discharges, upwelling) which inhibit the development of extensive coral formations. Nevertheless, biodiversity on Caribbean reefs is comparable to other Western Atlantic regions, and Brazilian reefs contain an unusually high proportion of endemic coral species.

SIMAC: The first nation wide reef monitoring programme in the STA region

As in many parts of the world, a considerable decline in coral reef health was observed in Colombia during the 1980s. Nevertheless, coral reef monitoring in Colombia did not start until 1992 when the "Instituto de Investigaciones Marinas y Costeras" (INVEMAR) joined the CARICOMP programme and implemented one monitoring station along the Caribbean coast. Based on this experience an expanded nation-wide reef monitoring programme was launched by INVEMAR in 1997 with the organisation of SIMAC (Sistema Nacional de Monitoreo de Arrecifes Coralinos en Colombia), with the support of COLCIENCIAS, other Colombian institutions and UNEP-RCU/CAR. A basic sampling programme (Level 1) was implemented in 1998 at 4 important reef areas (3 in the Caribbean, 1 in the Pacific). The Level 1 protocol consists of water quality measurements and yearly estimates of benthic reef cover, coral disease incidence, and fish diversity and abundance, at 2 sampling sites in each area. This protocol was formulated during national workshops in 1998 and 1999, to determine strategies and institutional coordination to maintain SIMAC in the long term. All the involved institutions have agreed to continue supporting SIMAC until after the end of COLCIENCIAS funded project in 2000, through the coordination of INVEMAR. However, current economic problems in Colombia may interrupt some of the SIMAC activities or preclude its expansion to other reef areas (the system is covering only 5 of nearly 30 Colombian reef localities). SIMAC has also promoted community monitoring by providing organizational, technical support and funding to Reef Check activities in several Colombian reef sites during the last 3 years. The SIMAC experience has been also essential for the organisation of the GCRMN STA regional Node in 2000.

The best-developed coral reefs occur along the coast of Panamá and associated with islands off Colombia and Venezuela in the Caribbean. Coral formations are comparatively poor in the Pacific and occur principally along the Costa Rica-Panamá coasts.

Most coral reefs in the region have undergone major changes during the last 30 years, but particularly during the 1980s. There have been considerable losses of live coral cover in many reef areas, while algae have become dominant. Diseases of corals are now commonly seen on Caribbean reefs. Nevertheless, high coral cover can still be found at numerous reef locations of both Caribbean (means between 20-40%) and Pacific (means above 40%) coasts.

Some of these changes were caused by natural agents (ENSO events, bleaching, disease outbreaks, phytoplankton blooms), but others are clearly related to human impacts (deforestation, increased sedimentation, sewage pollution, overfishing).

Bleaching events apparently increased in frequency, but decreased in severity, throughout the 1990s in the region. The second strong ENSO event (1997/98) generated only moderate bleaching and low coral mortality in the Pacific side. Mild bleaching, but insignificant coral mortality, occurred also in 1997/98 in several Caribbean and Atlantic localities. As global warming increases, it is predicted that coral bleaching will become a more frequent event during the next decade, although the level of bleaching and related mortality during each episode will probably continue to be low.

There are about 40 protected areas and reserves that include coral reefs in the region, but most of them remain as 'paper parks' because of limited funding to implement management plans and enforce regulations.

Despite the importance and accelerated degradation of coral reefs in the region, there are few specific government policies and laws to promote conservation of these ecosystems. Most protection is intended through the inclusion of reef areas within national parks and other reserves.

Reef monitoring in the region has been carried out for 20 years; however, geographic coverage and regional expansion are still very low due of funding limitations.

RECOMMENDATIONS

- It is necessary to complete the baseline characterisation of coral reefs in the region, including: mapping, biodiversity evaluations, population assessments of important species or groups of species, coral health assessments, and socioeconomic assessments.
- Reef monitoring in this region must be maintained in the long term and significantly expanded through the development of national monitoring programmes in each country, including the integration of these into the regional and global initiatives. National programmes need to address both protected and non-protected reef areas.
- Extraction of all reef organisms, including the so-called 'live rock' must be completely prohibited within the protected areas.

- Tourism in coral reef areas needs to be regulated, including the designation of dive sites based on scientific criteria to establish carrying capacity limits.
- New reef conservation programmes must include the surrounding areas, in particular adjacent and upstream large watersheds in order to minimise the input of sediments and pollutants into reef areas.
- Specific government policies and laws for coral reef sustainable management need to be developed during the next few years, as well as effective protection of natural parks and reserves through implementation of management plans and law enforcement.
- National governments need to increase regular budgets for institutions that have responsibilities for monitoring and conservation of coral reefs, but also the funding for scientific research on this ecosystem. At the same time, international programmes and funding agencies must provide important matching support for developing countries and for regional cooperative initiatives.

ACKNOWLEDGEMENTS

This work and the organisation of the GCRMN Node for Southern Tropical America (STA) was developed through an agreement (CR/0401/94-15-2218) between the 'Instituto de Investigaciones Marinas y Costeras' (INVEMAR) and the Regional Coordinating Unit for the Caribbean of the United Nations Environment Programmes (UNEP-RCU/CAR). Additional support has been received from the fund COLCIENCIAS-BID (project 2105-09-327-97). Helpful collaboration and information have been provided by the following colleagues: Juan M. Díaz, Jaime A. Rojas, Diego L. Gil, Maria C. Reyes, Nadia Santodomingo, Fernando Zapata and Pilar Herron from Colombia; Ana Fonseca from Costa Rica; Estrella Villamizar, J.M. Posada, F.J. Losada, Shiela Pauls and Pablo Penchaszadeh from Venezuela; Ernesto Weil from Puerto Rico; and Ruy Kikuchi and Viviane Testa from Brazil.

Jaime Garzón-Ferreira (jgarzon@invemar.org.co) and Alberto Rodríguez-Ramírez (betorod@invemar.org.co) are the Regional and Colombian Coordinators respectively of the GCRMN Node for Southern Tropical America at the Instituto de Investigaciones Marinas y Costeras (INVEMAR) in Colombia; Jorge Cortés (jcortes@cariari.ucr.ac.cr) is a National Coordinator of the STA Node at the CIMAR-Univ. Costa Rica; Aldo Croquer (croquer@telcel.net.ve) is a National Coordinator of the STA Node at the INTECMAR-Univ. Simón Bolívar in Venezuela; Héctor Guzmán (guzmanh@naos.si.edu) is a National Coordinator of the STA Node at the Smithsonian Tropical Research Institute in Panamá; and Zelinda Leao (zelinda@ufba.br) is a National Coordinator of the STA Node at the Instituto de Geociencias-Univ. Federal da Bahia in Brazil.

SUPPORTING DOCUMENTATION

Cortés, J. 2000. The status of coral reefs in Costa Rica. Unpublished report, CIMAR, Universidad de Costa Rica, San José, Costa Rica.

Cortés, J. (Ed.). 2000. Coral reefs of Latin America. Elsevier Science, Amsterdam: in press.

Croquer, A. 2000. The status of coral reefs in Venezuela. Unpublished report, INTECMAR, Universidad Simón Bolívar, Caracas, Venezuela.

Díaz, J.M. (Ed.). 2000. Areas coralinas de Colombia. Publicación Especial No. 5, INVEMAR, Santa Marta, Colombia, 171 p.

Garzón-Ferreira, J. & J.M. Díaz. 2000. Assessing and monitoring coral reef condition in Colombia during the last decade: 51-58. In Done, T. & D. Lloyd (eds.): Information Management and Decision Support for Marine Biodiversity Protection and Human Welfare: Coral Reefs. Australian Inst. Mar. Sci. (AIMS), Townsville, Australia.

Garzón-Ferreira, J. & J.M. Díaz. 2000. The Caribbean coral reefs of Colombia. In J. Cortés (ed): Coral reefs of Latin America. Elsevier Science, Amsterdam: in press.

Guzmán, H. 2000. Status of coral reefs in Panamá for 2000. Unpublished report, Smithsonian Tropical Research Institute, Balboa, Panamá.

Leao, Z.M.A.N.; R.K.P. Kikuchi & V. Testa. 2000. Corals and coral reefs of Brazil. In J. Cortés (ed): Coral reefs of Latin America. Elsevier Science, Amsterdam: in press.

Rodríguez, A. & J. Garzón-Ferreira. 2000. Status of the coral reefs of Colombia. Unpublished report, INVEMAR, Santa Marta, Colombia.

Zapata, F. & B. Vargas. 2000. Corals and coral reefs of the Pacific coast of Colombia. In J. Cortés (ed): Coral reefs of Latin America. Elsevier Science, Amsterdam: in press.

19. Sponsoring Organisations, Coral Reef Programmes and Monitoring Networks

AGRRA - Atlantic and Gulf Rapid Reef Assessment
AGRRA is an international collaboration of researchers and managers designed to evaluate reef condition throughout the Caribbean and Gulf of Mexico using a rapid assessment protocol. AGRRA was specifically designed for the western Atlantic to assess the regional condition of hard corals, composition of algal communities, and fish abundance and size. Principal indicators assessed are coral cover, coral mortality, coral recruitment, macro algal index, urchin density, abundance and size of key fish families; protocols are largely based on visual measurements (identification, size, abundance, partial mortality). Consistency between observers in measuring and recording data is essential, and training workshops for scientists and managers were held in Bonaire, Mexico, Jamaica and Florida to achieve this. AGRRA assessments have been in over 22 regions of the Caribbean (including Bahamas, Belize, Bonaire, Cayman Islands, Costa Rica, Cuba, Curaçao, Honduras, Jamaica, Mexico, Puerto Rico, St. Vincent, Turks and Caicos, US Virgin Islands, US Florida Keys and Flower Gardens, Venezuela) since June 1998. AGRRA has developed an ACCESS database for the data and long-term plans are to put survey results on the world-wide web. By examining many reefs, it is possible to develop scales of reef condition and perform regional comparisons, but these do not attempt to distinguish between cause and effect relationships on reef condition. These data will establish a present-day Caribbean-wide baseline and to develop hypotheses on regional patterns of reef condition.

For additional information on contact: Robert Ginsburg or Philip Kramer MGG/RSMAS, University of Miami, 4600 Rickenbacker Cswy, Miami, FL 33149.
E-mail: agrra@rsmas.miami.edu or rginsburg@rsmas.miami.edu;
Web site: http://coral.aoml.noaa.gov/agra/

AIMS - Australian Institute of Marine Science
AIMS is one of Australia's key research agencies and the only one committed primarily to marine research, with an emphasis on tropical marine science. It undertakes research and development to generate new knowledge in marine science and technology, and to promote its application in industry, government and environmental management. The research programme involves medium- to long-term research that is geared towards improved understanding of marine systems and the development of a capability to predict the behaviour of complex tropical marine systems. In the past 20 years the Institute has established a sound reputation for high quality research on coral reef and mangrove ecosystems, and on the water circulation around our coasts and continental shelf. Researchers have not only published extensively in scientific journals but have also written field guides, books and monographs for regional use. This work supports a wide range of studies for effective coral reef management.
Contact: Tel: + 61 7 4753 4444; Fax: + 61 7 4772 5852; Web site: www.aims.gov.au

CARICOMP - CARIBBEAN COASTAL MARINE PRODUCTIVITY PROGRAM

This is a regional scientific program that has been studying land-sea interaction processes in the Caribbean coastal zone since 1992. A Data Management Centre (now the Caribbean Coastal Data Centre) in the University of the West Indies in Jamaica archives data on mangrove forests, seagrass meadows and coral reefs. The initial focus was on monitoring relatively undisturbed sites to distinguish natural from anthropogenic disturbance. CARICOMP contributes coral reef data to REEFBASE, and has initiated the GCRMN in areas of the Caribbean. CARICOMP is a regional network of observers, able to collaborate on studies of region-wide events and there are institutions in 21 countries in the network: Bahamas, Barbados, Belize, Bermuda, Bonaire, Cayman Islands, Colombia, Costa Rica, Cuba, Curaçao, Dominican Republic, Haiti, Jamaica, México, Nicaragua, Panama, Puerto Rico, Saba, Trinidad and Tobago, USA, Venezuela. More details of the Program can be found at: www.uwimona.edu.jm/centres/cms/caricomp/

CORAL - THE CORAL REEF ALLIANCE

CORAL is a member-supported, non-profit organisation based in California that works with the dive community and others to promote coral reef conservation around the world. CORAL is creating a new constituency for coral reef conservation by building support from divers, snorkelers, and other concerned individuals. CORAL supports community-based organisations by providing financial and technical support for coral reef conservation in communities throughout the world. In addition, CORAL builds public awareness about coral reefs through various education programmes. CORAL's mission is to keep coral reefs alive. E-mail: Coralmail@aol.com; Web site: www.coral.org/

CORDIO – COral Reef Degradation in the Indian Ocean

CORDIO is a regional, multi-disciplinary program developed to investigate the ecological and socioeconomic consequences of the mass bleaching of corals during 1998 and subsequent degradation of coral reefs in the Indian Ocean. CORDIO is an operating program within ICRI. The general objectives are to determine the: biophysical impacts of the bleaching and mortality of corals and long term prospects for recovery; socioeconomic impacts of the coral mortality and options for mitigating these through management and development of alternative livelihoods for peoples dependent on coral reefs; and prospects for restoration and rehabilitation of reefs to accelerate their ecological and economic recovery. CORDIO assists and coordinates with the GCRMN in the Indian Ocean with monitoring and running the Node in East Africa. The participating countries are: Kenya, Tanzania, Mozambique, Madagascar, Seychelles, India, Maldives, Sri Lanka, Reunion, Comores, Mauritius and Chagos.
Program co-ordination contacts: Olof Lindén, Dept. of Zoology, Stockholm University, 106 91 Stockholm, Sweden
Tel: +46-156-310 77; Fax: +46-156-31087; E-mail: olof@timmermon.se
In South Asia: Dan Wilhelmsson, c/o SACEP, 10 Anderson Road, Colombo 5, Sri Lanka
Tel: +94-1-596 442; Fax: +94-1-589 369; E-mail: dan.wilhelmsson@cordio.org
In East Africa: David Obura, WIOMSA, P.O. Box 10135, Bamburi, Kenya
Tel/fax: + 254-11-486292; E-mail: dobura@africaonline.co.ke
In Island States: Jean Pascal Qoud, CloeCoop, Cellule Locale pour l'Environment, 16, rue Jean Chatel, 97400 Saint Denis, Reunion, E-mail: cloecoop@runtel.fr

GCRMN - GLOBAL CORAL REEF MONITORING NETWORK

The GCRMN was formed at the 1st ICRI international workshop in the Philippines in 1995 with a contribution from US Department of State to IOC-UNESCO, who invited UNEP, IUCN, and the World Bank as co-sponsors. The Global Coordinator is hosted at AIMS in partnership with ICLARM, which runs ReefBase, the global coral reef database.

The GCRMN seeks to encourage and coordinate three overlapping levels of monitoring:

- Community - monitoring by communities, fishers, schools, colleges, tourist operators and tourists over broad areas with less detail, to provide information on the reef status and causes of damage using Reef Check methodology and approaches.
- Government–monitoring by predominantly tertiary trained personnel in Government environment or fisheries departments, and universities for moderate coverage of reefs at higher resolution and detail using methods developed in Southeast Asia or comparable methods.
- Research –high resolution monitoring over small scales by scientists and institutes currently monitoring reefs for research.

Equal emphasises is placed on monitoring to gather biophysical and socioeconomic data, with a manual being issued for the latter. A major objective is to produce regular national, regional and global Status of Coral Reefs Report, such as those produced as the background for this report. The GCRMN functions as a network of independent Regional Nodes that will coordinate training, monitoring and databases within participating countries and institutes in regions based on the UNEP Regional Seas Programme:

- Middle East – Nodes being formed with the assistance of the Regional Organisation for the Conservation of the Environment of the Red Sea and Gulf of Aden (PERSGA) and the Regional Organisation for the Protection of the Marine Environment (ROPME), Contact: Fareed Krupp (fareed.krupp@persga.org);
- South-west Indian Ocean Island States – coordinating Comoros, Madagascar, Mauritius, Reunion and Seychelles with assistance from the European Union, the Indian Ocean Commission and the Global Environment Facility. Contact: Lionel Bigot, ARVAM La Reunion (lionelbigot.arvam@guetali.fr);
- East Africa – assisting Kenya, Mozambique, South Africa and Tanzania operating through the CORDIO network in Mombasa, in association with the Kenya Wildlife Service and the Kenya Marine and Fisheries Research Institute. Contact: David Obura in Mombasa (dobura@africaonline.co.ke);
- South Asia – for India, Maldives and Sri Lanka with support from the UK Department for International Development (DFID) and coordination through IOC-UNESCO. Contact: the Regional Coordinator Emma Whittingham, in Colombo (reefmonitor@eureka.lk);
- South East Asia or East Asian Seas – the countries prefer to act independently without the formation of a regional node. Contact: Chou Loke Ming, National University of Singapore (dbsclm@nus.edu.sg), or Hugh Kirkman of the UNEP offices in Bangkok (kirkman.unescap@un.org);

- Southwest Pacific and Melanesia, 'IOI-Pacific Islands Node' covers Fiji, Nauru, New Caledonia, Samoa, Solomon Islands, Tuvalu and Vanuatu. Contacts: Robin South (south_r@usp.ac.fj), Posa Skelton (skelton_p@student.usp.ac.fj), Ed Lovell for Reef Check (lovell@suva.is.com.fj);
- Southeast and Central Pacific, the 'Polynesia Mana Node' for the Cook Islands, French Polynesia, Kiribati, Niue, Tokelau, Tonga and Wallis and Futuna coordinated in French Polynesia from the CRIOBE-EPHE Research Station on Moorea. Contact: Bernard Salvat (bsalvat@uni-perp.fr);
- Northwest Pacific and Micronesia, the 'MAREPAC Node' for American Samoa, the Republic of the Marshall Islands, the Federated States of Micronesia (FSM), the Commonwealth of the Northern Mariana Islands (CNMI), Guam (an unincorporated Territory) and the Republic of Belau (Palau). Contact Robert Richmond, Marine Laboratory, University of Guam (richmond@uog9.uog.edu);
- Hawaiian Islands – for US islands in the Pacific. Contact: Dave Gulko Hawaii Department of Land & Natural Resources (david_a_gulko@exec.state.hi.us);
- US Caribbean – for US territories and states of Florida, Flower Garden Banks, Navassa, Puerto Rico, and US Virgin Islands. Contact Donna Turgeon, NOAA (donna.turgeon@noaa.gov) or (http.//coralreef.gov/);
- North Central America, Mesoamerican Barrier Reef System for Mexico, Belize, Guatemala, Honduras. Contact: Philip Kramer, University of Miami (pkramer@rsmas.miami.edu), Ernesto Arias-Gonzalez, CINVESTAV-University in Mexico (earias@mda.cinvestav.mx), and Marea Hatziolos of the World Bank in Washington DC (Mhatziolos@worldbank.org);
- Eastern Caribbean, for the islands of the Organisation of Eastern Caribbean States (OECS), Trinidad and Tobago, and Barbados, and including the French and Dutch Caribbean Islands. It is coordinated by CANARI, with support from UNEP-CAR/RCU (Caribbean Regional Coordinating Unit) from St Lucia. Contact: Allan Smith (smitha@candw.lc);
- Southern Tropical America Node for Costa Rica, Panama, Colombia, Venezuela and Brazil via the 'Instituto de Investigaciones Marinas y Costeras' (INVEMAR) with support from UNEP-CAR/RCU. Contact: Jaime Garzón-Ferreira (jgarzon@invemar.org.co) and Alberto Rodríguez-Ramírez (betorod@invemar.org.co).

Contact: Clive Wilkinson Global Coordinator at the Australian Institute of Marine Science, in Townsville (c.wilkinson@aims.gov.au); or Jamie Oliver at ICLARM in Penang Malaysia (j.oliver@cgiar.org); or home pages: www.environnement.gouv.fr/icri and www.coral.noaa.gov/gcrmn.

ICLARM - THE WORLD FISH CENTER
ICLARM is committed to contributing to food security and poverty eradication in developing countries. ICLARM efforts focus on benefiting poor people, and conserving aquatic resources and the environment. The organisation aims for poverty eradication; a healthier, better nourished human family; reduced pressure on fragile natural resources; and people-centered policies for sustainable development. ICLARM is an autonomous, nongovernmental, nonprofit organisation, established as an international centre in 1977, with headquarters in the Philippines. ICLARM headquarters relocated to Penang, Malaysia, in early 2000. The new headquarters will be a focal point for international efforts to tackle the major aquatic challenges affecting the developing world and to demonstrate solutions to resources managers worldwide. Contact: PO Box 500 GPO, 10670 Penang, Malaysia Tel: (604) 641 4623; Fax: (604) 643 4463; Web site: www.cgiar.org/iclarm/

ICRI - INTERNATIONAL CORAL REEF INITIATIVE
ICRI was launched in 1994 as a response to the UNCED Rio Earth Summit in 1992 and the concerns of Small Island Developing States expressed at a meeting in Barbados in 1993 about the declining status of coral reef resources. ICRI was built as a partnership of 8 founding countries under the initiative of the US State Department (Australia, France, Jamaica, Japan, Philippines, Sweden, UK and USA) along with CORAL, IOC-UNESCO, IUCN, UNDP, UNEP, the World Bank, with the objectives of mobilising global support for coral reef actions, often using diplomatic and international flora. The USA chaired the first ICRI Secretariat in 1995-96, followed by Australia in 1997-98, and France in 1999-2000. It is anticipated that the Secretariat will pass to the Philippines assisted by Sweden for 2001-02. ICRI held an international workshop in the Philippines in May 1995 with participation of over 40 countries, NGOs, researchers and private sector to develop **A Call to Action** and a **Framework for Action**. The major themes were: Integrated Coastal Management; Capacity Building; Research and Monitoring; and Reviewing Progress. The GCRMN was initiated at that meeting to meet the 3rd theme. Eight Regional ICRI workshops followed to assemble the concerns and priorities for action from over 100 countries. These action priorities were revised and consolidated in 1998 at the International Tropical Marine Ecosystems Management Symposium (ITMEMS) in Townsville Australia to produce a **Renewed Call to Action**.

ICRI represents coral reef partners on international fora, such as: the UNEP and IOC governing councils, Conventions on Sustainable Development, Biological Diversity, Framework Convention on Climate Change; Global Meetings of Regional Seas Conventions and Action Plans, Global Plan of Action to Protect the Marine Environment from Land-based Activities, Global Conference on Sustainable Development in Small Island Developing States, CITES. The GCRMN was the first operational unit of ICRI, and two more are being formed: ICRIN – the Information Network; and ICRAN – the Action Network. ICRAN was initiated by ICLARM and UNEP to create a global network of successful integrated coastal management and marine protected area models for expanded action. Membership of the ICRI Coordination and Planning Committee includes the countries and agencies above plus all Regional Seas Units of UNEP and the South Pacific Regional Environment Programme (SPREP), ICLARM, WWF, CBD, Ramsar, US Coral Reef Task Force, the French national committee (IFRECOR), International Society for Reef Studies (ISRS), and the Marine Aquarium Council (MAC).

Contacts: Bernard Salvat (bsalvat@univ-perp.fr) and Genevieve Verbrugge (genevieve.verbrugge@environnement.gouv.fr) of the ICRI Global Secretariat
www.environnement.gouv.fr/icri

ICRIN - INTERNATIONAL CORAL REEF INFORMATION NETWORK
This is a global public awareness initiative coordinated by the Coral Reef Alliance (CORAL) to educate about why and how to protect coral reefs; ICRIN will serve as a worldwide communication hub for those working on coral reef education and conservation. ICRIN is designed to: draw public attention to vital coral reef issues and promote coral reef conservation, research and monitoring; strengthen partnerships among coral reef groups and provide information and tools to support their work; and reach the mainstream media in order to create and maintain a common concern among the general public for the health of coral reefs. ICRIN is a component of the ICRI and the International Coral Reef Action Network (ICRAN), delivering information to the public, key government decision-makers, businesses, coral reef conservation groups, potential funders, and the media. ICRIN has coral reef information reference centre, available to the public and accessible via the Internet, with coral reef public awareness materials (brochures, pamphlets, videos, posters and books) from non-governmental organisations (NGOs), aquariums, schools, scientists, community groups and businesses around the world. ICRIN also distributes copyright-free photographs, videos and other images for use in coral reef public awareness programmes. ICRIN is a network for coral reef outreach coordinators, linking groups working on similar projects and channelling the latest coral reef information to the media and general public using press releases, the Internet, public service announcements, advertisements and videos.
E-mail: Coralmail@aol.com; Web site: www.coral.org/

IOC-UNESCO
The Intergovernmental Oceanographic Commission (IOC) in Paris has promoted marine scientific investigations and related ocean services for more than 30 years, with a view to learning more about ocean resources, their nature and sustainability. IOC, with United Nations Environment Programme (UNEP), World Conservation Union (IUCN) and the World Meteorological Organisation formed the Global Task Team on Coral Reefs in 1991 to select methods and develop plans to monitor the world's coral reefs. This Task Team was the immediate precursor to the Global Coral Reef Monitoring Network (GCRMN), and after the ICRI Dumaguete meeting in 1995, the IOC, UNEP and IUCN were invited to co-sponsor the GCRMN, with the World Bank joining as a co-sponsor in 1998. The GCRMN is coordinated and administered through the IOC and is a part of the Global Ocean Observing System, to which it contributes data on coral reef health and resources.
Tel: +33 1 45 68 41 89; fax: +33 1 45 68 58 12; Web site: http://ioc.unesco.org

IUCN-THE WORLD CONSERVATION UNION

Founded in 1948, IUCN brings together States, government agencies and a diverse range of non-governmental organisations in a unique world partnership: over 900 members in all spread across nearly 140 countries. As a Union, IUCN seeks to influence, encourage and assist societies throughout the world to conserve the integrity and diversity of nature and to ensure that any use of natural resources is equitable and ecologically sustainable.

Through its network of regional and country offices, expert Commissions and member organisations, and Marine Program, IUCN supports a variety of coral reef conservation and management efforts that promote: protection of critical habitats and threatened species; design of management plans with stakeholder participation; development of ecological and socioeconomic assessment methodologies; enhanced capacity building for sustainable management; and support of effective governance structures. IUCN is an original partner of the International Coral Reef Initiative (ICRI) and a founding co-sponsor of the Global Coral Reef Monitoring Network (GCRMN).

Tel: +1 202 387 4826; Fax: +1 202 387 4823
E-mail: oceans@iucn.org; Web site: www.iucn.org

REEF CHECK

Reef Check was designed to meet the need for a simple, rapid monitoring protocol that could be used to measure the health of coral reefs globally. Since 1997, over 1000 reefs in more than 50 countries have been surveyed with results documenting the global extent of overfishing, and the bleaching and mortality event of 1998. There are 3 principal goals: education, monitoring and management. The education is achieved by training government staff, volunteers and NGOs in coral reef monitoring. These people are encouraged to monitor reefs they visit each year to provide data for the annual global report and to assist management. In addition, Reef Check organises publicity events such as press conferences, beach clean-ups and group dive-ins to focus attention on the plight of coral reefs. Reef Check methods are designed to be carried out by teams of experienced recreational divers, snorkelers, or even village fishermen, lead by a scientists to produce statistically comparable results. Training requires only 1 day or less and surveys are conducted over 1 day as data are based on counting key indicator species, rather than measuring organisms.

Contact: Gregor Hodgson, Reef Check Foundation, Institute of the Environment, 1652 Hershey Hall, Box 149607
University of California at Los Angeles, CA 90095-1496 USA; www.ReefCheck.org; rcheck@ucla.edu

REEFS AT RISK IN SOUTHEAST ASIA

The World Resources Institute, in collaboration with regional partners, is currently analysing and consolidating data on threats to coral reefs in Southeast Asia, as a more detailed regional project following the global *Reefs at Risk* analysis completed in 1998. Reefs at Risk in Southeast Asia (RRSEA), aims to compensate for the lack of detailed information on the status of coral reefs, and has four primary goals:

- Improve the base of information available for examining the threats, status, value, and protection of coral reefs within Southeast Asia, through collecting, revising and integrating information;

- Model threats to coral reefs based upon population and development patterns, landuse change, and the location and intensity of certain activities known to degrade coral reefs (including watershed-based modelling of sediment potentially impacting coral reefs);
- Develop a geographic information system (GIS)-based tool for more local-level evaluation of development scenarios and related implications for coral reef health and associated economic implications;
- Raise awareness about human threats to coral reefs through wide dissemination of integrated data sets, model results, a published report, and the GIS planing tool.

The development and dissemination of an integrated, spatially-referenced base of information is the first step towards identifying causes of reef degradation. Regional-extent data sets being assembled and improved, include data on coral reef and mangrove locations, observations on coral condition (percent live coral cover, coral bleaching, coral diseases, and incidence of destructive fishing), and location and effectiveness of marine protected areas. In addition, land cover, elevation, bathymetry, hydrology, population density, and infrastructure data sets are assembled as background data sets for the threat modelling. The RRSEA project is implemented by WRI, in collaboration with over 20 international, national and more local partner institutions. A workshop on Reefs at Risk in Southeast Asia was held in April 2000 near Manila, with over 40 participants who reviewed and improved the base data sets and refined the model of threats to coral reefs. Although the main product of RRSEA is a standardised indicator of human pressure on coral reefs, the integrated database is an important related product. All project results and data will be available from on the Internet, www.wri.org, upon completion of the project in mid-2001. For additional information on the Reefs at Risk project, contact: Lauretta Burke, World Resources Institute, 10 G Street NE, Washington, DC 20009, USA; Tel: +1 202 729 7774; Fax: +1 202 729 7775; e.mail: lauretta@wri.org

UNEP - UNITED NATIONS ENVIRONMENT PROGRAMME
The mission of the United Nations Environment Programme (UNEP) is to provide leadership and encourage partnership in caring for the environment by inspiring, informing, and enabling nations and peoples to improve their quality of life without compromising that of future generations. UNEP makes a particular effort to nurture partnerships and enhancing the participation of civil society – the private sector, scientific community, NGOs and youth - in the achievement of sustainable development. Today, the challenge before UNEP is to implement an environmental agenda that is integrated strategically with the goals of economic development and social well-being – an agenda for sustainable development.
Contacts: P.O. Box 30552 Nairobi, Kenya
Tel: (254 2) 621234; Fax: (254 2) 623927/623692/623404; Cpiinfo@unep.org: www.unep.org

UNEP - WORLD CONSERVATION MONITORING CENTRE
There is a major focus on reef-related issues at UNEP-WCMC, including reef mapping, maintaining a global database on marine protected areas, monitoring trade in corals and other marine ornamentals, and looking at the global distribution of threats, including coral disease and coral bleaching. A major product from this work will be the *World Atlas of*

Coral Reefs, to be released in early 2001, which represents the culmination of six years of coral reef mapping with the latest information on the distribution of coral reefs world-wide. Colour maps will show the location of reefs, along with major population centres, forests, including mangrove forests, marine protected areas, and also dive-centres (defined as centres offering certified training courses). Maps will be accompanied by text describing the distribution and status of coral reefs in each country. Data tables will provide national statistics, including information on biodiversity, socioeconomics, and protected areas. A global section of the Atlas will address general issues of reef distribution and biodiversity, as well as human use of reefs, using maps wherever appropriate. Historically and modern reef maps will also be included. Following a major report on the legal (CITES regulated) trade in corals, UNEP-WCMC has started a major assessment of the trade in other 'marine ornamentals' including fish for the aquarium trade, in collaboration with the Marine Aquarium Council and members of the industry. The aim is foster a better understanding of this trade, and ascertain the impacts with the goal of developing a well regulated and sustainable trade that operates through certification schemes. UNEP-WCMC continues to maintain a global database on marine protected areas, with 3600 listings of which only 560 contain coral reefs. These and other data are increasingly being used by agencies and individuals world-wide. UNEP-WCMC is committed to increasing the accessibility of its coral reef information, both through collaboration with partners such as ICLARM, WRI, GCRMN and through increasing presentation of information on the internet. For additional information on the UNEP-WCMC reef-related activities, contact: Mark Spalding (mark.spalding@unep-wcmc.org) or Ed Green (edmund.green@unep-wcmc.org), UNEP-World Conservation Monitoring Centre, 219 Huntingdon Road, Cambridge, CB3 0DL, UK Tel: +44 (0)1223 277314; Fax: +44 (0)1223 277136; web site: www.unep-wcmc.org

WORLD BANK – ENVIRONMENT DEPARTMENT

The World Bank is an international financial institution dedicated to the alleviation of poverty. The Environment plays a crucial role in determining the physical and social well being of people. While poverty is exacerbated by deteriorating conditions in land, water and air quality, economic growth and the well being of communities in much of the developing world, continues to depend on natural wealth and the production of environmental goods and services. As a result, the Bank is committed to integrating environmental sustainability into its programmes, across sectors and regions and through its various financial instruments. Reducing vulnerability to environmental risk, improving people's health, and enhancing livelihoods through safeguarding the environment are the hallmarks of the Bank's emerging Environment Strategy. Support for coral reef conservation and sustainable use is consistent with this theme, as it potentially affects millions of people around the world. The challenge for the Bank and its many partners in coral reef conservation, such as ICRI and GCRMN, will be to help communities capture the benefits from the sound management of coral reefs to meet immediate needs, while at the same time ensuring the sustainability of these vital systems for generations to come. For information on the Environment Department, contact: Marea E. Hatziolos, Environment Department MC5-845, The World Bank, 1818 H. St., N.W., Washington, D.C. 20433 USA Tel: +1 202 473-1061; Fax: +1 202 522-0367; E-mail: Mhatziolos@worldbank.org

Suggested Reading

The following are some useful references to follow up the summaries in the Chapters above, however most Chapters include source references. Other useful material includes:

Birkeland, C. (1997). Life and Death of Coral Reefs. Chapman & Hall, New York, 536pp.

Birkeland, C. (1997). Symbiosis, fisheries and economic development on coral reefs. Trends in Ecology and Evolution 12:364-367.

Brown, B.E. (1997). Integrated Coastal Management: South Asia. Department of Marine Sciences and Coastal Management, University of Newcastle, Newcastle upon Tyne, United Kingdom

Bryant, D., Burke, L., McManus, J. and Spalding, M. (1998). Reefs at Risk: A Map-Based Indicator of Potential Threats to the World's Coral Reefs. World Resources Institute, Washington D.C, 56pp.

English, S., Wilkinson, C. and Baker, V. (1997). Survey Manual for Tropical Marine Resources. 2nd Edition. Australian Institute of Marine Science, Townsville, 390pp.

Ginsburg, R. N. (ed) (1993). Global Aspects of Coral Reefs: Health Hazards and History, 7-11 June 1993. University of Miami, Miami, 420pp. (This volume has a valuable series of papers on problems facing reefs around the world).

Hatziolos, M. E., Hooten, A. J. and Fodor, F. (1998). Coral Reefs: Challenges and Opportunities for Sustainable Management. In Proceedings of an associated event of the fifth annual World Bank Conference on Environmentally and Socially Sustainable Development. World Bank, Washington D.C., 224pp.

IOC/UNEP/IUCN. (1997). IOC/UNEP/IUCN Global Coral Reef Monitoring Network Strategic Plan. UNESCO, Paris, 10pp.

Johannes, R. E. (1998). The case for data-less marine resource management: examples from tropical nearshore finfisheries. Trends in Ecology and Evolution 13:243-246.

Kelleher, G. (1999). Guidelines for Marine Protected Areas. World Conservation Union, Washington D.C., 107pp.

Kelleher, G., Bleakley, C. and Wells, S. (1995). The Global Representative System of Marine Protected Areas. Volume II Wider Caribbean, West Africa and South Atlantic. Great Barrier Reef Marine Park Authority, Townsville ; World Bank, Washington D.C. ; World Conservation Union, Washington D.C., 93pp.

Kelleher, G., Bleakley, C. and Wells, S. (1995). The Global Representative System of Marine Protected Areas. Volume III Central Indian Ocean, Arabian Seas, East Africa and East Asia Seas. Great Barrier Reef Marine Park Authority, Townsville ; World Bank, Washington D.C. ; World Conservation Union, Washington D.C., 147pp.

Kelleher, G., Bleakley, C. and Wells, S. (1995). The Global Representative System of Marine Protected Areas. Volume IV South Pacific, Northeast Pacific, Southeast Pacific and Australia/New Zealand. Great Barrier Reef Marine Park Authority, Townsville ; World Bank, Washington D.C. ; World Conservation Union, Washington D.C., 212pp.

Maragos. J. E., Crosby, M. P. and McManus, J. W. (1996) Coral reefs and biodiversity: a critical and threatened relationship. Oceanography 9:83-99.

McClanahan, T. R., Sheppard, C.R., Obura, D.O. (eds) 2000. Coral reefs of the Indian Ocean: Their ecology and conservation. Oxford University Press, N.Y.

Salm, R.V., Clark, J.R. and Siirila, E. (2000). Marine and Coastal Protected Areas: A Guide for Planners and Managers. IUCN Washington DC 371 pp.

Sheppard, C. R. C. and Sheppard, A. L. S. (1991). Corals and Coral Communities of Arabia. Fauna of Saudi Arabia 12.

Wells, S. and Hanna, N. (1992). The Greenpeace Book of Coral Reefs. Sterling Publishing Co., New York, 160pp.

Wells, S.M. (ed) (1988). Coral Reefs of the World. Volume 1: Atlantic and Eastern Pacific. UNEP, Nairobi ; International Union for Conservation of Nature and Natural Resources, Switzerland.

Wells, S.M. (ed) (1988). Coral Reefs of the World. Volume 2: Indian Ocean, Red Sea, and Gulf. UNEP, Nairobi ; International Union for Conservation of Nature and Natural Resources, Switzerland.

Wells, S.M. (ed) (1988). Coral Reefs of the World. Volume 3: Central and western Pacific. UNEP, Nairobi; International Union for Conservation of Nature and Natural Resources, Switzerland.

Wilkinson, C. R. (1998). Status of Coral Reefs of the World: 1998. Global Coral Reef Monitoring Network and Australian Institute of Marine Science, Townsville, Australia 184pp. (The bulk of the essays were drawn from papers presented at the 8th International Coral Reef Symposium, held in Panama City, June 1996 and published in Volume 1 of the Proceedings. The reference for these essays is:

Lessios, H.A. and Macintyre, I.E. (eds) (1997). Proceedings of the 8th International Coral Reef Symposium, Panama, June 24-29 1996, Volume 1. Smithsonian Tropical Research Institute, Balboa, Panama.

Wilkinson, C. R. and Buddemeier, R. W. (1994). Global Climate Change and Coral Reefs: Implications for People and Reefs. Report of the UNEP-IOC-ASPEI-IUCN Global Task Team on Coral Reefs. IUCN, Gland, 124pp.

LIST OF ACRONYMS

AGRRA	Atlantic and Gulf Reef Assessment
AIMS	Australian Institute of Marine Science
ASEAN	Association of South East Asian Nations
BAPPEDA	Council for Provincial Planning and Development
CATIE	Centro Agronomico Tropical de Investigacion y Ensenanza
CFC	Chlorofluorocarbon
CITES	The Convention on International Trade in Endangered Species of Wild Fauna and Flora
COTS	Crown-of-thorns starfish (*Acanthaster planci*)
CNA	Clean Nigeria Association
CSD	Convention for Sustainable Development
DNA	Deoxyribonucleic Acid
EIA	Environmental Impact Assessment
GBR	Great Barrier Reef
GBRMP	Great Barrier Reef Marine Park
GBRMPA	Great Barrier Reef Marine Park Authority
GEF	Global Environment Facility
ICRI	International Coral Reef Initiative
ICAM	Integrated Coastal Area Management
ICM	Integrated Coastal Management
ICZM	Integrated Coastal Zone Management
IMO	International Maritime Organisation
IUCN	World Conservation Union
MARPOL	International Convention of the Prevention of Pollution from Ships
MPA	Marine Protected Area
NGO	Nongovernmental organisation
NOAA	National Oceanic and Atmospheric Administration (of USA)
Ramsar	International Convention on Wetlands
RAP	Rapid Assessment Protocol
SCUBA	Self-contained underwater breathing apparatus
SIDS	Small Island Developing States
UNEP	United Nations Environment Programme
UNESCO	United Nations Educational Scientific and Cultural Organisation
UV	Ultraviolet radiation